Lecture Notes in Physics

For information about Vols. 1–131, please contact your bookseller or Springer-Verlag.

Vol. 132: Systems Far from Equilibrium. Proceedings, 1980. Edited by L. Garrido. XV, 403 pages. 1980.

Vol. 133: Narrow Gap Semiconductors Physics and Applications. Proceedings, 1979. Edited by W. Zawadzki. X, 572 pages. 1980.

Vol. 134: γγ Collisions. Proceedings, 1980. Edited by G. Cochard and P. Kessler. XIII, 400 pages. 1980.

Vol. 135: Group Theoretical Methods in Physics. Proceedings, 1980. Edited by K. B. Wolf. XXVI, 629 pages. 1980.

Vol. 136: The Role of Coherent Structures in Modelling Turbulence and Mixing. Proceedings 1980. Edited by J. Jimenez. XIII, 393 pages. 1981.

Vol. 137: From Collective States to Quarks in Nuclei. Edited by H. Arenhövel and A. M. Saruis. VII, 414 pages. 1981.

Vol. 138: The Many-Body Problem. Proceedings 1980. Edited by R. Guardiola and J. Ros. V, 374 pages. 1981.

Vol. 139: H. D. Doebner, Differential Geometric Methods in Mathematical Physics. Proceedings 1981. VII, 329 pages. 1981.

Vol. 140: P. Kramer, M. Saraceno, Geometry of the Time-Dependent Variational Principle in Quantum Mechanics. IV, 98 pages. 1981.

Vol. 141: Seventh International Conference on Numerical Methods in Fluid Dynamics. Proceedings. Edited by W. C. Reynolds and R. W. MacCormack. VIII, 485 pages. 1981.

Vol. 142: Recent Progress in Many-Body Theories. Proceedings. Edited by J. G. Zabolitzky, M. de Llano, M. Fortes and J. W. Clark. VIII, 479 pages. 1981.

Vol. 143: Present Status and Aims of Quantum Electrodynamics. Proceedings, 1980. Edited by G. Gräff, E. Klempt and G. Werth. VI, 302 pages. 1981.

Vol. 144: Topics in Nuclear Physics I. A Comprehensive Review of Recent Developments. Edited by T.T.S. Kuo and S.S.M. Wong. XX, 567 pages. 1981.

Vol. 145: Topics in Nuclear Physics II. A Comprehensive Review of Recent Developments. Proceedings 1980/81. Edited by T. T. S. Kuo and S. S. M. Wong. VIII, 571-1.082 pages. 1981.

Vol. 146: B. J. West, On the Simpler Aspects of Nonlinear Fluctuating. Deep Gravity Waves. VI, 341 pages. 1981.

Vol. 147: J. Messer, Temperature Dependent Thomas-Fermi Theory. IX, 131 pages. 1981.

Vol. 148: Advances in Fluid Mechanics. Proceedings, 1980. Edited by E. Krause. VII, 361 pages. 1981.

Vol. 149: Disordered Systems and Localization. Proceedings, 1981. Edited by C. Castellani, C. Castro, and L. Peliti. XII, 308 pages. 1981.

Vol. 150: N. Straumann, Allgemeine Relativitätstheorie und relativistische Astrophysik. VII, 418 Seiten. 1981.

Vol. 151: Integrable Quantum Field Theory. Proceedings, 1981. Edited by J. Hietarinta and C. Montonen. V, 251 pages. 1982.

Vol. 152: Physics of Narrow Gap Semiconductors. Proceedings, 1981. Edited by E. Gornik, H. Heinrich and L. Palmetshofer. XIII, 485 pages. 1982.

Vol. 153: Mathematical Problems in Theoretical Physics. Proceedings, 1981. Edited by R. Schrader, R. Seiler, and D.A. Uhlenbrock. XII, 429 pages. 1982.

Vol. 154: Macroscopic Properties of Disordered Media. Proceedings, 1981. Edited by R. Burridge, S. Childress, and G. Papanicolaou. VII, 307 pages. 1982.

Vol. 155: Quantum Optics. Proceedings, 1981. Edited by C.A. Engelbrecht. VIII, 329 pages. 1982.

Vol. 156: Resonances in Heavy Ion Reactions. Proceedings, 1981. Edited by K.A. Eberhard. XII, 448 pages. 1982.

Vol. 157: P. Niyogi, Integral Equation Method in Transonic Flow. XI, 189 pages. 1982.

Vol. 158: Dynamics of Nuclear Fission and Related Collective Phenomena. Proceedings, 1981. Edited by P. David, T. Mayer-Kuckuk, and A. van der Woude. X, 462 pages. 1982.

Vol. 159: E. Seiler, Gauge Theories as a Problem of Constructive Quantum Field Theory and Statistical Mechanics. V, 192 pages. 1982.

Vol. 160: Unified Theories of Elementary Particles. Critical Assessment and Prospects. Proceedings, 1981. Edited by P. Breitenlohner and H.P. Dürr. VI, 217 pages. 1982.

Vol. 161: Interacting Bosons in Nuclei. Proceedings, 1981. Edited by J.S. Dehesa, J.M.G. Gomez, and J. Ros. V, 209 pages. 1982.

Vol. 162: Relativistic Action at a Distance: Classical and Quantum Aspects. Proceedings, 1981. Edited by J. Llosa. X, 263 pages. 1982.

Vol. 163: J. S. Darrozes, C. Francois, Mécanique des Fluides Incompressibles. XIX, 459 pages. 1982.

Vol. 164: Stability of Thermodynamic Systems. Proceedings, 1981. Edited by J. Casas-Vázquez and G. Lebon. VII, 321 pages. 1982.

Vol. 165: N. Mukunda, H. van Dam, L.C. Biedenharn, Relativistic Models of Extended Hadrons Obeying a Mass-Spin Trajectory Constraint. Edited by A. Böhm and J.D. Dollard. VI, 163 pages. 1982.

Vol. 166: Computer Simulation of Solids. Edited by C.R.A. Catlow and W.C. Mackrodt. XII, 320 pages. 1982.

Vol. 167: G. Fieck, Symmetry of Polycentric Systems. VI, 137 pages, 1982.

Vol. 168: Heavy-Ion Collisions. Proceedings, 1982. Edited by G. Madurga and M. Lozano. VI, 429 pages. 1982.

Vol. 169: K. Sundermeyer, Constrained Dynamics. IV, 318 pages. 1982.

Vol. 170: Eighth International Conference on Numerical Methods in Fluid Dynamics. Proceedings, 1982. Edited by E. Krause. X, 569 pages. 1982.

Vol. 171: Time-Dependent Hartree-Fock and Beyond. Proceedings, 1982. Edited by K. Goeke and P.-G. Reinhard. VIII, 426 pages. 1982.

Vol. 172: Ionic Liquids, Molten Salts and Polyelectrolytes. Proceedings, 1982. Edited by K.-H. Bennemann, F. Brouers, and D. Quitmann. VII, 253 pages. 1982.

Lecture Notes in Physics

209

Mathematical and Computational Methods in Nuclear Physics

Proceedings of the Sixth Granada Workshop
Held in Granada, Spain, October 3–8, 1983

Edited by J. S. Dehesa, J. M. G. Gomez and A. Polls

Springer-Verlag
Berlin Heidelberg GmbH 1984

Editors

J. S. Dehesa
Departamento de Física Nuclear, Facultad de Ciencias
Universidad de Granada, 18001 Granada, Spain

J. M. G. Gomez
Departamento de Física Nuclear, Facultad de Ciencias
Universidad de Salamanca, Salamanca, Spain

A. Polls
Departamento de Fisica Teórica, Facultat de Ciencias
Universitat de Barcelona, Barcelona, Spain

ISBN 978-3-540-13392-6 ISBN 978-3-540-38943-9 (eBook)
DOI 10.1007/978-3-540-38943-9

Originally published by Springer-Verlag Berlin Heidelberg New York Tokyo in 1984

2153/3140-543210

PREFACE

The present proceedings contain the talks given at the Sixth International Granada Workshop on "Mathematical and Computational Methods in Nuclear Physics", held in Granada (Spain), October 3rd-8th, 1983. The meeting was organized by the Departamento de Física Nuclear, Universidad de Granada, and the Departamento de Física Nuclear, Universidad de Salamanca.

The invited lecturers were O. Bohigas and M.J. Giannoni (Orsay), J.B. French (Rochester), K. Goeke (Julich), M.H. Kalos (New York), and A. Zuker (Strasbourg). The meeting was attended by some 50-odd participants of different nationalities, in an informal atmosphere of spontaneous and constructive scientific discussion.

The lectures, covering various aspects of the many-body problem in nuclei, review present knowledge and include some unpublished material as well. Bohigas and Giannoni discuss the fluctuation properties of spectra of many-body systems by means of random matrix theories, and the attempts to search for quantum mechanical manifestations of classical chaotic motion. The role of spectral distributions (expressed as explicit functions of the microscopic matrix elements of the Hamiltonian) in the statistical spectroscopy of nuclear systems is analyzed by French. Zuker, after a brief review of the theoretical basis of the shell model, discusses a reformulation of the theory of effective interactions and gives a survey of the linked cluster theory. Goeke's lectures center on the mean-field methods, particularly TDHF, used in the investigation of the large-amplitude nuclear collective motion, pointing out both the successes and failures of the theory.

In addition the present volume also contains the seminars on related topics given by J.M.G. Gomez (Salamanca), J. Martorell (Palma de Mallorca), E. Moya de Guerra (Badajoz), A. Poves (Madrid), and D. Provoost (Julich). Unfortunately, Kalos' lectures on Green's Function Monte-Carlo Methods, and M. Zirnbauer's seminar have not been received in time for the publication of the present proceedings.

The organization of this conference would not have been possible without the invaluable help of many members of the host institution, particularly Professor R. Guardiola. The organizing committee acknowledges the financial support of the Fundación Juan March, the Grupo Interuniversitario de Física Teórica (GIFT), the Academia de Ciencias Matemáticas, Físico-Químicas y Naturales de Granada, the Ayuntamiento de Granada, the Instituto de Ciencias de la Educación of the Universidad de Granada and the Asociación Cultural Hispano-Alemana , and the kind cooperation of the Director of the "Colegio Mayor Cardenal Cisneros", where the participants were lodged during the conference.

J.S. Dehesa
J.M.G. Gomez
A. Polls

TABLE OF CONTENTS

CHAOTIC MOTION AND RANDOM MATRIX THEORIES

Oriol BOHIGAS and Marie-Joya GIANNONI
Division de Physique Théorique*, Institut de Physique Nucléaire,
91406 ORSAY CEDEX, France

"Cada ciència ha mester los vocables per los quals mills sia manifestada ; e car a aquesta ciència demostrativa sien mester vocables escurs e que los hòmens lecs no han en ús, e car nós façam aquest llibre als hòmens lecs per açò breument e ab plans vocables parlarem d'esta ciència".
Ramon Llull (Libre de Gentil, 1273 ?)

I - INTRODUCTION

In the early usage, the word chaos referred to the darkness, the vacuum. John Milton, paraphrasing the Old Testament Genesis, wrote in the Paradise Lost

In the beginning how the heavens and earth
Rose out of Chaos

Since Milton, the word chaos has altered this meaning, to indicate confusion, mixing, complexity, disorder. One usually opposes chaos to order, or complexity to simplicity and one can imagine different sorts of evolutions or transitions [DH-80] : order to order, chaos to chaos, order to chaos and chaos to order. The first two are not unexpected and easy to illustrate. The third fits with everyday experience and looks consistent with the Second Principle of Thermodynamics. But the last one is surprising and intriguing. Most of the questions we will address ourselves in these lectures are of the type "How is the order diluted when going from order to chaos" and "What is the underlying order behind chaos". And the appropriate language, in most cases, will of course be the one of probability theory. In particular, it will emerge that probabilistic and deterministic points of view are not as irreconciliable as they may appear at first sight. Indeed, we will go through systems governed by causal equations of motions although their future motion does not depend on their past, i.e. we will discuss memory-loosing or unpredictable systems.

*Laboratoire associé au C.N.R.S.

The kind of systems we shall deal with are of several types :

i)- Points distributed on the real axis : $x_1, x_2, x_3, \ldots$. The axis may be the energy, the points corresponding to the discrete energy levels of a quantum system (an atomic nucleus, an atom or a molecule) ; or the frequency axis, the points corresponding, for instance, to the normal frequencies of a vibrating membrane ; or the time axis, the points corresponding to successive epochs of occurrence of a given event (times of arrivals on a telephone line in queuing problems, for instance).

- Points distributed on a segment or on the circle (for instance, eigenvalues of unitary matrices).

- For illustrative purposes, we shall also briefly describe some number-theoretic functions f(n) defined on the positive integers 1,2,3,...

ii)- Sequences of points $\underset{\sim}{x}_1, \underset{\sim}{x}_2, \ldots, \underset{\sim}{x}_k, \ldots$ on an N-dimensional space defined, for instance, through some transformation $\underset{\sim}{x}_{n+1} = T\underset{\sim}{x}_n$. For instance, snapshots of the classical trajectory of a point particle of certain dynamical systems, or sequences of values of some variables obtained through a sequential algorithm.

For the first kind of systems (discussed in this Section and in Section II), before a study of the fluctuations -one of our main goals- can be attempted, one must make a clean separation between the smooth (or average) behaviour and a fluctuating (or oscillating) part. In some cases this step will be obvious. For the second kinds of systems (treated in Section III), the attention will be focused on the structure of phase space, how it is filled by trajectories and special emphasis will be given to the question of stability of orbits.

In order to get some flavour about general ideas and concepts, let us now give a few examples coming from very different fields.

Prime Numbers (Number Theory)

The theory of numbers is at the same time one of the most elementary branches of mathematics, because it deals with the arithmetic properties of integers 1,2,3,.. and one of the most difficult branches insofar as it leads to new difficult problems and techniques. Which sort of techniques ? Some titles of monographs or specialized articles reveal it : "Statistical independence in probability, analysis and number theory" [Ka-59], "On the density of certain sequences of integers", etc.. The list could be made very long. Notice the presence of words like probability, density.

Let us now consider the sequence of prime numbers 2,3,5,7,.. We want to emphasize two facts [Za-77] : "The first is that, despite their simple definition and role as the building blocks of the natural numbers, the prime numbers belong to the most arbitrary and ornery objects studied by mathematicians... The second fact is even more astonishing, for it states just the opposite : that the prime numbers exhibit stunning regularity, that there are laws governing their behaviour, and that they obey these laws with almost military precision".

The first aspect is illustrated on Table I-1. In the interval [1-100] there are 25 primes, the largest gap between two successive primes being 97-89=8. In the interval

[1,100]	2,3,5,7,11,13,17,19,23,29,31,37,41,43,47,53,59,61,67, 71,73,79,83,89,97
$[10^7-100,10^7]$	9999901, 07,29,31,37,43,71,73,91
$[10^7,10^7+100]$	10 000 019, 079

Table I-1 : List of primes in three different intervals of length 100

$[10^7-100,100^7]$ there are 9 primes, the largest gap being 28. In the interval $[10^7,10^7+100]$ there are only two primes, which differ by 60. Despite this in appearance completely erratic behaviour, are there systematic patterns ? The first important feature, already known by Euclid (ca. 300 b.C.) is that the sequence of primes is infinite. But, obviously, there are much less primes than integers. For instance, the harmonic series $\sum(1/n)$ summed over integers n is known to diverge. More precisely

$$\lim_{x\to\infty}\left(\sum_{n\leq x}\frac{1}{n} - \ln x\right) = \gamma = 0.577.... \qquad \text{(Euler constant)} \qquad \text{(I-1)}$$

What about summing over primes p only, i.e. $\sum(1/p)$? If one performs the "experiment", for instance by summing for $p \leq 10^9$ one finds 3.3, whereas summing up to $p \leq 10^{18}$ one would find $\simeq 4$. In fact, the exact asymptotic result is

$$\sum_{p\leq x}\frac{1}{p} = \ln\ln x + C + \eta(x) , \qquad \text{(I-2)}$$

where $C \simeq 0.261497$ and $\eta(x) \to 0$ when $x \to \infty$; thus, the series diverges. This example clearly shows that in some cases, the empirical (numerical) observation is unable, not only to produce, but even to guess, the exact result. But it is not always like that. For instance, let us consider the number $\pi(x)$ of primes which are less than or equal to x . The function $\pi(x)$ is an infinite staircase, which increases by one each time one "crosses" a prime. The first 25 steps of this irregular staircase are shown on Fig.I.1. Has this staircase function an average behaviour ? Gauss observed as early as 1792 that the density of prime numbers $d\pi(x)/dx$ appears on the average to be $1/\ln x$.

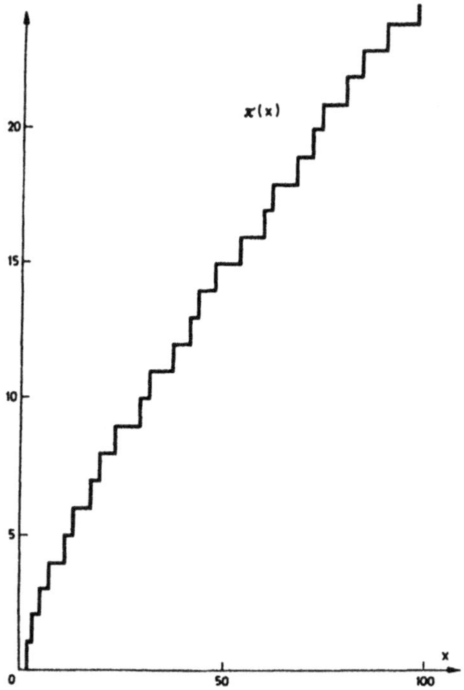

Fig.I.1 - Plot of the function $\pi(x)$ for $1 < x < 100$

He was thus led, from "empirical observation" of primes, to approximate $\pi(x)$ by the integral logarithm Li(x)

$$\pi(x) \simeq Li(x) = \lim_{\epsilon \to 0} \left[\int_0^{1-\epsilon} \frac{dt}{\ln t} + \int_{1+\epsilon}^{x} \frac{dt}{\ln t} \right]$$

$$= Li(2) + \int_2^x \frac{dt}{\ln t} = 1.04 + \int_2^x \frac{dt}{\ln t} \; . \qquad \text{(I-3)}$$

Li(X) admits the following expansion

$$Li(x) = \gamma + \ln \ln x + \sum_{n \geq 1} \frac{(\ln x)^n}{n.n!} \; . \qquad \text{(I-4)}$$

The quality of the approximation $\pi(x) \simeq \mathrm{Li}(x)$ is, for many purposes, very good. For instance, for $x \leq 10^7$, the relative error $|\pi(x) - \mathrm{Li}(x)|/\pi(x)$ is smaller than 5×10^{-5}. On Fig.I.2 is plotted, for $x \leq 10^7$, the difference $\mathrm{Li}(x) - \pi(x)$. It can be seen that although this difference is small, it is not featureless : for instance, for $x < 10^7$, it steadily increases.

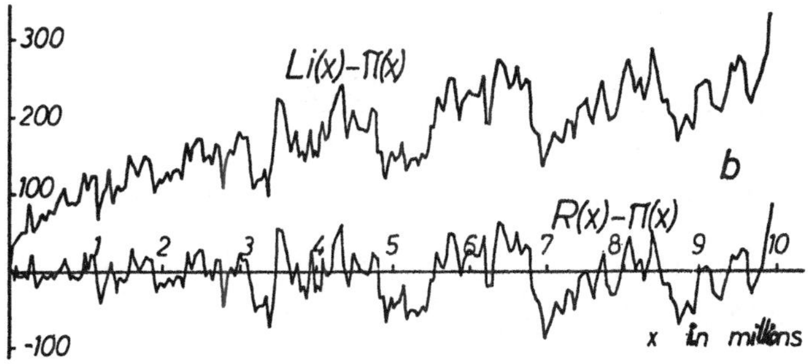

Fig.I.2 - Difference between Gauss (Riemann) approximation Li (x) ($R(x)$) to $\pi(x)$ and $\pi(x)$ for $x \leq 10^7$ (taken from Ref.[Za-77])

What about rigorous, "non-empirical", results ? One of the main questions, in the middle of the 19th century, was to prove the prime number theorem (PNT) namely

$$\pi(x) \underset{x\to\infty}{\sim} \frac{x}{\ln x} \quad . \tag{I-5}$$

Notice that Gauss approximation is consistent with the PNT. The first major result in the direction of the PNT was obtained by Chebyshev in 1850, who proved that

$$0.89 \frac{x}{\ln x} < \pi(x) < 1.11 \frac{x}{\ln x} \tag{I-6}$$

for sufficiently large x . Although the PNT is true, the approximation

$$\pi(x) \simeq \frac{x}{\ln x} \tag{I-7}$$

is much poorer than the approximation introduced by Gauss (see Fig.I.3).

Riemann, in his famous memoir "Uber die Anzahl der Primzahlen unter einer gegebenen Grösse" introduced, based on empirical evidence and intuition, a better approximation $R(x)$ to $\pi(x)$

$$\pi(x) \simeq R(x) = \sum_{n=1}^{\infty} \frac{\mu(n)}{n} \mathrm{Li}\left(x^{1/n}\right), \tag{I-8}$$

where $\mu(n)$ is the Möbius function defined as follows : $\mu(1) = 1$

$$\mu(n \neq 1) = \begin{cases} 0 & \text{if n is divisible by a prime square} \\ 1 & \text{if n is a product of an even number of distinct primes} \\ -1 & \text{if n is a product of an odd number of distinct primes} \end{cases} \qquad \text{(I-9)}$$

For instance, $\mu(2) = \mu(3) = \mu(5) = \mu(7) = -1$; $\mu(6) = \mu(2 \times 3) = 1$; $\mu(4) = \mu(2^2) = 0$.
R(x) is an entire function of $\ln x$ with the following expansion

$$R(x) = 1 + \sum_{n \geq 1} \frac{1}{n\zeta(n+1)} \frac{(\ln x)^n}{n!} , \qquad \text{(I-10)}$$

where ζ is the Riemann zeta function of $s = \sigma + it$, which is defined by

$$\zeta(s) = \sum_{n \geq 1} \frac{1}{n^s} = \prod_{p \geq 2} (1 - p^{-s})^{-1} \qquad \text{(I-11)}$$

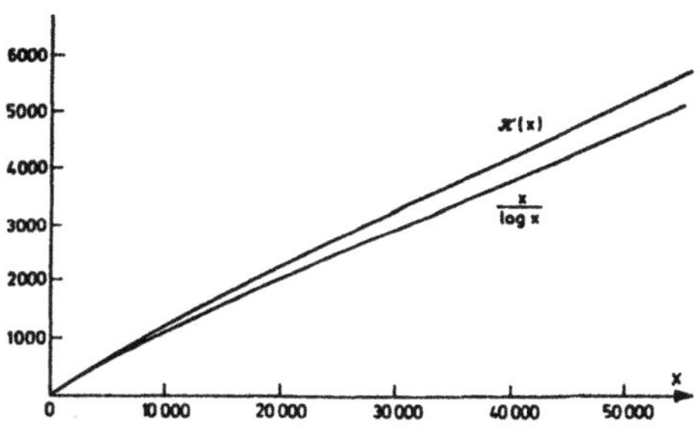

Fig.I.3 - Plot of $\pi(x)$ and $x/\ln x$ for $x \leq 5 \times 10^4$
(taken from Ref.[Za-77])

for $\sigma > 1$ and by analytic continuation for $\sigma \leq 1$, $s \neq 1$. Notice the connection between $\zeta(s)$ and the prime numbers p. For $x = 10^6$ one has $\pi(10^6) = 78498$ and the first two terms of (I-8) give $\mathrm{Li}(10^6) - \frac{1}{2}\mathrm{Li}(10^3) = 78628 - \frac{1}{2} \times 178 = 78539$.
On Fig.I.2 is shown the difference $R(x) - \pi(x)$. As can be seen, no structure is anymore present. It seems now that one can consider $R(x)$ to be the smooth behaviour of $\pi(x)$; after subtracting R to π only fluctuations are left out. In fact, Riemann, although unable to prove the PNT(*), did something even more astonishing. He derived an exact relation for $\pi(x)$:

$$\pi(x) = R(x) - \sum_{\rho} R(x^{\rho}) , \qquad \text{(I-12)}$$

(*)The PNT was proved simultaneously and independently by Hadamard and de la Vallée-Poussin in 1896. Hadamard was born in 1865 and de la Vallée-Poussin in 1866 and they died in 1963 and 1962 respectively !

where the sum runs over the "non trivial" ($\neq$ -2,-4,-6,...) roots ρ of $\zeta(s)$. The first term in (I-12) gives the approximation discussed above. Putting $\rho = \frac{1}{2} + i\alpha$ (one assumes the Riemann Hypothesis (RH) to be true ; see below), performing the summation over α in order of size, pairing terms of the sum corresponding to ρ and ρ^* (if ρ is a root of $\zeta(s)$, so is ρ^*), one can write

$$\pi(x) = R(x) + \sum_{k=1}^{\infty} T_k(x) \quad , \tag{I-13}$$

where

$$T_k(x) = -\left[R(x^{\rho_k}) + R(x^{\rho_k^*}) \right] \qquad k=1,2,\dots \, . \tag{I-14}$$

The terms $T_k(x)$ are oscillating ones. The first few are shown on Fig.I.4 and on Fig.I.5 is reproduced the result obtained by adding 10 and 29 oscillating terms to $R(x)$ [Za-77, Si-79]. As can be seen, by the time when $\sim$ 30 terms are added, one is obtaining an almost exact representation of $\pi(x)$ in the range $1 < x < 100$ (compare to Fig.I.1).

Some remarks are in order. We have mentioned that $Li(x) - \pi(x) > 0$ for $x < 10^7$ (see Fig.I.2). How incredibly weak may the position of the physicist be when looking for asymptotic results is illustrated by the following fact : it has been shown (Littlewood) that there exist numbers for which $Li(x) - \pi(x)$ is negative and Skewes proved that there is one smaller than

$$10^{10^{10^{34}}} \quad ,$$

a number of which Hardy said that it was surely the biggest that had ever served any definite purpose in mathematics. So, although one knows no number for which $Li(x) - \pi(x)$ is negative, one knows that this difference cannot increase steadily, as suggested by the enormous available "empirical data" coming from all the presently known primes. On the opposite, we have also seen (at least in the hands of Gauss and Riemann !) how extremely powerful and far-reaching approximations can be derived from empirical observation.

In the same vein, let us illustrate how empirism and heuristic arguments are used and differently appreciated by pure mathematicians. We have already mentioned the Riemann Hypothesis (RH), which is by universal agreement the out-standing unsolved problem in mathematics. It states that all the zeros $\rho = \sigma + it$ of the Riemann zeta function lie on the critical line $\sigma = 1/2$, except for the "trivial zeros" lying on the real axis at the values $\sigma = -2,-4,-6,\dots$ It has been proved that there are an infinite number of zeros on the critical line (Hardy) but not that all of them are on it. It has also been proved that the RH is equivalent to

$$\lim_{x\to\infty} M(x)\, x^{-(1/2)-\epsilon} = 0 \quad \text{for all } \epsilon > 0 . \qquad \text{(I-15)}$$

In (I-15) $M(x)$ is

$$M(x) = \sum_{n<x} \mu(n) , \qquad \text{(I-16)}$$

the summation is over integers and μ (n) is the Möbius function defined above (I-9). People have tried, using heuristic arguments having a probabilistic basis, to reinforce the belief that the RH is in fact true. We reproduce one of them for its simplicity [DH-80., GC-68]. If we make a plot of the Möbius function $\mu(n)$ it looks random, in the sense that it shows no discernible regularity, except for the fact that μ is just as likely to be equal 1 or -1. Now, what is the chance for $\mu(n) \neq 0$? This will happen if n is not a multiple of 4 or a multiple of 9, or a multiple of 25 or of any other square of a prime. The probability that a number chosen at random is not a multiple of 4 is 3/4, that it is not a multiple of 9 is 8/9, that it is not a multiple of p^2 is $(p^2-1)/p$. These conditions being all independent we obtain, for the probability that $\mu(n) \neq 0$

$$\prod_{p\geq 2} \frac{p^2-1}{p^2} = \frac{6}{\pi^2} . \qquad \text{(I-17)}$$

Therefore, the probability that : i) $\mu(n) = 1$ is $3/\pi^2$, ii) $\mu(n) = -1$ is $3/\pi^2$, iii) $\mu(n) = 0$ is $1 - 6/\pi^2$. Let us now examine $M(x)$

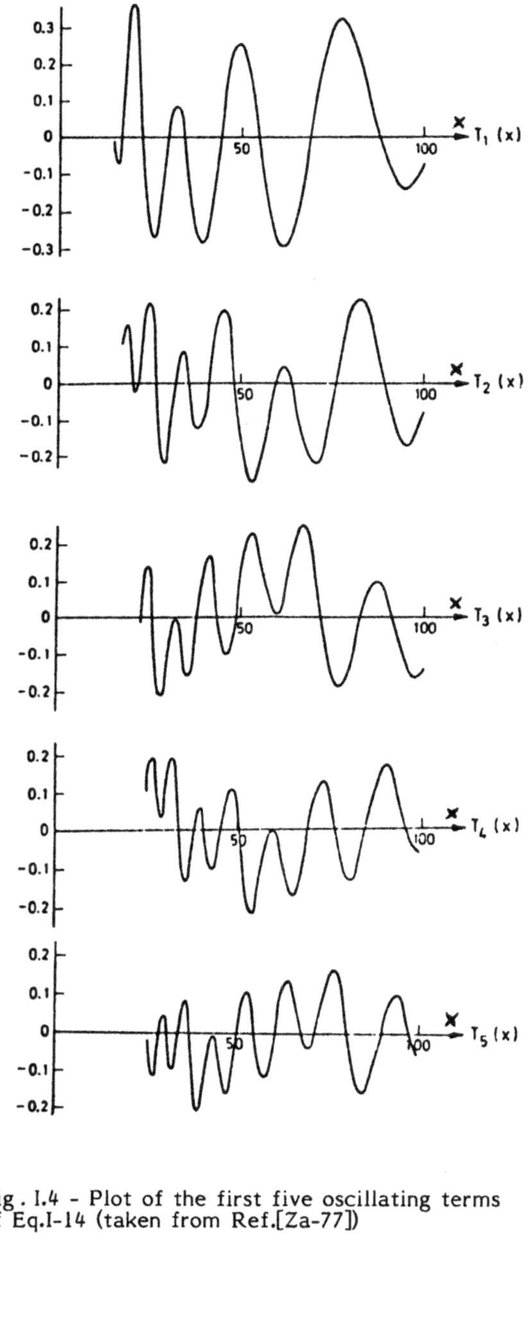

Fig. I.4 - Plot of the first five oscillating terms of Eq.I-14 (taken from Ref.[Za-77])

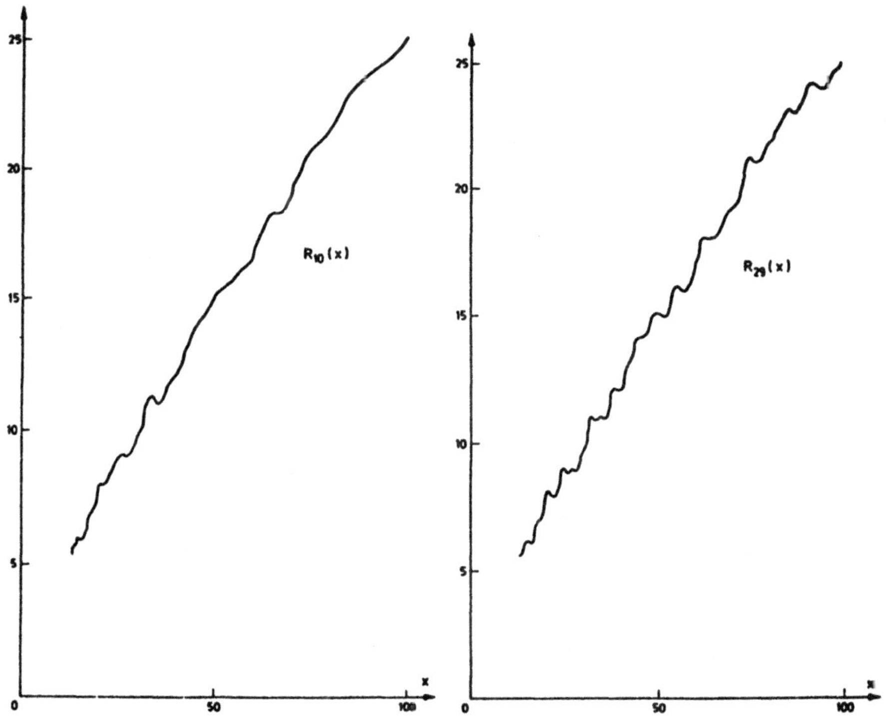

Fig.I.5 - Value of $\pi(x)$ including 10 and 29 oscillating terms in Eq.I-13 (taken from Ref. [Za-77])

and assume that each term in the summation in (I-16) can be considered as an independent random variable with the probabilities just derived. Hausdorff's inequality says then that, with probability 1, $M(x)$ grows no faster than a constant times $x^{(1/2)+\epsilon}$, exactly what is needed to proof RH ! However, we have made an unacceptable trick. Instead of adding the values of μ for the N values in the range 1 to N = Int(x), we have taken N integers at random. We have done this because we feel that the table of values of μ is "random" or "unpredictable" although we know that the Möbius function is completely deterministic. The author of the authoritative work on the zeta function, H.M. Edwards [Ed-74], calls this type of heuristic reasoning "quite absurd" and Littlewood wrote : "I should also record my feeling that there is no imaginable reason why the RH should be true". Nevertheless, if one persists with the "absurd" reasoning, one predicts that the number of zeros of $\mu(n)$ between 1 and 33×10^6 is [GC-68] $33 \times 10^6 \times (1-6/\pi^2)$= 12 938405.6 whereas the actual number is 12 938407, an 8 place accuracy result !

Zeros of the Riemann Zeta Function

Let N(T) be the number of zeros $\rho = 1/2 + i\gamma$ of $\zeta(s)$ with $0 < \gamma \leq T$. Then [Mo-76]

$$N(T) = \frac{T}{2\pi} \ln \frac{T}{2\pi} - \frac{T}{2\pi} + S(T) + \frac{7}{8} + O\left(\frac{1}{T}\right) . \tag{I-18}$$

The maximum order of S(T) remains unknown. Probably

$$S(T) = O\left(\sqrt{\frac{\ln T}{\ln \ln T}}\right) . \tag{I-19}$$

One knows the first 7×10^8 zeros of $\zeta(s)$ and stretches of 10^5 successive zeros have been computed around the 10^9-th zero and around the 10^{11}-th zero [Od-82].

There is a conjecture that the distribution of zeros of $\zeta(s)$ resembles that of the eigenvalues of a random complex hermitian or unitary matrix [Mo-73a,74] (see Section II). There are very strong numerical indications supporting this conjecture [Od-82].

Vibrations of a membrane (the drum)

A membrane is a perfectly flexible and infinitely thin lamina of solid matter, of uniform material and thickness which is stretched in all directions by a tension so great as to remain unaltered during its vibrations. The principal subject in this field is the investigation of the transverse vibrations of membranes of different shapes, whose boundaries are fixed. One considers then (Fig.I.6) a membrane streched over the area Ω included within a fixed, closed plane boundary Γ. Taking the plane of the membrane as that of $x-y$, let Ψ denote the small displacement perpendicular to its original plane. It has been known for all over a century that Ψ obeys the wave function

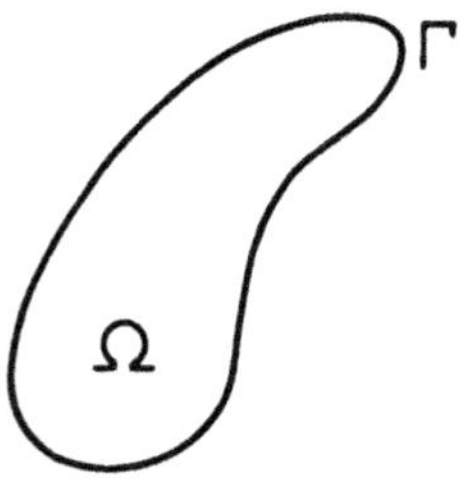

Fig.I.6

$$\frac{\partial^2 \Psi}{\partial t^2} = c^2 \nabla^2 \Psi , \tag{I-20}$$

where c is a constant with dimensions of a velocity, depending on the physical properties of the membrane as well as on the tension under which the membrane is held. Of special interest, both to mathematicians and to musicians, are solutions of the form

$$\Psi(x,y;t) = \psi(x,y)\, e^{i\omega t} \tag{I-21}$$

for, being harmonic in time with frequency ω, they represent the pure tones the membrane is capable of producing. These special solutions are known as normal modes. To find the normal modes, one substitutes (I-21) in (I-20) and one finds that ψ must satisfy the eigenvalue equation

$$\nabla^2 \psi = -\frac{\omega^2}{c^2}\psi = -k^2\psi = -E\psi \tag{I-22}$$

with the condition that ψ vanishes on the boundary Γ. In (I-22) ω is a frequency, k a wave number and E an energy. Equation (I-22) possesses an infinite number of eigenvalues E_n which are real and non-negative and have no accumulation point. One has therefore

$$0 \leqslant E_1 \leqslant E_2 \leqslant E_3 \leqslant \dots, \quad \lim_{n\to\infty} E_n = \infty \quad . \tag{I-23}$$

We are now interested in properties of the sequence of eigenvalues. For instance, on what characteristics of the boundary depends the number of modes per unit energy (or unit frequency), i.e. what is the density of eigenmodes. In these lectures the emphasis will be put on the 2-dimensional case, but the problem can also be considered in 3 dimensions, or more generally in N dimensions. In three dimensions, it corresponds to the study of the vibrations of an acoustical resonator or the vibrations of an elastic body or the free electromagnetic oscillations in the interior of a cavity, with adequate changes of the boundary conditions. In one dimension, the problem corresponds to the vibrations of a string, but in this case there is no room for the influence and regularity of the shape. In quantum physics, Eq.(I-22) represents the time-independent Schrödinger equation for a free particle of mass m moving in the interior of a box with infinite walls, with $\hbar k = \sqrt{2mE}$, where E is the kinetic energy.

The asymptotic number of modes per unit frequency was first established for the case of the rectangular parallepiped by Rayleigh (1905). Sommerfeld and H.A. Lorentz had drawn attention to the effect of the domain, or, following Kac to put it in a picturesque way, one may ask "Can one hear the shape of a drum ?" [Ka-66]. Johanna Reudler, a student of Lorentz, in a Leiden dissertation, verified that the asymptotic number of modes depends only on the volume for the special cases parallepiped, sphere and cylinder. The history of how the first important result in this field was obtained is worth to be briefly remembered [Ka-66]. Lorentz was invited to Göttingen in 1910 to deliver the Wolfskehl lectures (Wolfskehl awarded a prize for proving or disproving Fermat's last theorem[(*)] and in case the prize would not be awarded, stipulated that the

---

(*) In 1637 Fermat stated that the diophantine equation $x^n + y^n = z^n$ with integral $n > 2$ has no solutions in positive integers x, y, z. Fermat asserted to have a "truly marvelous proof" of his statement, but today it is generally believed that his argument, which was apparently never revealed, must have been incomplete.

proceeds from the principal should be used to invite eminent scientists to lecture in Göttingen). Lorentz gave some lectures under the general title "Alte und neue Fragen der Physik". At the end of one of them he asked ""In an enclosure with a perfectly reflecting surface there can form standing electromagnetic waves... The mathematical problem is to prove that the number of sufficiently high overtones which lie between ν and $\nu + d\nu$ is independent of the shape of the enclosure and is simply proportional to its volume ". If one believes an apocryphal report Hilbert predicted that the theorem would not be proved during his life. Less than two years later Hermann Weyl, who was present at the Lorentz's lecture, using the theory of integral equations which his teacher Hilbert developed only a few years before, proved the theorem, long before his death.

Let us now go back to the initial two-dimensional problem. Progress has been made since the pioneering work by Weyl. We are interested in extracting a smoothed eigenvalue distribution $N_{av}(E)$, i.e. the smoothed function giving the number of eigenvalues less than or equal to E in order to study the fluctuations or oscillations of the exact eigenvalue distribution around the averaged value $N_{av}(E)$. In the context of the previous example on prime numbers, we are searching the function $N_{av}(E)$ which has a similar relationship to $N(E)$ as $R(x)$ to $\pi(x)$. Reference [BH-76] gives a complete account of the results obtained so far in this field. The function $N_{av}(E)$ can be written

$$N_{av}(E) = \frac{\sigma}{4\pi} E - \frac{\gamma}{4\pi} \sqrt{E} + K + O\left(E^{-\eta/2} \ln\sqrt{E}\right), \tag{I-24}$$

where $0 < \eta \leq 1$. In (I-24) σ is the surface of the area Ω (Weyl's term) and γ is the perimeter of the boundary Γ(*). K is a constant term containing complex information on the geometrical and topological properties of the domain. The geometrical features contributing to the constant term are : i) Curvature contribution

$$(1/12) \int_\Gamma \kappa(l)\, dl , \tag{I-25}$$

where $\kappa(l)$ denotes the local curvature ; for instance, the curvature contribution for the circle is 1/6. ii) Corners contribution ; for a square (or a rectangle), it is 4 x (1/48). The topological features concern the connectivity of the surface ; for a multiply connected drum containing r holes, the contribution to the constant term is (1-r) x (1/6).

On Fig.I.7 are compared the exact function N(E) and the smoothed function $N_{av}(E)$ given by Eq.(I-24) for two different shapes, namely a quarter of a circle and a stadium (see Section III). It can be seen that $N_{av}(E)$ indeed reproduces perfectly the average behaviour of N(E), not only asymptotically but starting from the bottom of the spectrum.

(*) If one uses Neumann instead of Dirichlet boundary conditions, (I-24) is still valid except for the sign of the perimeter term

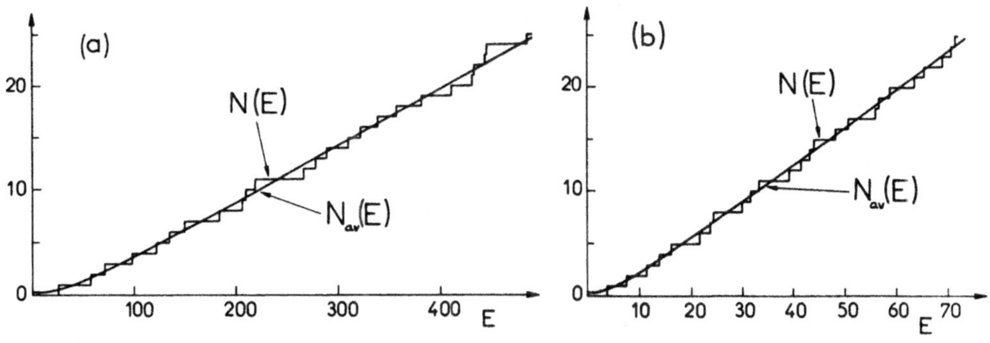

Fig.I.7 - Cumulative density of eigenvalues N(E) of Eq.(I-22) and its average $N_{av}(E)$ given by Eq.(I-24) for two different shapes of the boundary : (a) a quarter of a circle of radius unity ; (b) a stadium with straight line and radius of curvature equal to one (See Section IV)(taken from [Sc-84,BGS-84b]).

How to characterize fluctuations

The examples treated so far should have convinced the reader that it is possible in many cases, for a sequence of discrete points on the real line, to separate the staircase function $N(\xi)$ in a smooth part $N_{av}(\xi)$ and a fluctuating part $N_{fl}(\xi)$

$$N(\xi) = N_{av}(\xi) + N_{fl}(\xi) \quad . \tag{I-26}$$

Before studying fluctuations one wants to get rid of $N_{av}(\xi)$ in order to compare the fluctuation patterns of different systems whose corresponding average behaviours $N_{av}(\xi)$ are not the same. For that purpose, one "unfolds" the original spectrum ξ_i through the following mapping $\xi \mapsto x$

$$x_i = N_{av}(\xi_i) \qquad i=1,2,\dots \quad . \tag{I-27}$$

Consider now the sequence $\{x_i\}$ and its corresponding smooth behaviour $\hat{N}_{av}(x)$. The effect of (I-27) is that $\hat{N}_{av}(x) = x$, i.e., the sequence $\{x_i\}$ has on the average a constant mean spacing (or a constant density) equal to unity, as can be seen from

$$N_{av}(\xi) = \int_0^{\xi} \rho_{av}(\xi')\,d\xi' = \int_0^{x} dx' = x = \hat{N}_{av}(x), \tag{I-28}$$

where $\rho_{av}(\xi)$ is the average density of $\{\xi_i\}$. For instance, consider the (fluctuation-free) sequence $\xi_k = k^2$ (k = 1,2,...) ; then $N_{av}(\xi) = \sqrt{\xi}$ and $x_k = k$, a sequence of equally spaced points or picket fence. In summary, after unfolding, we shall study quantities related to

$$\hat{N}(x) = \hat{N}_{av}(x) + \hat{N}_{fl}(x) = x + \hat{N}_{fl}(x) \ . \tag{I-29}$$

When considering fluctuation properties of sequences $\{x_i\}$ we shall come across different situations : i) cases in which the system is known to be, from a statistical point of view, translational invariant or stationary, i.e., the fluctuation properties are the same irrespective of which region of the spectrum (of the sequence) one is considering ; ii) cases in which the system is not stationary but one is interested in asymptotic properties of the spectrum.

The question now is to discover the stochastic laws governing sequences having very different origins, as illustrated on Fig.I.8, which is inspired from a similar figure of Ref.[BFF-81]. There are displayed six spectra, each containing 50 levels.(*) Column (a) corresponds to a Poisson system : Take a random variable s whose probability density $p(x)$ is e^{-x}. Construct a sequence $\{x_i\}$

$$x_1 = 0 \ , \qquad x_{i+1} = x_i + s_i \qquad i = 1,2,3.. \ , \tag{I-30}$$

where s_i are outcomes of <u>independent</u> trials of the variable s. The resulting spectrum is what is called a <u>Poisson spectrum</u>, which is obviously stationary. For instance, if one studies the counting rate of a decaying source, the successive times of decay x_i will form a Poisson spectrum, the time being measured in units of the mean life of the source. Column (b) shows an example of a segment of prime numbers in the interval [7791097-7791877] Ref.[Si-79] ; column (c) the resonance energies $J^{\pi} = 1/2^+$ of the compound nucleus observed in the reaction n + ^{166}Er (see Section II) ; column (d) the eigenvalues (associated to eigenfunctions with a given symmetry) corresponding to the transverse vibrations of a membrane whose boundary is the Sinai's billiard (see Section IV) ; column (e) the positive imaginary part of the 1551-th to the 1600-th zero of the Riemann zeta function [HM-63] ; column (f) an equally spaced sequence of levels (picket fence). Columns (a) and (f) represent two limiting cases, maximum randomness and no randomness at all respectively.

Can one deduce some features just by inspecting Fig.I.8 ? Arrows indicate spacings $s_i = x_{i+1} - x_i$ which are smaller than 1/4. The Poisson spectrum shows 12 arrows out of 49 spacings, the prime number "spectrum" shows 9 arrows, the Er spectrum only 2 arrows, the frequencies of the membrane 3 arrows, the zeros of $\zeta(s)$ no arrow and, of course, the picket fence no arrow. One therefore sees a statistical

(*) The spectra have been rescaled to the same spectrum span [0,49], thereby introducing an artificial rigidity (see below).

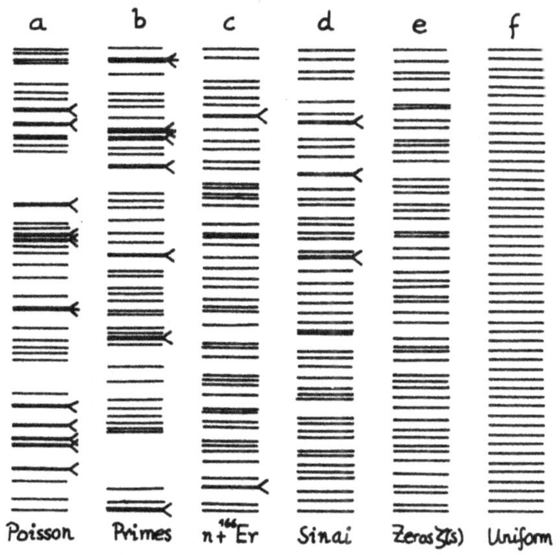

Fig.I.8 - Segments of "spectra", each containing 50 levels. The "arrowheads" mark the occurrence of pairs of levels with spacings smaller than 1/4. See text for further explanation.

similarity between (a) and (b) : large probability of small spacings and occurrence of some large spacings. On the contrary, (c) (d) and (e) show small probability of small and large spacings, the small probability of small spacings being usually referred to in the literature as the phenomenon of level repulsion. The spectrum (c), (d) and (e) deviate from (f) less strongly than (a) and (b). The picket fence (f) is a spectrum that we may qualify as absolutely rigid, in the sense that there is no departure at all from uniformity. Once the position of one level X_i is known, the position of any other level is determined, no matter how far it is from X_i . For this system the correlations between spacings are maximum and it shows perfect short and long range order. At the opposite extreme, the Poisson spectrum contains no correlations between spacings : the knowledge of a stretch of the spectrum puts no restriction on the behaviour of the spectrum beyond the interval considered (this is of course true irrespective of the form of the function $p(x)$ chosen in Eq.(I-30)). In intermediate situations between Poisson and the picket fence the degree of the spectral rigidity will depend on the nature and strength of the correlations between spacings.

Although this topic will be treated in greater detail in the next Section, let us already give some examples of characterization of fluctuation properties. We have mentioned before the spacing distribution $p(x)$ between adjacent levels. Let us reproduce a simple heuristic argument due to Wigner [Wi-56] that illustrates the presence or absence of level repulsion. Consider the probability $p(x)dx$ that, given a level at x_0, the next level is in the small interval $dI = [x_0+x , x_0+x+dx]$ (see Fig. I.9). It can be represented as the product of two factors

$$p(x)dx = \Pr(\text{one level in } dI/ \text{ no level in } I) \times \Pr(\text{no level in } I) \quad , \qquad \text{(I-31)}$$

where Pr means probability and Pr(a/b) is the conditional probability of having a if b

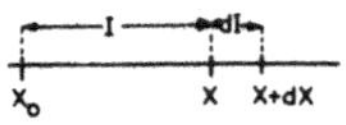

Fig.I.9

is true. One has

$$\Pr\,(\text{no level in I}) = \int_x^\infty p(x')\,dx' \tag{I-32}$$

and Pr (one level in dI/no level in I) = $\mu(x)\,dx$. Therefore

$$p(x) = \mu(x)\int_x^\infty p(x')\,dx', \tag{I-33}$$

which can be solved to give

$$p(x) = C\,\mu(x)\,e^{-\int^x \mu(x')\,dx'} \quad . \tag{I-34}$$

Now, for a Poisson sequence $\mu(x)$ is independent of x and one obtains $(\int p(x)\,dx = \int x\,p(x)\,dx = 1)$

$$p(x) = e^{-x} \qquad x \geqslant 0 \quad . \tag{I-35}$$

A linear level repulsion can be introduced by assuming $\mu(x) = \beta\,x$ and one obtains

$$P_W(x) = \frac{\pi}{2}\,x\,e^{-\frac{\pi}{4}x^2} \qquad x \geqslant 0 \; . \tag{I-36}$$

The result (I-36) for the spacing distribution is known in the literature o random matrices(*) as the Wigner surmise and it shows explicitly the level repulsion or tendency to avoid clustering of levels. Indeed, $p(x=0) = 0$ in constrast to the Poisson case, for which the corresponding spacing distribution (I-35) is maximum at the origin.

The spacing distribution $p(x)$ contains no information about spacing correlations. A simple measure of spacing correlations is the correlation coefficient C between, for instance, two adjacent spacings. Let $\{x_i\}$ be the sequence of levels and s_i the spacings $s_i = x_{i+1} - x_i$; C is given by

$$C = \sum_i (s_i - 1)(s_{i+1} - 1) \Big/ \sum_i (s_i - 1)^2 \;, \tag{I-37}$$

(*) (I-36) is a standard distribution in statistics, called a "Rayleigh distribution" : Consider on the plane a point of cartesian coordinates (x, y) and take x and y as independent random variables normally distributed with zero mean and with the same variance $\sigma^2 = 2/\pi$. The probability density of the radius vector $r = \sqrt{x^2 + y^2}$ is (I-36).

where we assume the mean spacing equal to unity. For a Poisson spectrum, C = 0.

A convenient variable which is often used is $\Delta_3(L)$ [DM-63]. It measures, given an interval $[\alpha, \alpha + L]$ of length L, the least square deviation of the staircase $\hat{N}(x)$ from the best straight line fitting it (see Fig.I.10) :

$$\Delta_3(\alpha;L) = (1/L) \min_{A,B} \int_{\alpha}^{\alpha+L} [\hat{N}(x) - Ax - B]^2 dx \quad . \tag{I-38}$$

Notice that we are dealing with spectra for which the average part $\hat{N}_{av}(x)$ of $\hat{N}(x)$ is a straight line $y = x$. However, when considering a given interval $[\alpha, \alpha + L]$, the best straight line fitting $\hat{N}(x)$ will not just be $y = x$ but another (presumably close lying) straight line $Ax + B$ which is determined by (I-38) The value of $\Delta_3(L)$ for $L \gtrsim 1$ for a picket fence spectrum is 1/12 whereas the average $\langle \Delta_3(L) \rangle$ of $\Delta_3(\alpha;L)$ (take many non overlapping adjacent segments of length L of the spectrum, compute the value of Δ_3 for each segment and perform the average(*)) for a Poisson spectrum is L/15. The departure of the average value of $\Delta_3(L)$ from the linear increase with L characteristic of the Poisson spectrum will give, in intermediate situations between Poisson and picket fence, information on the correlations between spacings and on the spectral rigidity or spectral stiffness.

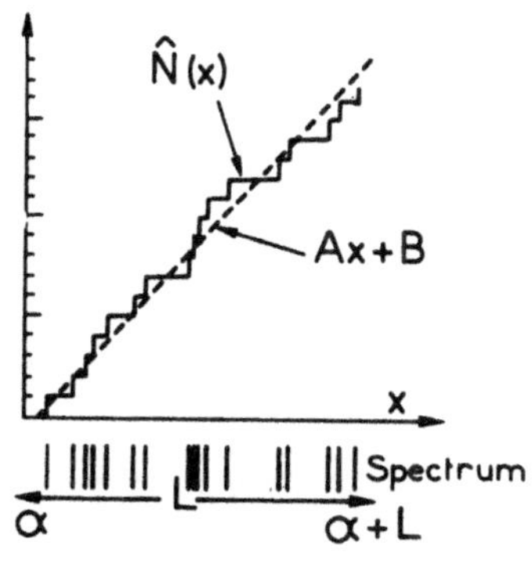

Fig.I.10- The Δ_3-statistic of Dyson-Mehta

In Ref.[BG-75] a convenient way to compute $\Delta_3(L)$ has been given, once the ordered sequece of points $x_1, x_2, \ldots, x_n$ in the interval $[\alpha, \alpha + L]$ is known. Take as origin the center of the interval, i.e., take $\tilde{x}_i = x_i - (\alpha + \frac{L}{2})$; $\Delta_3(L)$ can then be obtained from

$$\Delta_3(\alpha;L) = \frac{n^2}{16} - \frac{1}{L^2}\left[\sum_{i=1}^{n} \tilde{x}_i\right]^2 + \frac{3n}{2L^2}\left[\sum_{i=1}^{n} \tilde{x}_i^2\right] - \frac{3}{L^4}\left[\sum_{i=1}^{n} \tilde{x}_i^2\right]^2 + \frac{1}{L}\left[\sum_{i=1}^{n} (n-2i+1)\tilde{x}_i\right] . \tag{I-39}$$

Another useful variable to be discussed is the "number statistic n(L)"(**): given an interval $[\alpha, \alpha + L]$ of length L, it counts the number of levels contained in the interval. It is a discrete variable which can take the values 0 (no level in the interval), 1 (one level in the interval), 2,3,... The average value of n(L) is L, if the mean spacing is unity. We will consider higher moments or cumulants of n(L) (variance $\Sigma^2(L)$, skewness $\gamma_1(L)$, excess $\gamma_2(L)$). Qualitatively we expect that if the

(*) when the spectrum is stationary, the average $\langle \Delta_3(L) \rangle$ does not depend on α .
(**) For the sake of simplicity and when no confusion is possible, we shall omit in the notation the α-dependence

spectrum is stiff, the variance of n(L) will be small (in most cases the actual number of levels found in an interval of length L will differ only slightly from L) whereas for a non-rigid or compressible spectrum like Poisson the variance of n(L) will be comparatively large. For a Poisson spectrum one has $\Sigma^2(L) = L$ which tells nothing but the familiar result that if one takes an interval of length L one expects to find $L \pm \sqrt{L}$ levels in the interval. At the opposite extreme, for the picket fence, one will have $L \pm 0$. Again we will be interested in what happens in intermediate situations.

BIBLIOGRAPHY

- For some general ideas and reflections on mathematical practice we recommend ref.[DH-80], a beautiful book addressed to the layman
- On prime numbers : ref.[Za-77], addressed to non mathematicians
- On the Riemann Zeta Function : ref.[Ed-74], a classic
- On properties of vibrating membranes : refs.[Ka-66,BH-76]
- On general ideas on fluctuations : Introductory Section of ref.[BFF-81].

II - RANDOM MATRIX THEORIES

The random matrix theories we shall describe have their origin in the following physical problem. In the low energy region of the excitation spectrum of a nucleus, the level density ρ (E) is small and one expects to describe most of the states in a detailed way using nuclear models. However, due to the rapid increase of the level density with the excitation energy E

$$\rho(E) \simeq \frac{c}{(E-\Delta)^{5/4}} \exp\left(a\sqrt{E-\Delta}\right) \tag{II-1}$$

where c, Δ and a are for a given nucleus constants, by the time one reaches the region, for instance, of the neutron threshold ($E \sim 6$ MeV), the number of levels is so high that one must give up a description of microscopic detail, a description of the individual levels. The aim of nuclear models at this and higher excitation energies is rather to describe special states, like giant resonances, analogue states, etc.., which have a peculiar structure. But the detailed description of the sea of background states around the the collective ones must and should be abandoned. Twenty years ago, Dyson made an eloquent resumé of the situation [Dy-62a] :

"Recent theoretical analyses have had impressive success in interpreting the detailed structure of the low-lying excited states of complex nuclei. Still, there must be a point beyond which such analyses of individual levels cannot usefully go. For example, observations of levels of heavy nuclei in the neutron-capture region give precise information concerning a stretch of levels from number N to number (N+n), where N is an integer of the order of 10^6. It is improbable that level assignments based on shell structure and collective or individual-particle quantum numbers can ever be pushed as far as the millionth level. It is therefore reasonable to inquire whether the highly excited states may be understood from the diametrically opposite point of view, assuming as working hypothesis that all shell structure is washed out and that no quantum numbers other than spin and parity remain good. The results of such an inquiry will be a statistical theory of energy levels. The statistical theory will not predict the detailed sequence of levels in any one nucleus, but it will describe the general appearance and the degree of irregularity of the level structure that is expected to occur in any nucleus which is too complicated to be understood in detail.

"In ordinary statistical mechanics a comparable renunciation of exact knowledge is made. By assuming all states of a very large ensemble to be equally probable, one obtains useful information about the over-all behaviour of a complex system when the observation of the states in all its detail is impossible. This type of statistical mechanics is clearly inadequate for the discussion of nuclear energy levels. We wish to make statements about the fine detail of the level structure, and such statements cannot be made in terms of an ensemble of states. What is required is a new kind of statistical mechanics, in which we renounce exact knowledge not of the state of a system but of the nature of the system itself. We picture a complex nucleus as a "black box" in which a large number of particles are interacting according to unknown laws. The problem is then to define in a mathematically precise way an ensemble of systems in which all possible laws of interaction are equally probable".(*)

At the end of these lectures the reader will judge whether this programme, initiated by Wigner, has been successfully accomplished. And it will appear that partial justification of the theory may be found in concepts elaborated only recently.

The appropriate language to define an ensemble of systems is provided by random matrix theory. One considers the Hamiltonian matrix H as an $N \times N$ stochastic matrix (its matrix elements are random variables) and the question is to specify the probability density $\mathcal{P}(H)dH$. In order to proceed one must answer the following questions [Wi-67a,Dy-72]: i) what are the admissible Hamiltonians and what is the proper measure in the ensemble of these Hamiltonians ; ii) given the ensemble of Hamiltonians, are the properties in which we are interested common to the vast majority of them. Question i) is answered on the basis of general symmetry principles as well as of physical plausibility. Question ii) should be answered by the affirmative and one can then perform ensemble averages which are equivalent to averages over a generic matrix of the ensemble (ergodic property). A proper ensemble of random matrices should also fulfil the requirement that the mathematical problems involved should be tractable, in order to avoid situations in which one must exclusively rely on results obtained numerically (Monte Carlo). Of course, the final goal of the theory will be to successfully predict the empirical observations.

The systems we shall deal with are characterized by their Hamiltonians which can be represented by Hermitian matrices. When there are some exact quantum numbers corresponding to exact integrals of motion, like angular momentum and parity (J^{π}), and if the basis states are labelled by these exact quantum numbers, the Hamiltonian matrix will split into blocks (Fig.II.1), and the matrix elements connecting different blocks will vanish. We shall assume that such a basis has already been chosen and restrict our attention to one of the diagonal blocks, an $N \times N$ Hermitian matrix in which N is a large integer, for the systems we want to describe contain many levels. The theoretical results are

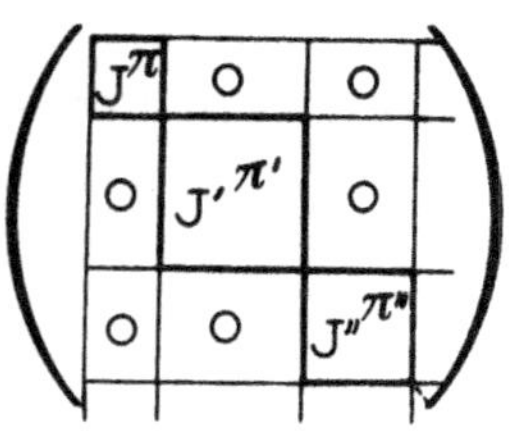

Fig.II.1

(*)The underlining is ours.

in most cases derived in the limit of large N. If this asymptotic limit is reached very fast (N $\simeq$ 100, for instance), the dimensionality will introduce no uncertainty when comparing to data.

II.1 GAUSSIAN ENSEMBLES [Po-65a,Me-67,Dy-62a,b]

The underlying space-time symmetries obeyed by the system put important restrictions on the admissible matrix ensembles. If the Hamiltonian is time-reversal invariant and invariant under rotations, the Hamiltonian matrices can be chosen real symmetric. If the Hamiltonian is not time-reversal invariant then, irrespective of its behaviour under rotations, the Hamiltonian matrices are complex Hermitian. Finally, if the system is time-reversal invariant but not invariant under rotations, and if it has half-odd-integer total angular momentum, the matrices are "quaternion real". In this last case all energy levels are doubly degenerate (Kramer's degeneracy).

Time Reversal Invariance	Angular Momentum	Rotation Invariance	Hamiltonian Matrix	Canonical Group	β	Number of Independent Real param.
Yes	Integer	-	Real Symmetric	Orthogonal	1	$\frac{1}{2}N(N+1)$
	$\frac{1}{2}$-Odd-Integer	Yes				
	$\frac{1}{2}$-Odd-Integer	No	Quaternion Real	Symplectic	4	N(2N-1)
No	-	-	Complex Hermitian	Unitary	2	N^2

Table II.1

Due to its importance in physical applications, we shall mainly concentrate in the case $\beta = 1$ (see Table II.1). Notice that the real symmetry property is preserved under orthogonal transformations, but not under a larger subgroup of the unitary transformations. The case $\beta = 4$ is included in Table II.1 for completeness, but no further mention will be made to it.

In order to introduce a proper measure dH in the space of matrices one can proceed as follows. A metric is defined in the matrix space to which H belongs by expressing

$$ds^2 = \mathrm{Tr}\ \delta H \delta H^+ \tag{II-2}$$

in terms of the independent variables x_μ of H, as

$$ds^2 = \sum_{\mu,\nu=1}^{M} g_{\mu\nu}\, \delta x_\mu\, \delta x_\nu \,, \tag{II-3}$$

where M is the number of independent variables (matrix elements). The basic measure dH is then the measure induced by this metric, namely

$$dH = (\det g_{\mu\nu})^{1/2} \prod_{\mu=1}^{M} dx_\mu \,. \tag{II-4}$$

If one considers ($\beta = 1$) real symmetric matrices $H(=H^*=H^T)$ there are $(1/2)N(N+1)$ independent variables and one has

$$ds^2 = \sum_{1\leq i\leq N} (\delta H_{ii})^2 + 2 \sum_{1\leq i<j\leq N} (\delta H_{ij})^2 \tag{II-5}$$

so that

$$dH = 2^{N(N-1)/4} \prod_{1\leq i\leq N} dH_{ii} \prod_{1\leq i<j\leq N} dH_{ij} \,. \tag{II-6}$$

For the case ($\beta = 2$) of Hermitian matrices $H(=H^+)$ one has N^2 independent variables. Writing $H = H' + iH''$ where $H' = H'^* = H'^T$, $H'' = H''^* = -H''^T$, one has

$$ds^2 = \sum_{1\leq i\leq N} (\delta H'_{ii})^2 + 2 \sum_{1\leq i<j\leq N} \left[(\delta H'_{ij})^2 + (\delta H''_{ij})^2\right] \tag{II-7}$$

so that

$$dH = 2^{N(N-1)/2} \prod_{1\leq i\leq N} dH'_{ii} \prod_{1\leq i<j\leq N} dH'_{ij}\, dH''_{ij} \,. \tag{II-8}$$

Any automorphism of the studied matrix space which leaves the metric (II-2) invariant will leave the associate measure dH invariant. For instance, in the case of real symmetric matrices, the invariance of (II-2) (II-5) under a real orthogonal transformation implies the invariance of (II-6).

Let us now define the <u>Gaussian Orthogonal Ensemble (GOE)</u> in the space of real symmetric matrices by demanding two requirements :

1. The ensemble is invariant under every orthogonal transformation

$$H' = W^T H W \quad , \tag{II-9}$$

where W is any real orthogonal matrix, i.e., the probability $\mathcal{P}(H)dH$ that a matrix H will be in the volume element dH (Eq.(II-6)) is invariant under orthogonal transformations(*)

$$\mathcal{P}(H')dH' = \mathcal{P}(H)dH \tag{II-10}$$

2. The various elements $H_{ij}(i \leqslant j)$ are independent random variables .

We make the first requirement because we don't want that any given state plays a particular role : all basis states, and therefore all states, should behave in the same way. The second requirement has no special physical origin. It is put for the sake of simplicity with the hope of leading to a mathematically soluble problem.

Similarly, the Gaussian Unitary Ensemble (GUE) in the space of Hermitian matrices is defined by the properties

1'. The ensemble is invariant under every unitary transformation

$$H' = U^+ H U \tag{II-11}$$

where U is any unitary matrix, i.e., the probability $\mathcal{P}(H)dH$ that a matrix H will be in the volume element dH (Eq.(II-8)) is invariant under unitary transformations

$$\mathcal{P}(H')dH' = \mathcal{P}(H)dH \tag{II-12}$$

2'. The various elements H'_{ij} $(i \leqslant j)$, $H''_{ij}(i < j)$ are independent random variables, i.e., $\mathcal{P}(H)$ is a product of N^2 functions.

These two requirements (1. and 2. or 1'. and 2') determine uniquely the ensembles. The function $\mathcal{P}(H)$, which will also be invariant under the corresponding automorphism, can be written

$$\mathcal{P}_{N\beta}(H) = K_{N\beta} \exp\left\{-\mathrm{Tr}(H^2)/4\sigma^2\right\} , \tag{II-13}$$

(*) Notice that from (II-10) and from the invariance of the measure dH, one has that $\mathcal{P}(H)$ must also be invariant under orthogonal transformations.

where $K_{N\beta}$ is a normalization constant (N refers to the dimensionality of the matrix and β = 1,2,4 labels the different cases, see Table II.1). In (II-13)

$$\mathrm{Tr}\, H^2 = \sum_{1\leq i\leq N} H_{ii}^2 + 2 \sum_{1\leq i<j\leq N} H_{ij}^2 \qquad \text{for } \beta=1 \tag{II-14}$$

and

$$\mathrm{Tr}\, H^2 = \sum_{1\leq i\leq N} H_{ii}'^2 + 2 \sum_{1\leq i<j\leq N} (H_{ij}'^2 + H_{ij}''^2) \quad \text{for } \beta=2\,. \tag{II-14'}$$

For the GOE one therefore has that each matrix element H_{ij} is distributed normally (or Gaussian distributed, from where the name Gaussian ensembles) with zero mean

$$\overline{H_{ij}} = 0 \qquad i\leq j \tag{II-15}$$

and variance

$$\overline{H_{ij}^2} = (1+\delta_{ij})\,\sigma^2 \qquad i\leq j\quad . \tag{II-15'}$$

We use the notation : for any quantity W, $\overline{W}$ is its ensemble average. For GUE one also has normal distribution of the different variables with

$$\overline{H_{ii}'} = \overline{H_{ii}''} = 0 \quad , \quad \overline{H_{ii}'^2} = 2\sigma^2 \qquad i=j \tag{II-16}$$

and

$$\overline{H_{ij}'} = \overline{H_{ij}''} = 0 \quad , \quad \overline{H_{ij}'^2} = \overline{H_{ij}''^2} = \sigma^2 \qquad i<j\,. \tag{II-16'}$$

We have thus defined two ensembles -GOE and GUE- depending on two parameters : a trivial scale factor σ^2 and the dimensionality N. Equation (II-13) clearly exhibits the statistical independence of the matrix elements.

Balian [Ba-68] has used a different and more general approach to derive (among others), the Gaussian Ensembles. He uses concepts borrowed from information theory, in particular the amount of information $\mathcal{I}$ associated to the probability $\mathcal{P}(H)$

$$\mathcal{I} = \int \mathcal{P}(H)\, \ln \mathcal{P}(H)\, dH\,. \tag{II-17}$$

One looks for the function $\mathcal{P}(H)$ that minimizes $\mathcal{I}$, which is equivalent to assuming the least possible knowledge about $\mathcal{P}(H)$. Before minimizing one has to face the following problem : in the cases we are studying, the range of variation of the matrix elements H_{ij} (or (H_{ij}', H_{ij}'')) is infinite. In order to confine the eigenvalues of H to a finite range, one has to impose a condition on its norm $(\mathrm{Tr}\, H^2)^{1/2}$. That is, one asks that $\mathcal{P}(H)$ should minimize $\mathcal{I}$ subject to the constraints

$$\int \mathcal{P}(H)\, dH = C_1 \qquad \text{(normalization)} \tag{II-18}$$

$$\int \mathcal{P}(H)\, \mathrm{Tr}(H^2)\, dH = C_2 \quad . \tag{II-18'}$$

This leads to the result (II-13). When following this derivation of Gaussian Ensembles, the only input one needs is the invariant measure dH (plus the constraints) and the statistical independence of the matrix elements follows as a result.

Making use of the invariance properties, the joint probability density of the eigenvalues E_i can be extracted from (II-13) and one obtains :

$$P_{N\beta}(E_1, E_2, \dots, E_N) = C_{N\beta} \exp\left\{-\frac{1}{4\sigma^2} \sum E_i^2\right\} \prod |E_i - E_j|^\beta , \tag{II-19}$$

where $C_{N\beta}$ is a normalization constant determined by

$$\int P_{N\beta}(E_1, E_2, \dots, E_N)\, dE_1\, dE_2 \cdots dE_N = 1 \quad . \tag{II-19'}$$

$P_{N\beta}(E_1, \dots, E_N)\, dE_1 \cdots dE_N$ gives the probability, regardless of labelling ($P_{N\beta}$ is a symmetric function under the interchange $E_i \longleftrightarrow E_j$), of having one eigenvalue at E_1 , another at E_2 ..., another at E_N within each of the intervals [E_j , $E_j + dE_j$]. Equation (II-19) contains all the information concerning the eigenvalue distribution and the correlations among eigenvalues. By performing suitable integrations, one can in principle derive all the quantities related to eigenvalues which are of interest. Notice the last factor on the r.h.s. of (II-19) : when $E_i = E_j$, $P_{N\beta}(E_1, \dots, E_i, \dots, E_j, \dots, E_N)$ becomes zero (there is level repulsion !). By integrating (II-19) over all the variables but one, one finds, in the limit of large N for the three cases β = 1,2,4, the ensemble averaged eigenvalue density $\bar{\rho}$ (E)

$$\bar{\rho}(E) = \int \cdots \int P_{N\beta}(E, E_2, \dots, E_N)\, dE_2 \cdots dE_N$$

$$= \begin{cases} \dfrac{1}{2\pi} \dfrac{1}{N\sigma^2} (4N\sigma^2 - E^2)^{1/2} & \text{for } |E| \leqslant 2\sqrt{N\sigma^2} \\ 0 & \text{for } |E| \geqslant 2\sqrt{N\sigma^2} \end{cases} \tag{II-20}$$

which, from (II-19'), is normalized to unity. If one takes $2(N\sigma^2)^{1/2} = 2(\overline{\mathrm{Tr}(H^2)}/N)^{1/2}$ as energy unit, $(\pi/2)\bar{\rho}$ is a semi-circle and (II-20) is the so called <u>Wigner semi-circle law</u>

for the level density of a random matrix. Although it has been derived for large N, it is followed rather closely even for low values of N ($N \simeq 20$). One should point out that it can be obtained under more general conditions than the ones fulfilled by the Gaussian Ensembles.(*) Let us finally remark that (II-20) does not reproduce the experimental level density (II-1). Whether this is a serious weakness of the theory or not will be discussed later in this Section.

II.2 FLUCTUATION PROPERTIES

Once the global behaviour of the eigenvalues has been established (the ensemble averaged eigenvalue density $\bar{\rho}(E)$(**)) we turn now our attention to the level fluctuations or departures of the spectrum from its average behaviour. For that purpose one introduces the k-level correlation functions $\tilde{R}_k(E_1, E_2, \dots, E_k)$ defined as follows

$$\tilde{R}_k(E_1, \dots, E_k) = \frac{N!}{(N-k)!} \int \cdots \int P_N(E_1, \dots, E_N)\, dE_{k+1} \cdots dE_N \quad (1 \leq k \leq N), \tag{II-21}$$

where P_N is given by (II-19). It follows that $\tilde{R}_1(E_1) = N\bar{\rho}(E_1)$, where $\bar{\rho}(E_1)$ is given by (II-20) and $\tilde{R}_N = N!\, P_N$ with P_N given by (II-19). $\tilde{R}_k(E_1, \dots, E_k) \times dE_1 \dots dE_k$ is the probability of finding one level, regardless of labelling, within each of the intervals $[E_j, E_j + dE_j]$. Like in the previous Section, to get rid of $\bar{\rho}(E)$, one introduces a new set of k-level correlation functions $R_k(x_1, \dots, x_k)$ and one considers the case of interest $N \gg k \geq 1$

$$R_k(x_1, \dots, x_k) = \lim_{N \to \infty} \frac{\tilde{R}_k(E_1, \dots, E_k)}{\tilde{R}_1(E_1) \dots \tilde{R}_1(E_k)} \tag{II-22}$$

where x is obtained from the mapping $E \mapsto x$ defined by (see (I-28))

$$E \mapsto x = \int_{-\infty}^{E} \tilde{R}_1(E')\, dE' . \tag{II-22'}$$

The fluctuation properties of the levels are completely characterized by the set of functions $R_k(x_1, \dots, x_k)$. From the definition (II-22), one has $R_1(x_1) = 1$. This expresses the fact that the level density (or the average spacing) in the variable x

---

(*) For real symmetric matrices, for instance, one only needs the statistical independence of the matrix elements plus (II-15,15') for almost all the matrix elements.

(**) $N\,\bar{\rho}(E)$ plays in the context of matrix ensembles the role of $\rho_{av}(\xi) = \frac{d}{d\xi} N_{av}(\xi)$ (see (I-26,28) in the previous Section).

is unity everywhere.

It is convenient to introduce the k-level cluster functions Y_k which are obtained from R_k by subtracting out the lower-order correlation terms

$$Y_k(x_1,\dots,x_k) = \sum_G (-)^{k-m}(m-1)!\prod_{j=1}^{m} R_{G_j}(x_t, \text{ with } t \text{ in } G_j) . \qquad \text{(II-23)}$$

Here G stands for any division of the indices [1,2,...,k] into subgroups [$G_1, G_2, \dots, G_m$]. For instance

— $k=1$ $\quad m=1 \quad$ [(1)] and

$$Y_1(x_1) = R_1(x_1) = 1 \qquad \text{(II-24)}$$

— $k=2$ $\quad m=1 \quad$ [(1,2)]

$m=2 \quad$ [(1)(2)] and

$$Y_2(x_1,x_2) = -R_2(x_1,x_2) + R_1(x_1)R_1(x_2) \qquad \text{(II-24')}$$

— $k=3$ $\quad m=1 \quad$ [(1,2,3)]

$m=2 \quad$ [(1)(2,3)], [(1,3)(2)], [(1,2)(3)]

$m=3 \quad$ [(1)(2)(3)] and

$$\begin{aligned} Y_3(x_1,x_2,x_3) = {} & R_3(x_1,x_2,x_3) \\ & -[R_1(x_1)R_2(x_2,x_3)+R_1(x_2)R_2(x_1,x_3)+R_1(x_3)R_2(x_1,x_2)] \\ & +2\,R_1(x_1)R_1(x_2)R_1(x_3) . \end{aligned} \qquad \text{(II-24'')}$$

The inverse of (II-23) is

$$R_k(x_1,\dots,x_k) = \sum_G (-)^{k-m}\prod_{j=1}^{m} Y_{G_j}(x_t, \text{ with } t \text{ in } G_j) . \qquad \text{(II-25)}$$

Thus each set of functions R_k and Y_k is easily determined in terms of the other. The advantage of the cluster functions is that they have the property of vanishing when any one (or several) of the separations $|x_i - x_j|$ becomes large. The function Y_k describes the correlation properties of a single cluster of k levels, isolated from more trivial effects of lower-order correlations.

It has been shown by Pandey [Pa-79] that the Gaussian Ensembles are stationary and ergodic in the limit of large N (this is obvious by construction for a Poisson ensemble). For instance, for one-point measures, the ensemble density $\bar{\rho}(E)$ is equal to the spectral averaged density $\langle \rho(E) \rangle_S$. The mapping $E \mapsto x$ (Eq. (II-22')) (unfolding) is such that the ensemble-averaged local spacing $1/\bar{\rho}(x)$ is stationary (independent of x) and equal to unity. And the k-level cluster functions Y_k, functions of the variables x_j, are also stationary (they only depend on the relative coordinates $x_{ij} = x_i - x_j$). This means that the fluctuation properties of several segments of a spectrum located at different positions will be the same: from the point of view of fluctuations, the spectrum is translationally invariant. Furthermore, after unfolding, a spectral average is equal to an ensemble average.

An alternative way to characterize fluctuations consists to deal with spacing distributions(*) and related quantities. In (II-21), instead of integrating from $-\infty$ to $+\infty$ without any restriction, one integrates some of the variables outside the interval $[\alpha, \alpha + L]$ whereas the others are integrated inside it. Assume that the unfolding (mapping $E \mapsto x$) has been performed. One defines ($N \gg k \gg 0$)

$$E(k;L) = \lim_{N\to\infty} \frac{N!}{(N-k)!} \int_{in} \cdots \int dx_1 \cdots dx_k \int_{out} \cdots \int dx_{k+1} \cdots dx_N \, P_N(x_1,\dots,x_N) \,. \tag{II-26}$$

If the system is stationary, $E(k;L)$ will be independent of α. $E(k;L)$ is the probability that in a sequence $\{x_i\}$ of levels with mean spacing unity an interval of length L taken at random contains exactly k levels. One useful aspect of the functions $E(k;L)$ is that they are very directly connected to the spacing distributions [MdC-72]

$$p(k;L) = (d^2/dL^2) \sum_{j=0}^{k} (k-j+1)\, E(j;L) \quad . \tag{II-27}$$

In particular, for the nearest-neighbour spacing distribution $p(x)$

$$p(x) \equiv p(0;x) = (d^2/dx^2)\, E(0,x) \,. \tag{II-27'}$$

The probability $F(k;L)$ that in a sequence $\{x_i\}$ of levels with mean spacing unity, an interval $[x_\alpha, x_\alpha + L]$ of length L which starts at a level x_α contains exactly k levels is also given in terms of the functions $E(k;L)$:

(*)The distribution of nearest-neighbour spacings has been denoted, and will continue to be when no confusion is possible, by $p(x) \equiv p(0;x)$; $p(k;x)$ denotes the distribution of spacings $s = x_{i+k+1} - x_i$ between two levels x_i and x_{i+k+1} having k levels in-between ($k = 0,1,2,\dots$).

$$F(k;L) = -(d/dL)\sum_{j=0}^{k} E(j;L) \tag{II-28}$$

and, in particular

$$F(0;L) = -(d/dL)E(0;L) \; . \tag{II-28'}$$

From their meaning it is clear that

$$\sum_{k=0}^{\infty} E(k;L) = \sum_{k=0}^{\infty} F(k;L) = 1 \tag{II-29}$$

and if there are no degeneracies

$$E(k;0) = F(k;0) = \delta_{k0} \; . \tag{II-30}$$

When comparing a theoretical model with experiment, due to the limited number of high quality data available, it is impossible to make a complete comparison between the set of k-level cluster functions Y_k (II-23) and the corresponding quantities extracted from experiment. It is convenient to work with integrated quantities like

$$y_k(L) = \int_0^L \cdots \int_0^L Y_k(x_1, \ldots, x_k)\, dx_1 \ldots dx_k \; . \tag{II-31}$$

Consider, for instance, the random variable n(L) "number of levels contained in an interval of length L" or, in short, the number statistic (see end of previous Section). Its first two moments and shape parameters(*) are given by [Pa-78]

(*)Let ξ be a random variable whose probability density is $f(x)$. Its expectation value m_1 and central moments M_k $(k \geq 2)$ are defined by $m_1 = \int f(x)\, dx$ and $M_k = \int f(x)(x-m_1)^k\, dx$. The characteristic function φ (t) is the expectation value of the random variables $\exp(it\xi)$: $\varphi(t) = \int f(x)\, e^{itx}\, dx$ . . Consider the function $\psi(t) = \ln \varphi(t)$ called the cumulant generating function, because its Mac Laurin expansion is given by $\psi(t) = \sum_{k=1}^{\infty} (K_k/k!)(it)^k$, where K_k are the cumulants. They are related in a simple way to the centered moments M_k of ξ : $K_1 = m_1$, $K_2 = M_2$ (variance), $K_3 = M_3$, $K_4 = M_4 - 3M_2^2, \ldots$ The shape parameters γ_k $(k \geq 1)$ are defined by $\gamma_k = K_{k+2}/(K_2^{(k+2)/2})$.
γ_1 is called the excess and γ_2 the asymmetry of $f(x)$. For a Gaussian distribution $f(x) = (1/\sqrt{2\pi M_2}) \exp\{-(x-m_1)^2/2M_2\}$ one has $K_1 = m_1$, $K_2 = M_2$ and $K_\nu = 0$ for $\nu \geq 3$, i.e., $\gamma_k = 0$.

$$m_1(L) = \bar{n}(L) = \sum_{k=0}^{\infty} k\,E(k;L) = y_1(L) = L \tag{II-32}$$

$$\Sigma^2(L) = M_2(L) = \overline{(n(L)-L)^2} = \sum_{k=0}^{\infty} (k-L)^2 E(k,L) = y_1(L) - y_2(L) \tag{II-32'}$$

$$\gamma_1(L) = \left\{\sum_{k=0}^{\infty} (k-L)^3 E(k;L)\right\} / M_2^{3/2}(L) = (y_1(L) - 3y_2(L) + y_3(L)) / M_2^{3/2}(L) \tag{II-32''}$$

$$\gamma_2(L) = \left\{\sum_{k=0}^{\infty} (k-L)^4 E(k;L) - 3M_2^2(L)\right\} / M_2^2(L) \tag{II-32'''}$$
$$= (y_1(L) - 7y_2(L) + 6y_3(L) - y_4(L)) / M_2^2(L) .$$

Thus, Σ^2, γ_1 and γ_2 are given in terms of the functions $E(k;L)$ and all values of k appear. However, they are in fact (2)-, (2+3)- and (2+3+4)- point measures respectively, as can be seen from the last equalities in (II-32). When dealing with the 2-level cluster function $Y_2(x_1, x_2)$ one uses the notation $Y_2(x)$ (*). Y_2 is related to the spacing distributions by

$$1 - Y_2(x) = \sum_{k=0}^{\infty} p(k;x) \tag{II-33}$$

and $(1 - Y_2(x))dx$ gives the probability of observing a level in an infinitesimal interval dx at a distance x from a given level. An alternative form of (II-32') is

$$\Sigma^2(L) = L - \int_0^L (L-r)\,Y_2(r)\,dr \quad . \tag{II-34}$$

Finally, consider the least-square statistic $\Delta_3(L)$ introduced in the previous Section (I-38). It can be shown that its ensemble average $\bar{\Delta}_3(L)$ can be obtained as follows [Pa-79]

$$\bar{\Delta}_3(L) = (2/L^4) \int_0^L (L^3 - 2L^2 r + r^3)\,\Sigma^2(r)\,dr \quad , \tag{II-35}$$

Therefore, like $\Sigma^2(L)$, it is also a 2-point measure (some particular integral of Y_2).

(*) Take as coordinates the center of the interval $X = (x_1 + x_2)/2$ and the relative coordinate $x = x_1 - x_2$. Then $Y_2(x_1, x_2) = Y_2(X, x)$ but as Y_2 does not depend on X, one simply writes $Y_2(x)$.

Its variance

$$Var\ \Delta_3(L) = \overline{(\Delta_3(L) - \bar{\Delta}_3(L))^2} \qquad \text{(II-36)}$$

is a (2+3+4)- point measure.

Let us now briefly describe the results obtained from different theoretical models. We shall mainly consider the GOE and also, for the sake of comparison and because it constitutes the limiting case of maximum randomness, the Poisson ensemble (an ensemble of sequences of points, not of eigenvalues of matrices).

Poisson Spectrum

- Correlation coefficient between adjacent spacings (I-37)

$$C = 0$$

- Functions E,F and spacing distributions (II-26,27,28)

$$E(k;L) = F(k;L) = p(k;L) = (L^k/k!)\,e^{-L} \qquad \text{(II-37)}$$

- k-level correlation functions (II-22)

$$R_k(x_1, \dots, x_k) = 1 \qquad k \geqslant 1$$

- k-level cluster functions (II-23)

$$Y_1(x) = 1, \quad Y_k(x_1, \dots, x_k) = 0 \qquad k \geqslant 2$$

and their integrals (II-31)

$$y_1(L) = L, \qquad y_k(L) = 0 \qquad k \geqslant 2$$

- Cumulants $K_\nu(L)$

$$K_\nu(L) = L$$

In particular (II-34)

$$\Sigma^2(L) = L$$

- Shape parameters (II-32)

$$\gamma_k(L) = L^{-k/2} \qquad k \geqslant 1,$$

which means that for large L $\gamma_k \to 0$, i.e., n(L) tends to be normally distributed.

- Average value of Δ_3(L) (II-35) [DM-63]

$$\bar{\Delta}_3(L) = L/15 \qquad \text{(II-38)}$$

and its variance (II-36) [BHP-83]

$$\mathrm{Var}\,\Delta_3(L) = \left(1+\frac{11}{30}L\right)\frac{L}{210} \tag{II-39}$$

Gaussian Orthogonal Ensemble

The derivation of the different quantities of interest, mainly due to Mehta, Gaudin and Dyson, needs beautiful classical mathematical techniques. The principal initial difficulty was encountered when trying to perform integrals of the type (II-21), difficulty that was solved by the method of integration over alternate variables invented by Mehta [Me-60]. One knows now, in closed analytical form, the different cluster functions. The functions related to the spacing distributions are only known (except for their behaviour for small and large values of the arguments) in numerical form and are connected to the spheroidal functions. For their relationship to the k-level correlation and cluster functions, see Ref.[MP-83].

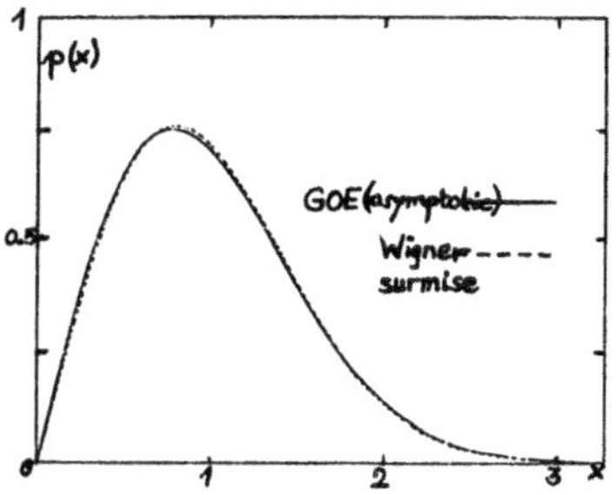

Fig.II.2 - The nearest neighbour-spacing distribution (taken from [Ga-61]).

Let us now give a brief account of the main results. On Fig.II.2 is reproduced the nearest-neighbour spacing distribution $p(x)$ (large N limit) and compared to the Wigner surmise $p_W(x)$ (I-36) derived in the previous Section. They are extremely close to each other. It is worth mentioning that $p_W(x)$ is the result corresponding to a GOE of $N \times N$ matrices of dimension N=2. The slope at the origin of $p(x)$ is $\pi^2/6$ to be compared to $\pi/2$ for $p_W(x)$. The two values differ by a factor of $\pi/3$, which is close to 1.

On Fig.II.3 are reproduced the functions E(0;L) and F(0;L) (II-26,28'). Notice that $F(0;L)-E(0;L) > 0$ for all $L > 0$. This means that the probability that in an interval of length L which starts at a level there is no level is larger than the probability that in an interval of the same length taken at random there is no level. This is actually what level repulsion means. This less conventional definition of level repulsion is sensible

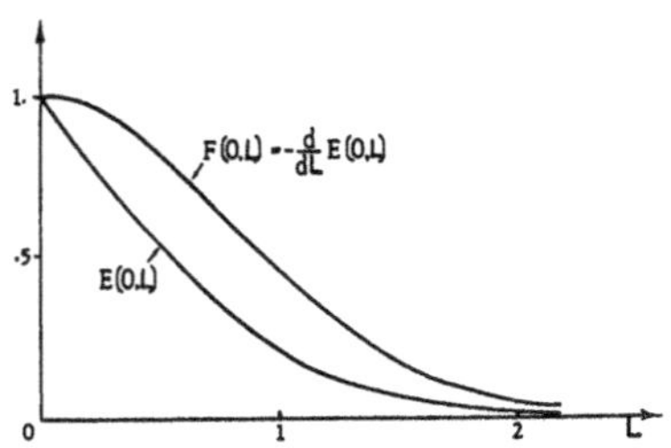

Fig.II.3 - The functions E(0;L) and F(0;L) for GOE

not only for small values of L, but for any value of L. For a Poisson spectrum one has F(0;L)=E(0;L) : it is irrelevant whether the interval L starts or not at a level.

On Fig.II.4 are reproduced the functions E(k;L) in the interval $0 \lesssim L \lesssim 5$. They can be used, for instance, to compute via (II-32) the values of γ_1 (L) and γ_2 (L) to be discussed later (this is a more practical method than to use the general expressions of k-level cluster functions). For adjacent spacings, the correlation coefficient (I-37) is C = -0.271, to be compared to 0 for a Poisson spectrum.

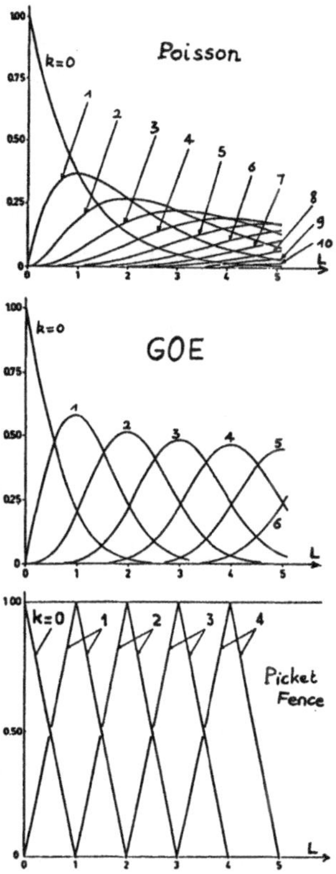

Fig.II.4 - The functions E(k;L) (II-26) of a Poisson spectrum, a GOE and a picket fence (taken from [MdC-72])

A closed expression has been given for the k-level cluster functions Y_k [Me-71,Dy-70]

$$Y_1(x) = 1$$

$$Y_k(x_1,\ldots,x_k) = \frac{1}{2}\mathrm{Tr}\sum_P\left[\sigma(x_{12})\sigma(x_{23})\cdots\sigma(x_{k1})\right] \qquad k \geqslant 2 \tag{II-40}$$

where $\sum_P$ denotes a sum over the (k-1)! distinct cyclic permutations of the indices (1,2,..., k), where $x_{ij} = x_i - x_j$, and where σ is a 2-dimensional matrix given by

$$\sigma(x) = \begin{pmatrix} s(x) & Ds(x) \\ Js(x) & s(x) \end{pmatrix} \tag{II-41}$$

In (II-41) $s(x)$, $Ds(x)$ and $Js(x)$ are given by

$$s(x) = \sin \pi x / \pi x \tag{II-42}$$

$$Ds(x) = (d/dx)\, s(x) \tag{II-42'}$$

$$Js(x) = Is(x) - \epsilon(x), \text{ where } \begin{cases} Is(x) = \int_0^{x'} s(x')\,dx' \\ \epsilon(x) = \begin{cases} 1/2 & x>0 \\ 0 & x=0 \\ -1/2 & x<0 \end{cases} \end{cases} \tag{II-42''}$$

As a particular case one has

$$Y_2(r) = (s(r))^2 - Ds(r)\,Js(r) \qquad \text{(II-43)}$$

with $r = x_1 - x_2$. For small and large values of r one has

$$Y_2(r) \underset{r\to 0}{\longrightarrow} 1 - (1/6)\pi^2 r + (1/60)\pi^4 r^3 + \cdots \qquad \text{(II-44)}$$

$$Y_2(r) \underset{r\to\infty}{\longrightarrow} 1/\pi^2 r^2 - (1+\cos^2\pi r)/\pi^4 r^4 + \cdots \qquad \text{(II-45)}$$

The behaviour of $Y_2(r)$ for small values of r ($Y_2(0)=1$) is responsible for the level repulsion, whereas the large-r behaviour ($Y_2(r)$ tends to zero as $1/\pi^2 r^2$) determines the presence of long range order. On Fig.II.5 is displayed the function $1-Y_2(r)$ in the range $0 \leqslant r \leqslant 2$.

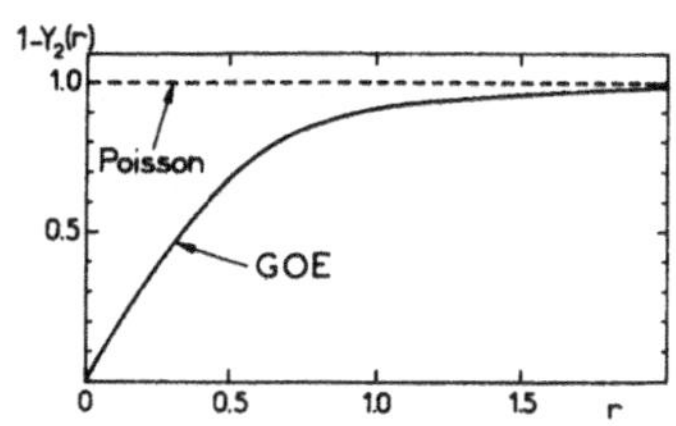

Fig.II.5 - Two-level cluster function for GOE

The value of $\Sigma^2(L)$ can be obtained by performing the integral in (II-34) with Y_2 given by (II-43). One obtains [DM-63,BFF-81]

$$\begin{aligned}\Sigma^2(L) = (2/\pi^2)\Big\{ &\ln(2\pi L) + \gamma + 1 + \tfrac{1}{2}[Si(\pi L)]^2 \\ &- \tfrac{\pi}{2}\,Si(\pi L) - \cos(2\pi L) - Ci(2\pi L) \\ &+ \pi^2 L\,[1-(2/\pi)\,Si(2\pi L)]\Big\}\end{aligned} \qquad \text{(II-46)}$$

$$\begin{aligned}&\underset{L\gtrsim 1}{\longrightarrow} (2/\pi^2)\{\ln(2\pi L) + \gamma + 1 - \pi^2/8\} \\ &= \frac{2}{\pi^2}\ln L + 0.44\end{aligned} \qquad \text{(II-46')}$$

In (II-46) γ is the Euler constant and Si and Ci are the sine and cosine integrals. Note the effect of the spectral rigidity : the value of $\Sigma^2(L)$ is only of the order of unity for L=100 and even for $L \sim 10^6$ the fluctuation is not more than a couple of levels. This should be compared to $\Sigma^2(L) = L$ for the Poisson case.

The ensemble average $\bar{\Delta}_3(L)$ of $\Delta_3(L)$ can be obtained by integrating numerically (II-35) [HPB-82]. For $L \gtrsim 15$ one may use the asymptotic (large L) result [DM-63]

$$\bar{\Delta}_3(L) \sim \frac{1}{\pi^2}\left[\ln(2\pi L) + \gamma - \frac{\pi^2}{8} - \frac{5}{4}\right] = \frac{1}{\pi^2}\ln L - 0.007 \qquad \text{(II-47)}$$

with very good accuracy. Again, in comparison with the Poisson value L/15, the GOE spectrum is seen to exhibit long range order.

The variance of $\Delta_3(L)$ has been obtained by Monte Carlo calculations [HPB-82]. For large L, it approaches the asymptotic (large L) value [DM-63]

$$\mathrm{Var}\,\Delta_3(L) \sim \frac{1}{\pi^4}\left[\frac{4\pi^2}{45}+\frac{7}{24}\right] = 0.012\,, \qquad \text{(II-48)}$$

a small constant ! By comparing to the Poisson result (II-39), once again it is seen that the fluctuations are small for GOE.

Although in what follows we shall be interested in the GOE, let us briefly mention that the fluctuation properties of the eigenvalues of Hermitian matrices of the Gaussian Unitary Ensemble (GUE, $\beta = 2$) are also known(*). The level repulsion is stronger for GUE than for GOE : in both cases the spacing distribution $p(x)$ vanishes at the origin, but near the origin $p(x) \sim (\pi^2/6)x$ for GOE whereas for GUE it is $p(x) \sim (\pi^2/3)x^2$. The expression of the k-level cluster functions reads as in (II-40) except that the factor (1/2)Tr should be dropped and $\sigma(x)$ is given by $s(x)$ (II-42). The resulting two-level cluster function $Y_2(r)$ contains only the first term of the r.h.s. of (II-43). The variance Σ^2 (L) of n(L) is, for L $\gtrsim$ 1, half the variance corresponding to the GOE plus 1/8, i.e., a GUE spectrum is more rigid than a GOE spectrum.

We have mentioned in the previous Section the conjecture that the fluctuations of the imaginary part of the zeros of the Riemann Zeta function are asymptotically (large imaginary part) identical to the fluctuations of a GUE spectrum. Indeed, one can see that the spectrum displayed in column (e) of Fig.I.8 shows a stronger level repulsion than the spectra of columns (c) and (d) which, as will be explained in the rest of this Section and in Section IV, are well described by the GOE.

II.3 COMPARISON OF GOE PREDICTIONS WITH EXPERIMENT

We shall concentrate on fluctuation properties of spectra and shall leave completely untouched the important subject of statistical properties of eigenvectors and its consequences (strength and width fluctuations) [BFF-81, MW-79, We-84]. In what follows we summarize the most significant comparisons performed so far between GOE predictions and nuclear, atomic and molecular spectra.

A large experimental effort has been and is still currently devoted to the measurements of positions and widths of resonances of the compound nucleus, effort motivated to a large extent by technological reasons. However, only a very limited part of the existing data, on neutron cross sections for instance, can be used for our purpose. What we want here is not a set of resonance energies but rather a

(*) And also of the eigenvalues of matrices belonging to the Symplectic Ensemble (see Table II.1, $\beta = 4$). The statistical properties of these eigenvalues are identical to those of an alternate series from the GOE.

complete and pure sequence of resonance energies corresponding to states having the same quantum numbers J^{π} (see Fig.II.1), in a given energy range. The very characteristic features of GOE fluctuations are rapidly lost if the sequence of levels is incomplete (missing levels) and/or polluted by spurious levels due to erroneous spin-parity assignments. Obviously, in order to test long range order effects, the pure series should contain many levels.

In the atomic and molecular cases nothing comparable to the nuclear case has been achieved, although recent and promising progress is to be noticed.

Nuclei

The main source of experimental data is provided by neutron resonance spectroscopy [Ly-68]. When transmission measurements are performed by sending neutrons on a nuclear target A, the study of the cross section σ_T as a function of the neutron energy E_n reveals the presence of sharp resonances, all over the periodic table, resonances that correspond to quasi-bound states of the compound nucleus (A+1) (see Fig.II.6 for an example). The mean spacings D and average widths Γ of the resonances near neutron threshold (excitation energy of the compound nucleus $\sim$ 6-7 MeV) vary widely over the full range of nuclei. Very roughly D $\sim$ 100 keV and $\Gamma \sim$ 10 keV for light nuclei, D $\sim$ 1 keV and $\Gamma \sim$ 100 eV for medium nuclei and D $\sim$ 10 eV and $\Gamma \sim$ 1 eV for heavy nuclei. Most of the resonances observed in low-energy neutron cross sections are excited by s-waves, because for all but s-wave neutrons the penetration factor, which depends on the centrifugal barrier to be overcome by the entering neutron, is very small at low energies. Thus, if the target has $J^{\pi} = 0^+$ and E_n is very low, one populates only $(1/2)^+$ states of the compound nucleus, i.e., one produces pure sequences. The energy range that can be explored is severely limited for two reasons : i) by increasing E_n, the p-wave contribution increases and $(1/2)^-$ and $(3/2)^-$ states are also populated, the series becoming then polluted, ii) the experimental resolution being proportional to $E_n^{3/2}$, it rapidly becomes coarser than the resonance spacing. In practice this method has been extensively used, mainly by the group of Columbia University, to produce presumably pure series for a large variety of nuclei, each series containing typically of the order of 50-100 levels [HWR-78].

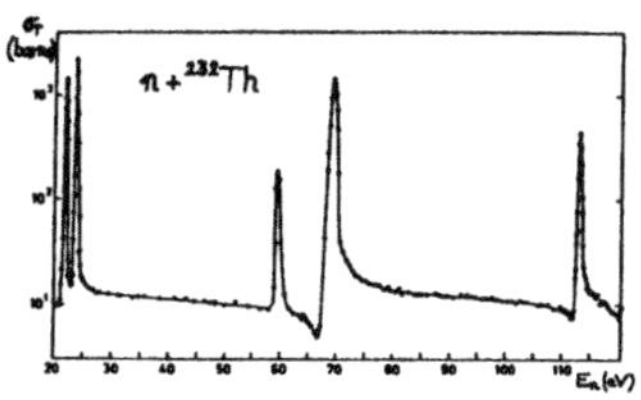

Fig.II.6 - Resonances for the reaction n+ ^{232}Th (taken from the compilation Neutron cross sections (1964))

High resolution proton scattering experiments on medium nuclei ($A \lesssim 65$) performed in Duke University provide additional information. In this case one uses the analogue state as an amplifying device : the fine structure states can be seen through the Coulomb force mixing of the analogue with the background sea of $T_<$ states. The experiments are performed with proton energies of the order of a few MeV and the energy resolution attained is of the order of 10^{-4} [BLM-76].

We shall not give a complete account of comparisons between GOE predictions and nuclear data (see [BFF-81,HPB-82,BHP-83,BHP-84]) but rather present a few typical examples. The combined set of nuclear resonance-energy data of different nuclei- in short, the nuclear data ensemble (NDE)- is treated as a sampling of eigenvalues of GOE matrices. The data analyzed consist of 1762 resonance energies corresponding to 36 sequences of 32 different nuclei. In Fig.II.7 are shown the spacing distribution $p(x)$ and the average value of Δ_3 as a function of L in the range $L < 25$. In Fig.II.8 are shown the variance Σ^2, asymmetry γ_1 and excess γ_2 of the distribution

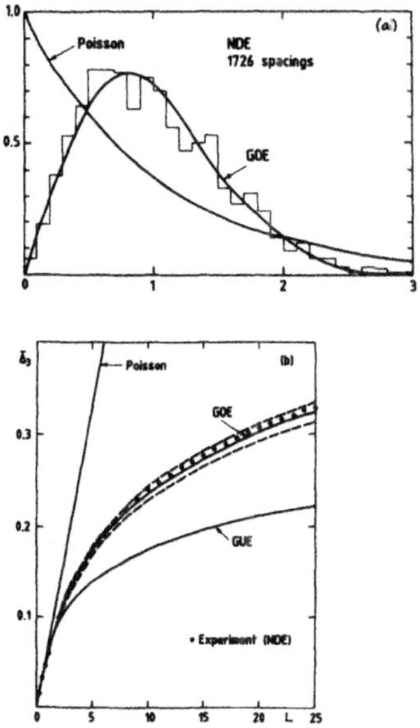

Fig.II.7 (a) Nearest-neighbour spacing histogram for NDE (experiment) ; (b) $\bar{\Delta}_3$ as a function of L ; dashed lines for GOE, to take into account finite sample size effects, correspond to one standard deviation from the average (taken from [HPB-82,BHP-83])

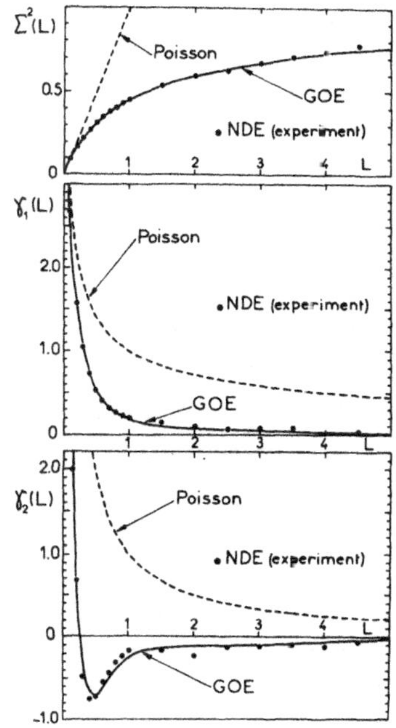

Fig.II.8 - (a) Σ^2, (b) γ_1, (c) γ_2 as functions of L (taken from [BHP-84])

of the number statistic n(L) in the range $L < 5$. The procedure for calculations is to evaluate for each of the 36 sequences the spectral-averaged measure, say $\langle \Delta_3(L) \rangle_S$ for $\bar{\Delta}_3(L)$, and then take their average, weighted according to the size of each sequence.

Remember that Σ^2 and $\bar{\Delta}_3$ are derived from the 2-level cluster function (see (II-34,35)), whereas $\gamma_1(\gamma_2)$ derives from (2+3)-((2+3+4)-) level cluster functions respectively (see (II-32',32")). However, the restrictions imposed by $y_2(L)$ on higher-order functions $y_k(L)$ are not well understood and therefore it is not easy to know how much information not contained already in y_2 is contained in γ_1 and γ_2.

Inspection of Figs.II.7 and II.8 illustrates the fact that all the fluctuation-measures considered so far, which include a thorough study of 2-point measures and to some extent more than 2-point measures as well, are fully consistent with the GOE predictions.

Atoms and Molecules

There exists a pioneering work by Rosenzweig and Porter [RP-60] in which atomic spectra were studied. They demonstrated that levels having the same quantum numbers do show level repulsion and that the spacing distribution $p(x)$ follows closely Wigner's prediction. It is only recently that Camarda and Georgopulos [CG-83] have tested more systematically energy-level-fluctuation predictions of GOE with atomic spectra. These authors have analyzed energy levels of neutral and ionized atoms in the rare-earth region. In contrast with the nuclear levels studied so far, which correspond to unbound or quasi-bound states lying above particle threshold, the atomic states are bound and can decay only by photon emission. Eight different stretches of spectra containing altogether 269 atomic energy levels have been studied. The results are reproduced in Table II.2 and Fig.II.9. One can see that the agreement

	Nd	Nd	Nd^+	Nd^+	Nd^+	Sm^+	Sm^+	Tb
J^π	4^-	6^-	$7/2^-$	$13/2^-$	$15/2^-$	$3/2^-$	$9/2^-$	$9/2^-$
L	35	38	34	28	32	26	31	45
$\bar{\Delta}_3(L)$ exp.	0.39	0.45	0.30	0.37	0.39	0.37	0.40	0.31
$\bar{\Delta}_3(L)$ GOE	0.35 ±0.11	0.36 ± 0.11	0.35 ±0.11	0.33 ±0.11	0.34 ± 0.11	0.32 ±0.11	0.34 ±0.11	0.38 ±0.11

Table II.2 - Values of Δ_3 for atomic energy levels. Each series is identified by the angular momentum and parity (J^π). For GOE the value of $\bar{\Delta}_3$ is followed by the square root of Var Δ_3 (II-48) (taken from [CG-83]).

between GOE predictions and experiment is good.

Finally let us mention very recent studies on molecular spectra. Haller et al. [HKC-83] have studied the statistical behaviour of molecular vibronic energy

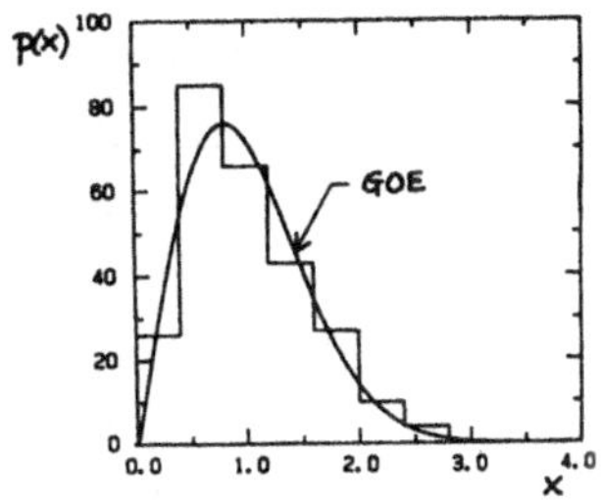

Fig.II.9- Histogram of nearest-neighbour spacing distribution $p(x)$ of eight sets of atomic energy levels containing 261 spacings (taken from [CG-83]).

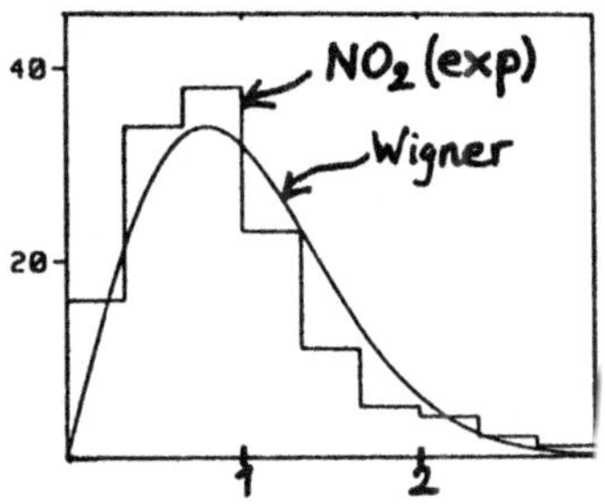

Fig.II.10 - Nearest-neighbour spacing histogram for energy levels of NO_2 (taken from [HKC-83])

levels. They have analyzed the interval 14900-19500 cm^{-1} above the ground state of the spectrum of the small polyatomic molecule NO_2. On Fig.II.10 is reproduced the spacing distribution $p(x)$ corresponding to the 140 measured levels in this energy range. It is qualitatively in agreement with the Wigner prediction. However, whereas 140 levels are found, calculations predict 201 levels of a given vibronic symmetry. In ref. [HKC-83] it is suggested that there are deficiencies in the data and that the sub-interval 16600-17300 cm^{-1}, which contains 46 levels, is most reliable. For this interval the value of Δ_3 computed from the measured spectrum is 0.38 to be compared with the GOE prediction (II-47,48) 0.38 $\pm$ 0.11 and the value of the correlation coefficient (I-37) C(exp) is -0.32 to be compared to the GOE prediction -0.27 $\pm$ 0.15.

We mention also the work of Mukamel et al. [MSP-84] who have studied level fluctuations from recent experimental data on highly vibrationally excited acetylene. The number of lines in the experiment is 65. However, due to insufficient spectral resolution, among other effects, one cannot attempt a sharp comparison between GOE fluctuations and data. The theory can, with the present quality of the data, be used to estimate the fraction of missing levels in the experiment. But one can expect that this type of analysis will become more useful when the spectral resolution will improve.

II.4 DISCUSSION

The conclusion that can be drawn from the previous subsection is that, when systematic and accurate data are available allowing a stringent comparison with the theory, the GOE describes level fluctuations remarkably well (nuclei). For atoms the agreement is significant and for molecules the comparison is still at a primitive stage. On the other hand, we know that GOE gives for the level density a semi-circle

(II-20) in contrast with the observed increase of the level density with excitation energy (II-1). In other words, local predictions (fluctuations) are in agreement with observation whereas global ones (average or smooth behaviour, see eq.(I-26)) are not. Is this a serious drawback of the theory ? The following statement provides a guideline : "In general, events distinguished by a great disparity in size have little influence on one another, they do not communicate and so phenomena associated with each scale can be treated independently" [Wi-79]. In our context, we have a density/fluctuation separation (I-26) ([BFF-81,FK-82,Dy-72]) and GOE should be considered as a model for fluctuations(*). Remember that all quantities (Eqs.(II-22,26)) used to describe fluctuations have been rescaled through the mapping $E \mapsto x$ (Eq.(II-22')), that we have been considering the behaviour of k points in the case $N \gg k \geq 1$, and that GOE-fluctuations have been proved to be stationary (translational invariant). We should also mention that methods have been proposed to construct ensembles of matrices with a prescribed eigenvalue density $\bar{\rho}$ (E) and the expectation is that local properties are independent of $\bar{\rho}$ (E) [Ba-68,Dy-72].

The question now is : are GOE-fluctuations specific of GOE or, on the contrary, are they really shared by other random matrix ensembles ? To answer this question, let us briefly describe other ensembles of random matrices and first the Circular Ensembles introduced by Dyson [Dy-62a]. Dyson wanted to avoid the following unsatisfactory feature of the Gaussian Ensembles (see (II-13)) : a uniform probability distribution on an infinite range being impossible to define, some arbitrary restriction of the magnitudes of the matrix elements is inevitable (II-18'). But then all interactions in GOE are not equally probable. By a small formal change, he introduced the following idea. Instead of thinking in terms of the eigenvalues E_i of the Hamiltonian matrix H, think in terms of eigenvalues $e^{i\theta_i}$ ($0 \leq \theta_i \leq 2\pi$) of a unitary matrix S connected to H, say, by

$$S = \exp(i\tau H) \quad . \tag{II-49}$$

The interest of this approach is that it leads to ensembles of unitary matrices S for which the probability is uniform. For instance, using Balian's information theory approach, one proceeds as follows : Minimize (II-17) in the space of symmetric unitary matrices with the constraint (II-18). One then obtains that $\mathcal{P}(S)$ is a constant. And instead of (II-19) one has(**)

$$P_{N\beta}(\theta_1,\dots,\theta_N) = C_{N\beta} \prod_{1 \leq i < j \leq N} |e^{i\theta_i - i\theta_j}|^\beta \tag{II-50}$$

The ensemble averaged density of points θ_i on the unit circle resulting from (II-50) is constant.

(*)In a different context, is Eq.(I-13) not contradicting the density/fluctuation separation ? In fact one needs of the order of x terms in (I-13) to "see" primes which are of the order of x.

(**)For $\beta = 2$, the ensemble is defined in the space of unitary matrices ; for $\beta = 4$, in the space of selfdual unitary quaternion matrices.

The fluctuation properties of the Circular Ensembles are stationary by construction and it has been shown that, as expected, in the limit of large N they give the same fluctuation properties as the corresponding Gaussian Ensembles (orthogonal, unitary and symplectic) [Dy-70,Me-71].

In the preceding, Eq.(II-49) was suggested as one possible device to define a relation between S and H. In fact the precise connection between S and H needs not be specified. Assume only that S is a function of H so that the angles θ_j are functions of E_j . Over a small range of angles, the relation between θ_j and E_j will be approximately linear and the basic statistical hypothesis is that the fluctuation properties of k consecutive levels of H will be the same as the ones of k consecutive angles on the unit circle.

There are other matrix ensembles having GOE-fluctuations. For instance the "random-sign ensemble", which is an ensemble of real symmetric matrices with matrix elements generated by coin tossing ($H_{ij} = \pm 1$), whose eigenvalue distribution is the semi-circle level density. Or Dyson's ensembles $\tilde{H}$ [Dy-71, BFF-81]

$$\tilde{H} = \sum_{i=1}^{M} A_i^T A_i \ , \tag{II-51}$$

where the A_i are independent real asymmetric random matrices whose matrix elements are normally and independently distributed with zero mean and the same variance. A particular case of H is an ensemble proposed by Wigner [Wi-72]

$$H = \mathrm{Re}\, M^+ M \ , \tag{II-52}$$

where M is an asymmetric complex matrix, the real and imaginary parts of the matrix elements being sampled normally and independently with zero mean and the same variance.

Shell-model nuclear spectra, calculated by diagonalizing a realistic (non-random) Hamiltonian in some finite subspace generated by putting n nucleons in Ω single-particle orbits, do also show GOE fluctuations [BFF-81]. In this case the average eigenvalue density is not a semi-circle but a Gaussian distribution. After adequately unfolding the spectrum, one can study the fluctuations and two examples of spacing distributions are given on Fig.II.11. In one case (Fig.II.11a) only levels having the same angular momentum and parity have been included and $p(x)$ is in agreement with the GOE-spacing law. When all levels (having different angular momenta) are included (Fig.II.11b), the spacing correlations are destroyed and the spacing distribution follows a Poisson law as expected(*). Does this mean that the GOE correctly

(*)Consider the following example : on the interval [0,1] take n points $x_i (i=1,\dots,n)$ at random uniformely distributed and construct an infinite spectrum by attaching to each point x_i a picket fence of unit spacing. The resulting spectrum, in the limit of large n , is a Poisson spectrum. So, by superposing most ordered spectra, one ends up with a Poisson spectrum.

models a shell model matrix ? This is certainly not true for in GOE, except for the symmetric nature of the matrices, all the matrix elements are <u>independent</u> random variables, whereas in the shell model, the Hamiltonian being a 2-body operator, the matrix elements of H between n -particle-states ($n > 2$) are linear combinations of 2-body matrix elements.

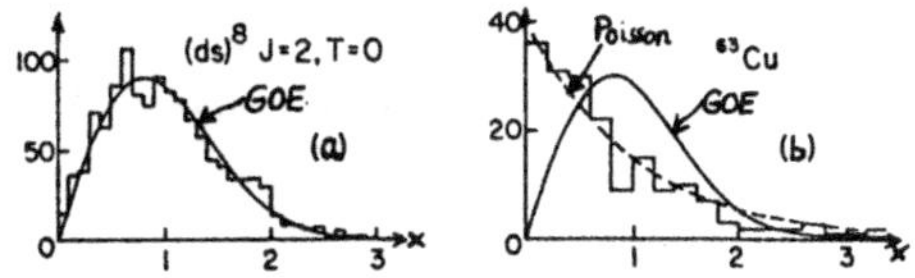

Fig.II.11 - Spacing histogram from two shell model spectra calculated with realistic interactions for ^{24}Mg [SZ-72] and ^{63}Cu [Wo-70] : (a) all states included have the same quantum numbers $J^{\pi} = 2^+$. (b) all states from a given spectrum span have been included ; it contains states with $J^{\pi} = 1/2^-$, $3/2^-$,...,$19/2^-$. (Taken from ref.[BFF-81]).

To follow closely the shell model, the microscopic model "par excellence" in nuclear physics, and to render matrix ensembles physically more plausible, a statistical extension of the shell model has been proposed : the "Two-Body Random Ensemble" (TBRE) introduced in refs. [FW-70,BF-71] and generalized by French and collaborators ("Embedded Gaussian Orthogonal Ensemble" (EGOE) [MF-75,BFF-81]). Consider a ν -body Hamiltonian acting in an n -particle space ($n \gg \nu$), the n identical fermions being distributed over Ω distinct single-particle states ($\Omega \gg n$). The ensemble of ν -body Hamiltonians is defined by

$$H = \sum_{\alpha \leq \beta} W_{\alpha\beta} A^+_\alpha(\nu) A_\beta(\nu) \ , \tag{II-53}$$

where $A^+_\alpha(\nu)$ is the operator creating ν particles in single-particle states labelled by $\alpha_1, \alpha_2, \ldots, \alpha_\nu$

$$A^+_\alpha(\nu) = C^+_{\alpha_1} C^+_{\alpha_2} \cdots C^+_{\alpha_\nu} \ , \tag{II-54}$$

and $A_\beta(\nu)$ is the Hermitian conjugate of $A^+_\beta(\nu)$. The ν -body matrix elements $W_{\alpha\beta}$ are taken as statistically independent random variables normally distributed with zero mean and variances

$$\overline{W^2_{\alpha\beta}} = (1 + \delta_{\alpha\beta})\sigma^2 \ . \tag{II-55}$$

When $n = \nu$ (II-53) reduces to GOE with a matrix dimension $N = \binom{\Omega}{\nu}$. For $n > \nu$, the dimensionality of the n -particle space is $\binom{\Omega}{n}$ and the matrix elements $\langle n\lambda | H | n\lambda' \rangle$ are linear combinations of the ν -body matrix elements $W_{\alpha\beta}$

$$\langle n\lambda | H | n\lambda' \rangle = \sum_{\alpha \leq \beta} C_{\alpha\beta}^{\lambda\lambda'} W_{\alpha\beta} \; , \qquad \text{(II-56)}$$

where

$$C_{\alpha\beta}^{\lambda\lambda'} = \langle n\lambda | A_\alpha^+(\nu) A_\beta(\nu) | n\lambda' \rangle \qquad \text{(II-57)}$$

and λ, λ' label the different n-particle states. The "realistic" case corresponds to $\Omega \gg n \gg \nu = 2$. It has been shown, using techniques which do not require the derivation of the joint probability density of eigenvalues $P(E_1, \dots, E_N)$, which is not known for EGOE, that, in the limit of a "dilute system" $\Omega \gg n \gg \nu \geqslant 1$, the eigenvalue density $\bar{\rho}(E)$ is Gaussian [MF-75,Ge-72]. For $n > \nu = 1$ the fluctuations are of Poisson-type but for $\nu \geqslant 2$ the methods used to derive $\bar{\rho}(E)$ fail in deriving fluctuation properties. However, there is strong numerical evidence coming from Monte-Carlo calculations that the fluctuation properties for $\nu \geqslant 2$ are identical to GOE-fluctuations(*). Very recently Verbaarschot and Zirnbauer, using the method of replicated variables combined with field-theoretic tools, have obtained the first analytical evidence that EGOE spectra show long-range order [VZ-84].

One can get some insight on the generality of GOE-fluctuations by the following argument. Consider a (non-random) Hamiltonian matrix H_0 corresponding, for instance, to a realistic nuclear Hamiltonian containing all sorts of non-statistical features (large pairing and quadrupole components, etc.) and perturb it with GOE matrices

$$H = H_0 + \lambda V_{GOE} \qquad \text{(II-58)}$$

It has been shown [Pa-81a] that there is a rapid transition from H_0-fluctuations to GOE-fluctuations when λ increases. Intermediate fluctuation patterns are to be expected only when the random-matrix elements are of the order of the local average spacing of the given (non-random) matrix H_0 (see also [ZVW-83]).

In summary, we have seen on the one hand that the spectra of very different systems (nuclei-light or heavy-, some atoms, maybe some molecules), when properly scaled, show identical fluctuation patterns. The scale (average spacing) covers five or six orders of magnitude, when going from a medium nucleus to a complex atom or molecule. Notice that one is considering extremely different systems, some of them governed by strong short range interactions and others by the Coulomb long range force. On the other hand, these characteristic fluctuation patterns, although

---

(*)If one deals with bosons instead of with fermions, there is numerical evidence that one also has GOE-fluctuations [Ma-83]

not specific of, are well reproduced by GOE. Presumably they are shared by broad classes of models. Thus, a simple picture emerges : there exists a universality of level fluctuation laws, as well from the experimental than from the theoretical point of view. Notice also that GOE is in some respects a disappointing model : although it predicts beautifully the observed level fluctuations, it does not provide a hint on the origin of its success. It rather looks like a mathematical device that mimics perfectly the observed level fluctuations. Wigner says : "The assumption is that the Hamiltonian which governs the behaviour of a complicated system is a random symmetric matrix, with no particular properties except for its symmetric nature" [Wi-67b]. But what is, then, a complicated system ? And are the laws that we have been discussing of purely quantal nature or do they have a classical counterpart ? With the hope of finding some clues on the origin of the success of the theory as well as on its domain of validity, let us do in the next section a long excursion in the realm of classical Hamiltonian systems, the emphasis being put on the concepts of simple and complicated, or of regular and irregular systems.

BIBLIOGRAPHY

- A good general review can be found in [Wi-67a]
- In [Po-65] are reprinted all the important papers on random matrix theory published before 1965. The volume is preceded by a good and rather detailed introduction by C.E. Porter
- [Me-67] is the classical text on the subject. The emphasis is on mathematical techniques and derivation of exact results
- [BFF-81] is a recent and complete review article, containing mainly developments since [Po-65] and [Me-67], with emphasis on both theory and applications
- A review article (Bohigas, Haq and Pandey) on the comparison with experiment of random matrix predictions for level fluctuations is in preparation.

III - AN INTRODUCTION TO CLASSICAL CHAOTIC MOTION

The aim of this section is to introduce the concept of chaos in classical mechanics of conservative systems.

Giving a rigorous presentation of the material would require a highly technical language, and elaborate mathematical tools. Our purpose is rather to present an elementary intuitive approach to the subject, in order to get a physical insight into the main ideas. The material of this section is far from being exhaustive, and many fundamental questions have been omitted (such as perturbation theory, bifurcating orbits, mechanism of destruction of tori, dissipative systems, etc.).

In what follows, we limit ourselves to the study of classical Hamiltonian systems which are conservative and time-reversal invariant ; moreover, we consider only initial conditions for which the motion can take place only in a bounded region of the phase space.

III.1 FROM REGULAR TO CHAOTIC MOTION

All conservative Hamiltonian systems with N degrees of freedom have in common three essential properties :

i) for a given set of initial conditions, the dimensionality of the accessible surface in phase space is less or equal to (2N-1) ; since the system is conservative, the energy is constant along this "energy surface".

ii) From Liouville's theorem, we know that the volume element in phase space is conserved. In other words, the Hamiltonian flow, which preserves the measure in phase space, is incompressible.

iii) Trajectories in phase space cannot cross.

Apart from these features which are shared by all systems, the motion in phase space can exhibit a great variety of behaviours. For instance, one may ask how does a given volume element evolve with time : does it tend to cover the whole energy surface S_E as time goes to infinity or does it remain in a restricted part of S_E ? Does it conserve approximately its initial shape, or does it display more or less dramatic deformations with time ? According to the answers to such questions, one can define a hierarchy of regularity for dynamical systems. As we shall see now, the most regular systems, lying at the bottom of this classification, can be used as clocks, whereas, at the opposite side, the most chaotic systems(*) can be used as random number generators.

(*) We speak here of dynamical systems in the enlarged sense of area-preserving mappings (see Sect.III.2).

Regular systems

Roughly speaking, a regular dynamical system is such that it can be integrated by quadratures. Let us be more precise, and define what are the conditions and the consequences of "extreme regularity". For a more rigorous account of the subject, see Refs. [Ar-76, AA-67].

We consider a time-independent Hamiltonian system with N degrees of freedom :

$$H = H(\underline{q}, \underline{p}) \quad , \quad \underline{q} = \{q_1, \dots, q_N\}$$
$$\underline{p} = \{p_1, \dots, p_N\} \, .$$

The equations of motion, written in Hamiltonian form, are :

$$\begin{cases} \dot{\underline{q}} = \vec{\nabla}_{\underline{p}} H(\underline{q}, \underline{p}) \\ \dot{\underline{p}} = -\vec{\nabla}_{\underline{q}} H(\underline{q}, \underline{p}) \, . \end{cases}$$

Definition

A time-independent Hamiltonian system with N degrees of freedom is said to be "integrable" if there exist N constants of motion $F_m(\underline{q}, \underline{p})$ (one of them being H itself) which are analytic functions of $\underline{q}$ and $\underline{p}$, single-valued, functionally independent, and in involution (for the Poisson bracket) : $\{F_n, F_m\} \equiv 0 \quad \forall \, n, m = 1, \dots, N$.

Theorem (Liouville-Arnold)

If a system is integrable, then

i) there exist a canonical transformation to action-angle variables :

$$(q_1, \dots, q_N \, ; p_1, \dots, p_N) \mapsto (\theta_1, \dots, \theta_N \, ; I_1, \dots, I_N)$$

such that the Hamiltonian, expressed in the new variables, depends only on the actions :

$$H(q_1, \dots, q_N \, ; p_1, \dots, p_N) \longmapsto \tilde{H}(I_1, \dots, I_N) = \tilde{H}(\underline{I}) \, .$$

The action variables are constants of motion :

$$\dot{\underline{I}} = 0 \quad ,$$

and the evolution of the angle variables is given by :

$$\dot{\underline{\theta}} = \frac{\partial \tilde{H}}{\partial \underline{I}} = \underline{\omega}(\underline{I}) ,$$
$$\underline{\theta}(t) = \underline{\omega}(\underline{I})\, t + \underline{\alpha} . \qquad \text{(III-1)}$$

ii) For each set of initial conditions, the accessible surface in phase space is a compact manifold (we consider only bounded trajectories) having the topology of an N-dimensional torus T^N (*). These tori are called "invariant tori", because a trajectory starting on one of them remains on it for ever.

According to this theorem, the motion of an integrable system is restricted to an N-dimensional surface, instead of a (2N-1)-dimensional energy surface for a generic system having no other constant of motion than the energy. Let us notice that all conservative systems with one degree of freedom are integrable (provided they satisfy the smoothness conditions on H) : the accessible phase space is of dimension one (tori are reduced to circles). For integrable systems with two degrees of freedom, the existence of a second integral of motion reduces the dimension of accessible phase space from three to two.

Each value of $\underline{I}$ defines the torus T^N, whereas the vector $\underline{\theta}(t)$ gives the position of the trajectory on the torus at each time t. If the frequencies ω_i are mutually commensurable i.e. if there exist (N-1) independent relations

$$\underline{\omega} \cdot \underline{m} = 0 , \qquad \text{(III-2)}$$

where the vector $\underline{m}$ has integer (positive or negative) components and $\underline{m} \neq 0$, then the trajectory is closed on the torus ; in this case, the motion in phase space takes place on a one-dimensional region, and the motion is periodic. Systems having this property exhibit therefore the strongest degree of regularity, as an "ideal clock" would do.

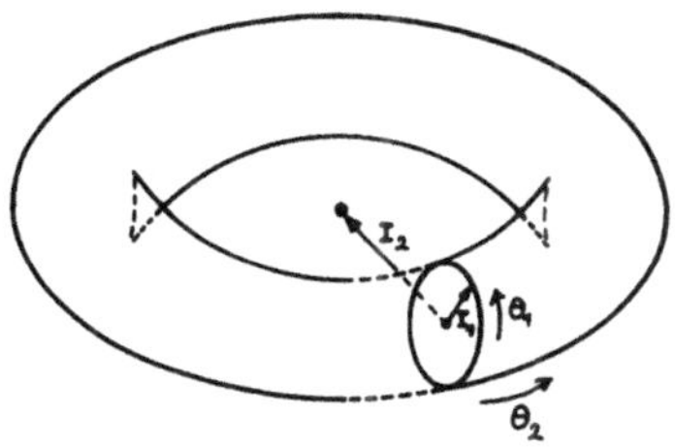

Fig.III.1 - An invariant torus of an integrable system with two degrees of freedom ; the actions I_1 and I_2 are the radii of the two circles defining the torus T^2, and the angles θ_1 and θ_2 define a point on T^2.

(*) An N-dimensional torus T^N is a direct product of n circles. A point on T^N can be defined by N angular coordinates ($\theta_1,\dots,\theta_N$) [see Fig.III.1]. The torus is often represented as an N-dimensional hypercube defined by $\{(\theta_1,\dots,\theta_N) : 0 \leqslant \theta_i \leqslant 2\pi\}$.

At the opposite, if no relation of the type (III-2) holds, the orbit never closes, but densely covers the torus after infinite time : such a spiralling orbit is "ergodic" on the torus (see footnote(**) p.67) -called "irrational", or "non-resonant" torus-, and exhibits a strong regularity. Intermediate cases where p independent relations ($0 < p < N-1$) like (III-2) hold can also occur ; orbits then lie on (N-p)-dimensional manifolds of T^N.

It should be noticed that in the generic integrable case (see below the harmonic oscillator and the Kepler motion as exceptions), the frequencies depend on the actions (which define the invariant tori), i.e. on the set of initial conditions fixing the values of the N first integrals of the motion. Consequently for a generic integrable system, there exist simultaneously -corresponding to different sets of initial conditions- non-resonant tori, covered by ergodic trajectories and resonant tori ; measure theoretic arguments show that for such systems, the irrational tori form a set which is almost everywhere dense [Ar-76]. In other words, almost all the tori of a generic integrable system are irrational, in the same way as almost all real numbers are irrational.

Examples of regular systems

i) for N=1, we already saw that all conservative systems are regular (therefore periodic)

ii) for N=3, all systems submitted to a central field force V(r) are regular : $\vec{L}^2$ and L_z (orbital moment) are conserved, together with H.

Two particular systems of this kind are of special interest :

- for the Kepler motion ($V(r) = -\frac{k}{r}$) :

$$H = \frac{p_r^2}{2m} + \frac{p_\theta^2}{2mr^2} + \frac{p_\varphi^2}{2mr^2\sin^2\theta} - \frac{k}{r} \quad ,$$

The Hamiltonian is well known to write, in terms of the action variables $I_i = \oint p_i \, dq_i$ (each integral is over the period corresponding to the q_i) as :

$$\tilde{H} = \frac{-2\pi^2 m k^2}{(I_r + I_\theta + I_\varphi)^2}$$

- for the isotropic harmonic oscillator ($V(r) = \omega^2 \frac{r^2}{2}$), the Hamiltonian takes the form

$$\tilde{H} = \omega\,(I_1 + I_2 + I_3) \quad .$$

In both cases, for all values of the actions, i.e. for all initial conditions, the three frequencies coincide; the orbits are closed, with period $\tau = \frac{2\pi}{\omega}$. This means that for these two particular systems, all tori are rational (no orbit covering densely a torus). To understand why such a situation is an exception, one has to remember Bertrand's theorem [Be-1873] : Consider the motion of a point-mass under the action of a spherically

symmetric potential V(r), supposed to be a C^3 function of r. Then a necessary and sufficient condition for all the bounded trajectories to be closed is that V(r) behaves like r^2 or $(-\frac{1}{r})$. As is well known, the origin of this particular behaviour of spherical harmonic oscillator and Kepler motion is the existence of a further first integral ("dynamical invariance"), due to the particular shape of the potential, whereas the other three independent first integrals are merely consequences of rotational invariance. For this reason, these systems are called "overintegrable" systems. For Kepler motion, the extra integral is the Runge-Lenz vector, which is carried by the focal axis. Another peculiarity of Kepler motion and harmonic oscillator lies in the fact that they are "separable" in orthogonal coordinate systems. Most of the integrable systems we are used to consider possess this property ; however, separability is a very special case of integrability.
In fact, no general method is available to know whether a system is integrable or not. Moreover, even if one system is known to be integrable, finding action-angle variable is not guaranted !

- The system of a particle in a parallelepipedic box of sides a,b,c is also a regular system ; the energy writes

$$E = \frac{\pi^2}{2m}\left(\frac{I_1^2}{a^2} + \frac{I_2^2}{b^2} + \frac{I_3^2}{c^2}\right)$$

iii) The N-dimensional (anisotropic or isotropic) oscillator of frequencies ω_i $(i=1,\dots,N)$ is also a separable regular system :

$$\tilde{H} = \underline{\omega}\cdot\underline{I}$$

iv) There are some other systems which are known to be regular (see below regular billiards) : these are all the systems which are solved by quadratures in textbooks or articles... but they are not so numerous !

Regular systems had, up to one or two years ago, a so strong monopole in textbooks, that one might believe their properties as being generic of any classical systems [Fo -83]. The fact is that regular systems are an <u>exception</u> among dynamical systems, and that, for other kinds of systems, many difficult questions remain open.

Ergodic systems

The most popular definition of ergodicity is the equality of time averages and phase space averages. In more abstract language, one often defines an ergodic system as a dynamical system whose phase space is metrically indecomposable under the Hamiltonian flow. Let us get in some more detail into these two equivalent definitions of ergodicity, and quote some important consequences.

Let us consider a conservative dynamical system, of phase space X. Let $x(t) = [q_1(t), \dots, q_N(t), p_1(t), \dots, p_N(t)]$ be any point of X. The evolution of x with time can be described by an operator T :

$$X \xrightarrow{T_t} X \ :$$

$$x \longmapsto T_t x = x_t \ .$$

The transformations T_t have trivially the property of forming a continuous group T depending on one parameter t. This group is often called the "dynamical group" or the "Hamiltonian flow".

The Liouville measure of a subset V of X (which is the volume of V) :

$$\mu(V) = \int_V dq_1 \dots dq_N \, dp_1 \dots dp_N$$

is, according to Liouville's theorem, invariant under the Hamiltonian flow :

$$\mu(T_t V) = \mu(V) \ . \tag{III-3}$$

For a conservative system, one can define a surface element of the energy surface S_E as

$$d\sigma = d\mu \ \| \overrightarrow{\mathrm{grad}}\, H \|_{H=E}$$

where $$\| \overrightarrow{\mathrm{grad}}\, H \|^2 = \sum_{i=1}^{N} \left[\left(\frac{\partial H}{\partial p_i} \right)^2 + \left(\frac{\partial H}{\partial q_i} \right)^2 \right] ,$$

and get from (III-3) the invariance of the induced measure σ under the group T. In the following, the measure σ will be normalized to one (i.e. $\sigma(S_E) = 1$).

The system is said to be ergodic, or metrically transitive (or indecomposable) if it is indecomposable into non trivial subsets which are invariant under T, i.e. :

$$\nexists \ A \subset S_E \quad (\sigma(A) \neq 0, \ \sigma(A) \neq 1) \quad \text{such that} \quad T_t A = A \ . \tag{III-4}$$

The fact that the only subset of S_E of non-zero measure which is invariant under T is the whole energy surface S_E (or any subset of S_E having the same area as S_E itself) intuitively suggests that a typical trajectory cannot be confined in a restricted region of phase space, but, on the contrary, explores the whole energy surface (except for a set of zero measure)(*). In fact, an even stronger property holds for ergodic systems : one can show that almost every orbit explores almost every point of the energy surface S_E (densely covers S_E), spending in any part A a time proportional to the area of A. Indeed, let $x \in S_E$ and $t(x, A, \tau)$ be the time that x spends in $A \subset S_E$ between instants 0 and τ. According to the ergodic theorem (Birkhoff-von Neumann), the limit

$$\lim_{\tau \to \infty} \frac{t(x, A, \tau)}{\tau}$$

does exist for almost all x's, and is equal to the area of A :

$$\lim_{\tau \to \infty} \frac{t(x, A, \tau)}{\tau} = \sigma(A)$$

if the system is ergodic.

The equality of phase averages and time averages constitutes an alternative definition of ergodicity (i.e. is equivalent to metrical transitivity) ; it can be formulated as follows : a system is ergodic iff, for any integrable function $f(x)$ ($x \in X$, $\int |f(x)| d\sigma < \infty$),

$$\lim_{t \to \infty} \frac{1}{t} \int_0^t dt\, f(T_t x) = \int_{S_E} f(x)\, d\sigma \qquad \text{(III-5)}$$

for almost all $x \in S_E$. Eq.(III-5) is often referred to as a version of the law of large numbers, the left hand side representing the infinitely many trials approaching the probability in the right hand side.

Equilibrium Statistical Mechanics is built on the ergodic hypothesis. But ergodicity is not sufficient for a system to reach an equilibrium state : one additional property -called mixing- is needed, which concerns the way any volume element evolves with time.

(*) It is clear that no trajectory can explore the whole energy surface (i.e. ergodicity in the sense of Boltzmann can never hold). Indeed, a trajectory which would pass through any point of S_E should intersect with itself, which is impossible. What we call ergodicity here is often referred to as the quasi-ergodic hypothesis in Statistical Mechanics.

Mixing systems

The concept of mixing was first introduced by Gibbs. Qualitatively, the underlying idea of mixing is the property, for any part A of the energy surface ($\sigma(A) \neq 0$), of getting spread uniformly over the energy surface as $t \to \infty$. To get an "experimental feeling" of this feature, let us imagine a shaker in which one puts 80 % of orange juice, and 20 % of vodka ; then, one continuously stirs the shaker. If the system is mixing, one will find, as time tends to infinity, 20 % of vodka in any volume of the liquid, i.e. the vodka will be uniformly mixed with the orange juice in any part of the cocktail. Of course, such a property implies for the initial volume of vodka (or of any part of the energy surface) being strongly deformed as time runs, i.e. transforming with time in a thinner and thinner filament which ultimately will be present in any volume of the liquid (or will densely cover the energy surface).

This intuitive description of mixing makes now the mathematical definition very easy to understand :

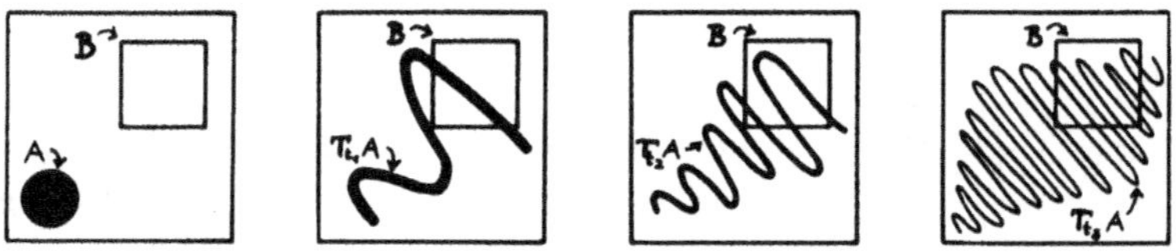

Fig.III.2 - Schematic representation of the mixing property : B is any fixed area of the surface energy, and A is a surface element considered at initial time t_0 , whose evolution is drawn for successive times $t_1 < t_2 < t_3$. As time is running, the Hamiltonian flow T_t tends to dilute the initial surface A uniformly in the whole energy surface. As time tends to infinity, the fraction of the test area B occupied by $T_t A$ is equal to the fraction of the energy surface S_E initially occupied by A .

A system is said to be mixing if

$$\forall A, B \subset S_E \quad (\sigma(A) \neq 0 \neq \sigma(B)), \quad \lim_{t \to \infty} \frac{\sigma(B \cap T_t A)}{\sigma(B)} = \sigma(A) . \qquad \text{(III-6)}$$

Let us look at the right hand side of Eq.(III-6). One can write it as :

$$\sigma(A) = \frac{\sigma(T_t A)}{\sigma(S_E)} ,$$

since $\sigma(T_t A) = \sigma(A)$ from Liouville's theorem, and $\sigma(S_E) = 1$. So the r.h.s. of Eq.(III-6) represents the fraction of the energy surface occupied by $T_t A$, while the l.h.s. represents the fraction of B occupied by $T_t A$. According to the uniform spreading of A over S_E, these two terms should be equal.

Some important consequences of mixing should be noticed. First, one can show that mixing implies ergodicity. Indeed, suppose that the mixing property holds ; take for A an invariant set of S_E, and choose $B \equiv A$. Then ($B \cap T_t A$) is simply A , and Eq.(III-6) implies that $\sigma(A)$ is zero or one, i.e. there does not exist any non-trivial subset A of S_E which is invariant by T_t . This is exactly the indecomposability property characterizing ergodic systems (see (III-4).

Notice that the converse is not true, i.e. ergodicity does not imply mixing ; indeed, one can imagine systems for which any surface elements explores the whole energy surface without being deformed with time (see Fig.III.3 and example III-3-b-1) ;

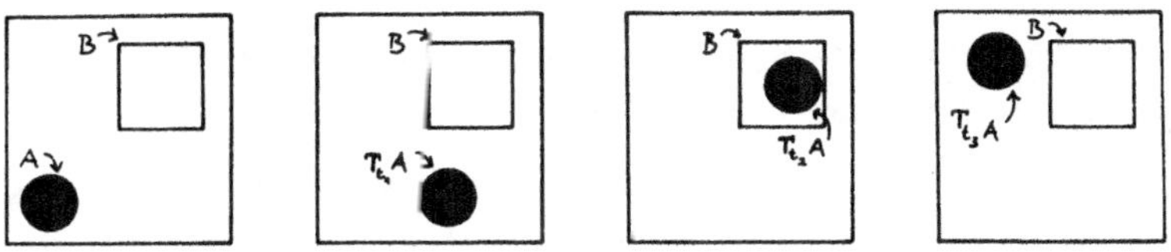

Fig.III.3 - Schematic representation of an ergodic system which does not have the mixing property. The surface element A tends to explore the whole energy surface as time is running, spending equal times in equal areas (ergodic property), but its shape remains unchanged, so that Eq.(III-5) (mixing property) cannot hold.

in such systems, the distance between two points of S_E remains of the same order of magnitude as time is running. On the contrary, in a mixing system, the spreading property allows two points which are initially close to each other to get as far from each other, in the limit $t \to \infty$, as is permitted by the constrain of staying on the energy surface. Thus, mixing implies instability with respect to initial conditions, or, in other words, loss of information with time, since a small error in measurement of $x(t=0)$ can propagate with time and eventually induce so large errors on $x(t)$ that long time predictions for the system are prohibited.

Finally, let us mention an important characteristic (i.e. which could be used as a definition) property of mixing, which is the decay of correlations between two functions : in mixing systems, any two square integrable function (for the measure σ) asymptotically become statistically independent, i.e.

$$\forall f, g \in \mathcal{L}^2_\sigma(S_E), \qquad \int f(x)\, g(T_t x)\, d\sigma \underset{t\to\infty}{\longrightarrow} \left[\int f(x)\, d\sigma\right]\left[\int g(x)\, d\sigma\right].$$

As a particular case, let us take for g a non-equilibrium normalized distribution $\rho(x)$:

$$\int f(x)\,\rho(T_t x)\,d\sigma \underset{t\to\infty}{\longrightarrow} \int f(x)\,d\sigma,$$

which means that, in average, the distribution $\rho(x)$ tends to a uniform equilibrium probability density.

We shall not discuss the subtle relations of mixing with irreversibility. Let us only emphasize in this respect that the equations of motion allow to recover any point $x(t=0)$ knowing $x(t>0)$; however, all memory of the initial state is lost as time tends to infinity, and it is only in this limit that one can speak of irreversibility.

K -systems

Without no further assumption than mixing, one cannot give any quantitative information about the separation of orbits with time.

The so-called K -systems are mixing systems which possess so strong instability that most orbits starting from close points separate, in the average, exponentially with time. For such systems, the knowledge of $x(t_i)$ for all discrete times $\{t_i\} = \{-\infty, \dots, t_0-1, t_0\}$ does no provide any useful information on the behaviour of the system for times $t>t_0$. Such a strange feature means that, though the system is completely deterministic -i.e. is governed by causal equations of motion-, the evolution of a generic point in phase space is very irregular, and exhibits some kind of random behaviour ; this is why one associates with K -systems, for which the motion does not depend on their distant past history, the idea of unpredictability.

The mathematical definition of a K -system would go beyond the scope of our "qualitative" introduction to chaotic phenomena (see for instance Refs.[AA-67,Sh-73,Or-74] for a rigorous approach). Let us only give the main ideas of characterization of K - systems, and try to understand roughly in which sense one can say that the past does not determine the future. All the concepts used are borrowed from information theory [SW-49,Bi-78,ME-81]. First, one introduces any finite ordered partition P of the energy surface S_E into cells (atoms), in order to define a measurement : the result of the measurement of the system associated with P at time t_0 is the n° of the cell of P which is crossed by the orbit at time t ; then, an experiment associated with P is a sequence of measurements for equally spaced times going from t_0 to ∞ . Now, one can define the entropy $h(T,P)$ of the hamiltonian flow T relative to the partition P, which represents the mean rate (averaged over the whole sequence of measurements of the experiment) of information[(*)] generated by the time running for the experiment associated with P. Consider, for instance, a very regular system, whose motion is periodic in phase space : for such a system, the knowledge of a finite set of sequential measurements is sufficient to determine the outcomes of all further measurements, i.e. the information produced by all but a finite set of measurements is zero, and therefore the average of information generated for a given experiment is zero. For strongly irregular systems, even the knowlecge of an infinite set of sequential measurements is not sufficient to know, with probability one, the outcomes of later measurements, which means that the average of information -or entropy- is positive.

K -systems are now defined as systems for which the entropy relative to any finite partition P is positive :

$$h(T,P) > 0 \quad \forall P .$$

These are therefore systems whose motion looks irregular and unpredictable whatever may be the experiments one can imagine. A quantity of interest to measure the degree of irregularity is the so-called metric entropy, or Kolmogorov-Sinai (K-S) entropy : it is defined as the maximum value of $h(T,P)$, taken over all the possible finite partitions P :

$$h_{K\text{-}S}(T) = \max_{P_i} h(T,P_i) \tag{III-7}$$

It is clear from their definition that K-systems have positive K-S entropy. Let us emphasize that this entropy does not depend on any dynamical state under consideration, contrarily to the entropy commonly used in Statistical Mechanics, but is an intrinsic quantity associated to the dynamical system considered as a whole at a given energy E. This K-S entropy provides a measure of the strength of mixing of the system, and is related to the mean rate of exponential separation of trajectories.

Bernouilli systems

These are, among the K-systems, the most unpredictable ones. Indeed, they have the property that there exist a partition P_0 such that the sequence of measurements of the associated experiment are completely uncorrelated, as would be the outcomes of games with a roulette wheel ! This partition P_0 must also satisfy another condition

(*) Notice that the amount of information gained by making a measurement is equal to the lack of information (i.e. to the uncertainty) before making the measurement, which is also called the entropy associated with the experiment.

(whose statement would require entering in abstract mathematical considerations : see for instance [Sh-73] or [Pa-81]) in order for the system to be Bernouilli.

For a Bernouilli system, the K-S entropy is simply equal to the entropy of the particular partition P_0 , i.e.

$$h_{K-S}(T) = -\sum_i \mu_i \ln \mu_i , \qquad \text{(III-8)}$$

where μ_i is the measure of the ith atom of P_0 (i.e. the area of the ith cell of $P_0 \subset S_E$), and the sum runs over all the atoms of P_0 .

The different types of systems we have just introduced illustrate the transition from regularity to chaos, i.e. randomness produced by causal equation of motion. Several remarks should be made at this stage :

i) Part of the vocabulary introduced above ("ergodicity", "mixing") sounds familiar, and might evoque Statistical Mechanics. We emphasize, however, that all the considerations made in the preceding are not restricted to systems with a great number of degrees of freedom, but really can apply to very simple systems, eventually having only two degrees of freedom (see example in the following).

ii) For physical systems which are suspected to belong to a given member of the hierarchy just presented (except for certain integrable systems), it is in general extremely difficult to prove that indeed they are ergodic, or mixing, etc.. The main rigorous results available so far concern billiards, and we shall devote a particular attention to this kind of systems. Also are known interesting results for "abstract" dynamical systems, many of them being very "educative". Several examples will be presented in Sect.III.3, as illustrations of the rather abstract ideas introduced up to now.

iii) One question of interest is the following : can "most" of the dynamical systems be classified according to the preceding scheme, i.e. are physical systems, in their majority, either integrable, or ergodic ? There is at least a very peculiar(*) kind of systems for which the answer to this question is trivially "no" : these are conservative systems having a number n of independent first integrals, with $n > 1$, but $n < N$; for such systems, the motion in phase space is restricted to a ($2N - n$)-dimensional surface. Thus they are non-integrable systems which are not ergodic on the whole energy surface S_E. We can therefore reformulate the question as : "are conservative systems having no other constant of motion than the energy E ergodic ? Much attention has been paid to this problem during the first half of the century, and we recall in a few words the revolution produced in the 1960's by the so-called Kolmogorov-Arnold-Moser theorem.

(*) Here, the word peculiar refers to the rarity of such systems among the family of all dynamical systems ; however, these are generic systems for the physicist ! (isolated physical system possess other constants of motion than the energy).

Quasi-integrable systems

Conservative quasi-integrable systems are, roughly speaking, obtained by inducing a slight perturbation preserving the energy conservation on an integrable system. In 1892, Poincaré showed that, for such systems, apart from exceptional kinds of perturbations, all the constants of motions other than the energy are destroyed. It was probably this theorem which encouraged people to try to demonstrate -or at least to convince themselves- that most of the integrable systems can be made ergodic by small perturbations. (For instance Fermi [1923] and Birkhoff [1927] attempted to develop arguments in favour of this assumption : see Ref.[Mo- 73b]). In view of this, the result of Kolmogorov, Arnold, and Moser, now known as the KAM theorem, was a major turning point in the history of dynamical systems. Indeed, the main information of this theorem is that, under rather general conditions, "most" of the irrational tori persist (though slightly deformed) under a small perturbation.

From this fundamental result, one can easily imagine that, for a "generic" dynamical system, the structure of phase space is very complicated, showing both regular and chaotic regimes (there is an enormous gap between integrability and ergodicity !). In this respect, the above classification, according to increasing degree of chaos, is irrelevant for most systems, and may appear somewhat "academic". However, to study open problems such as the characterization of quantum chaos, it seems natural to begin with systems whose classical analogues are reasonably well understood -for instance K-systems-, even though they are known to be rare among the whole family of chaotic dynamical systems.

III.2 - MEASURE-PRESERVING MAPPINGS

III.2.a) Surfaces of section

For the sake of simplicity, let us restrict for a moment to physical conservative systems with two degrees of freedom. As already mentioned, it is far from being easy, even for these "simplest" systems, to study analytically their degree of irregularity. However, one can get for them quite useful informations from very simple numerical experiments, as we shall see now.

For the systems considered here, the accessible portion of phase space at a given energy E -the surface energy S_E- is a three-dimensional bounded surface. The method of Poincaré surfaces of section consists in choosing an "appropriate" two-dimensional surface Σ in the phase space, and look at the figure generated by successive crossings of one given orbit with Σ ; by "appropriate", we mean that

i) the trajectory should cross the surface Σ an infinite number of times, as time goes from zero to infinity

ii) the area bounded by any closed curve on Σ should be conserved with time (see Eq. (III-9)).

Let us illustrate the interest of the method by considering a particular surface Σ_y, defined in the following way. Let us suppose that the surface $\{x = 0\}$ intersects with the energy surface $H(x, y, p_x, p_y) = E$; we call this intersection Σ_y, which is a two-dimensional surface. A point on Σ_y can be defined by the coordinates (y, p_y). The integral of motion $H = E$ defines p_x^2 as a function of (p_y, x, y) ; therefore, one can attach to a point on Σ_y a set of coordinates $(x = 0, y, \epsilon|p_x|, p_y)$, with the sign $\epsilon = \pm 1$ of p_x depending on the sense with which a given trajectory crosses Σ_y. To get this sign unambiguous, we choose a given side of Σ_y, and look at the successive intersections Q_n of a given trajectory with Σ_y which, for instance, go out of the chosen side of Σ_y :

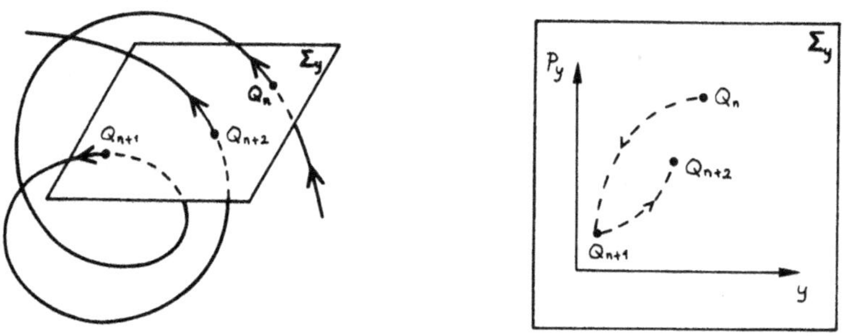

(a) (b)

Fig.III.4 - (a) A trajectory crossing the surface of section Σ_y ; (b) Points of the Poincaré map on Σ_y

Now, the knowledge of the coordinates of Q_n on Σ_y completely specifies the whole set of coordinates of Q_n in phase space, i.e. the knowledge of one pair of conjugate variables on Σ_y completely determines the dynamical state. What has been lost by reducing the initial three-dimensional problem to a two-dimensional one is the "story" of the orbit between two successive crossings with Σ_y, but this is not essential for our purpose.

The repeated crossings with Σ_y define a **mapping** M :

$$\Sigma_y \xrightarrow{M} \Sigma_y$$

Given an initial condition $\underline{\xi}_0 = (x_0 = 0, y_0, p_{x_0}, p_{y_0})$ for a point on Σ_y, the study of the dynamical system is now reduced to the study of the set of points

$\{ \underline{\xi}_n = M^n \underline{\xi}_0 \quad , \quad n = 0,1,2,\ldots \infty \}$.

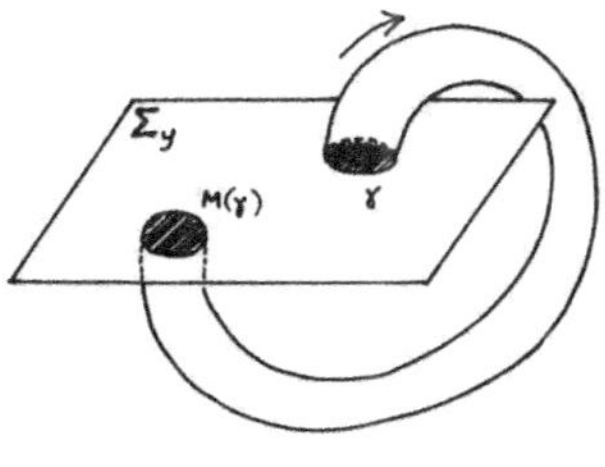

Fig.III.5 - Illustration of the property (III-9) : Σ_y is a surface of section , γ a closed curve on Σ_y , and $M(\gamma)$ the image of γ on Σ_y under the Hamiltonian flow. The areas bounded by γ and $M(\gamma)$ are equal.

The property ii) stated above means that the mapping M is area (or measure)-preserving, i.e. that, given any closed curve γ on Σ_y , the area bounded by the image curve $M(\gamma)$ of γ by the mapping M should be equal to the area bounded by the curve γ :

$$\oint_\gamma p_y \, dy = \oint_{M(\gamma)} p_y \, dy . \qquad \text{(III-9)}$$

This can be easily shown from the Hamiltonian character of the motion (see e.g. Ref.[AA-67, Appendix 31]).

Now, let us look at the pattern generated on Σ_y by the points $M^n \underline{\xi}_0$, ($n = 0,1,2,\ldots\infty$), and see which kind of information we can learn from it. First, let us consider a system which has, besides the energy, another constant of motion F in involution with H (i.e. an integrable system)

$$F(x,y,p_x,p_y) = C \quad , \quad H(x,y,p_x,p_y) = E . \qquad \text{(III-10)}$$

From (III-10) one can eliminate two of the canonical variables, say x and p_x , which yields

$$p_y = p_y(x,y) . \qquad \text{(III-11)}$$

Since $x = 0$ on Σ_y , Eq.(III-11), "projected" on to Σ_y is the equation of a curve Γ , which is invariant under the mapping M , and which is simply the intersection of the 2-dimensional invariant torus T^2 with Σ_y . Therefore, the successive crossings of the orbit with Σ_y all lie on a unique smooth curve ($M(\Gamma) \equiv \Gamma$), for any set of initial conditions, which is a signature of integrability. Two possibilities can occur in such a case :

i) the invariant torus T^2 is resonant (or rational). Then, the motion is periodic on the torus, i.e. the orbit is closed, and has only a finite number k of crossings with Σ_y , which are all fixed points for M, since $M^k \underline{\xi}_0 = \underline{\xi}_0$ (see Fig.III.6).

ii) if, on the contrary, the invariant torus T^2 is irrational, the orbit densely covers the torus (is ergodic on the torus), and the points generated by successive iterations of the mapping will ultimately densely cover the invariant curve Γ (see Fig.III.7).

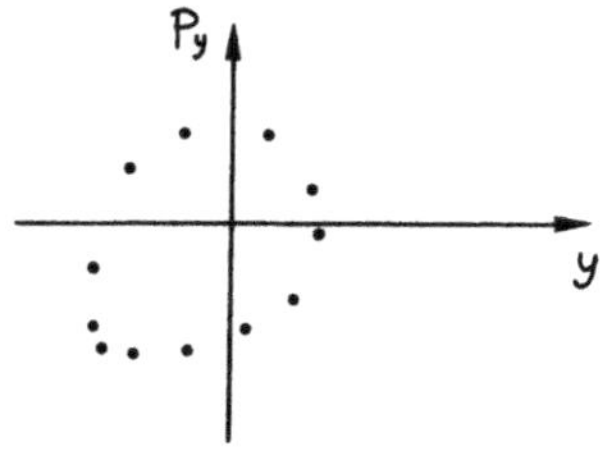

Fig.III.6 - Typical pattern of a Poincaré map of two-dimensional integrable system for initial conditions defining a rational torus

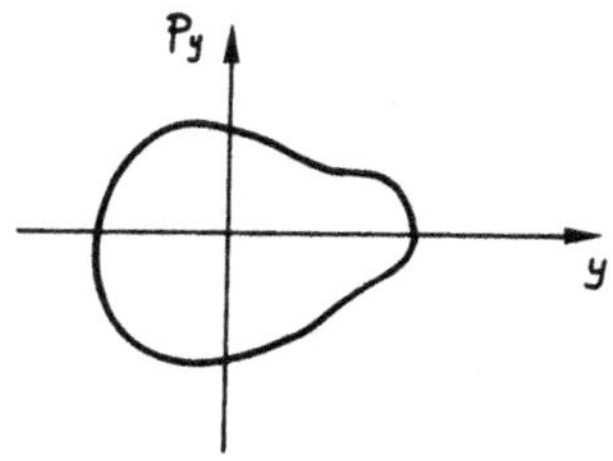

Fig.III.7 - Same as Fig.III.6, but for an irrational invariant torus

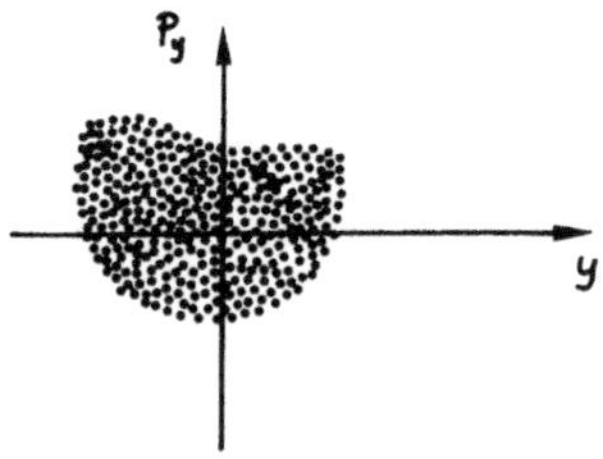

Fig.III.8 - Idealized pattern of a Poincaré map of an ergodic system

Let us next consider a dynamical system which is ergodic (on the energy surface). Then, almost all trajectories densely cover a three-dimensional region of phase space, and their intersections with Σ_y fill densely some area (see Fig.III.8).

We should notice that they may exist "non-typical" trajectories for ergodic systems such as, for instance, periodic trajectories (see examples for chaotic billiards in Sect.III.4); but these form a set of zero measure.

Between integrable and ergodic systems, there is a great variety of systems for which chaotic and regular regions coexist in phase space ; for such systems, the pattern of the Poincaré map shows both invariant curves and sparse points, as can be seen on the example of Fig.III.12. (It is usual to represent on the same Poincaré map impacts of several trajectories corresponding to different initial conditions).

Comments

i) The method of surfaces of section is highly efficient, and provides valuable informations on the structure of phase space. (The example of the Toda lattice presented below nicely illustrates the success of the method. Many other examples can be found in Ref.[He-83]). However, it should, in any case, remain no more than a guide for further theoretical studies, and can never be used as a proof, neither of integrability, nor of ergodicity.

ii) Of course, the pattern gets more and more informative as the number of iterations increases. But the number of iterations necessary to get a reasonable image of the structure of phase space highly depends of the nature of the system, and of the choice of initial conditions. This is particularly true for strongly chaotic systems, for which a trajectory may happen to be trapped in a given region of phase space, and to escape only after a long time ; therefore, the true pattern appears only after a great number of iterations. In this respect, an "abusive" use of Poincaré maps for such systems might suggest that the system is not ergodic, since a typical trajectory for an ergodic system must have spent equal times in equal areas after an infinite time ; however, if a trajectory happens to be confined for a long time in a restricted region of the energy surface S_E, it will visit again this region much less frequently than other regions of S_E, in order to preserve the ergodic property in the limit $t \to \infty$.

iii) Among the limitations of the method, one should quote its inefficiency to detect any kind of instability of the system ; for instance, the method is completely inadequate to distinguish simple ergodicity from mixing or K-property. These properties are indeed very delicate to study, even by numerical means.

iv) More "exotic" possibilities than those presented before for the pattern of Poincaré maps are open for dissipative systems and will not be discussed here. For instance, the points may be restricted to lie on "fractal" sets of the surface of section (as can happen in strange attractors).

v) We presented the method only for systems with $N = 2$ degrees of freedom. It can, in fact, be generalized to any number N of degrees of freedom ; in this case, the surface of section is (2N-2)-dimensional, which makes the study more difficult.

III.2.b) Abstract dynamical systems

An abstract dynamical system (X, μ, T) is usually defined given a space X, a measure μ, and a mapping T, in the following way. X is a measurable space with respect to the measure μ ; moreover, μ is a probability measure, i.e. $\mu(X) = 1$. T is a mapping $X \xrightarrow{T} X$, which is surjective, and which preserves the measure μ, i.e.

$$\mu(T^{-1}A) = \mu(A) \qquad \forall A \subset X : \mu(A) \neq 0 .$$

Moreover, the mapping T is often supposed to define a one parameter group of transformations T_t (the parameter may take continuous or discrete values). In this case,

T is one-to-one, and the measure-preserving property can also be expressed as

$$\mu(TA) = \mu(A) \,.$$

The most natural abstract dynamical system associated with a physical Hamiltonian system is such that X refers to the energy shell S_E, μ to the measure induced on S_E by the Liouville measure, and T is what we called the dynamical group, or Hamiltonian flow (see Sect.III.1) ; the property that μ is an invariant measure for T is simply the Liouville's conservation of surface on S_E. In this case the group T depends on a continuous parameter t , and the transformation T_t is a diffeomorphism on X (i.e. is a C^∞ automorphism on X). Another way of defining an abstract dynamical system given a physical Hamiltonian system is to take for X a surface of section Σ , for μ the normalized measure on it, and for T the mapping generating the successive crossings of an orbit with Σ . This time, the mapping is discrete, and its measure-preserving property is associated with the conservation of the area $\oint_{\Gamma \subset \Sigma} p\,dq$.

The discrete mapping associated with a Poincaré surface of section is in fact induced by the continuous dynamical flow, i.e. can be obtained by solving the differential equations of motion. But really abstract discrete mappings are defined by a "discrete time" transformation, which is simply a deterministic algorithm ; among them, the discrete area-preserving mappings of the plane are often considered. These are defined as follows : consider a compact surface X in the plane (x , y) with the measure $d\mu = dx\,dy$, and an application T :

$$X \xrightarrow{T} X$$

$$(x_n, y_n) \longmapsto T(x_n, y_n) = (x_{n+1}, y_{n+1}) \,.$$

The mapping T is area-preserving if

$$\det \begin{bmatrix} \dfrac{\partial x_{n+1}}{\partial x_n} & \dfrac{\partial x_{n+1}}{\partial y_n} \\ \\ \dfrac{\partial y_{n+1}}{\partial x_n} & \dfrac{\partial y_{n+1}}{\partial y_n} \end{bmatrix} = 1 \qquad \text{(III-12)}$$

Examples of such mappings will be presented in the next section.

These notions of dynamical measurable space and measure-preserving transformation have been used as key points in Sect.III.1 to introduce the hierarchy of irregularity for physical Hamiltonian systems. It should be clear now that the definitions given above of ergodicity, mixing, K-property and B-property (Bernouilli) trivially extend to abstract dynamical systems.

III.3 EXAMPLES

Let us now illustrate the concepts just introduced by concrete examples.

III.3.a) Physical Hamiltonian systems

There are now numerous Poincaré maps of physical systems available. Among them, the Toda lattice and the Hénon-Heiles systems are probably the most striking illustrations of the interest of surfaces of section.

The Toda lattice [To-67,To-70] is the system of three particles moving on a ring with exponential repulsive forces between them :

$$H = \frac{1}{2}(p_1^2+p_2^2+p_3^2) + e^{-(\phi_1-\phi_3)} + e^{-(\phi_2-\phi_1)} + e^{-(\phi_3-\phi_2)} - 3.$$

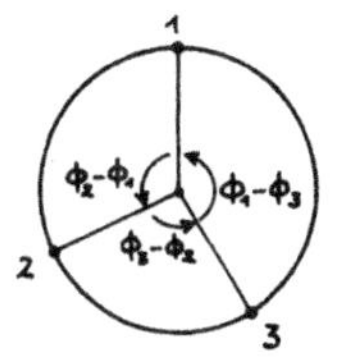

Fig.III.9 - The three-particle Toda lattice

This system possesses, besides the energy, another constant of motion :

$$P = p_1 + p_2 + p_3 = ct$$

associated with the invariance of H under a rigid rotation. Its true number of degrees of freedom is thus reduced from three to two(*). After some manipulation, it is possible to eliminate one irrelevant dynamical variable, and express the Hamiltonian in terms of two coordinates and their canonical momenta :

$$\bar{H} = \frac{1}{2}(p_x^2+p_y^2) + \frac{1}{24}\left[e^{2y+2\sqrt{3}x} + e^{2y-2\sqrt{3}x} + e^{-4y}\right] - \frac{1}{8}\,. \quad \text{(III-13)}$$

Now, the accessible phase space at a given energy $\bar{H} = E$ is located on the energy shell S_E, which is three-dimensional. To investigate the structure of the motion on

---

(*) In the given example, the existence of the extra-constant P reduces the dimension of the accessible phase space from five to three (i.e. the Toda lattice is truly a conservative system with two degrees of freedom). This is because the total momentum P can be considered -by means of an appropriate canonical transformation- as the conjugate momentum of an ignorable (or cyclic, or kinosthenic) variable, i.e. a coordinate which does not appear in the Hamiltonian function (see for instance Refs.[La-70,Go-51] for the general procedure of elimination of cyclic variables, and Ref.[LL-83] for the reduction of numbers of degrees of freedom of the Toda lattice).

S_E, Ford et al. [FST-73] studied the pattern of the motion on the Poincaré surface of section $\Sigma_y = (y, p_y)$:

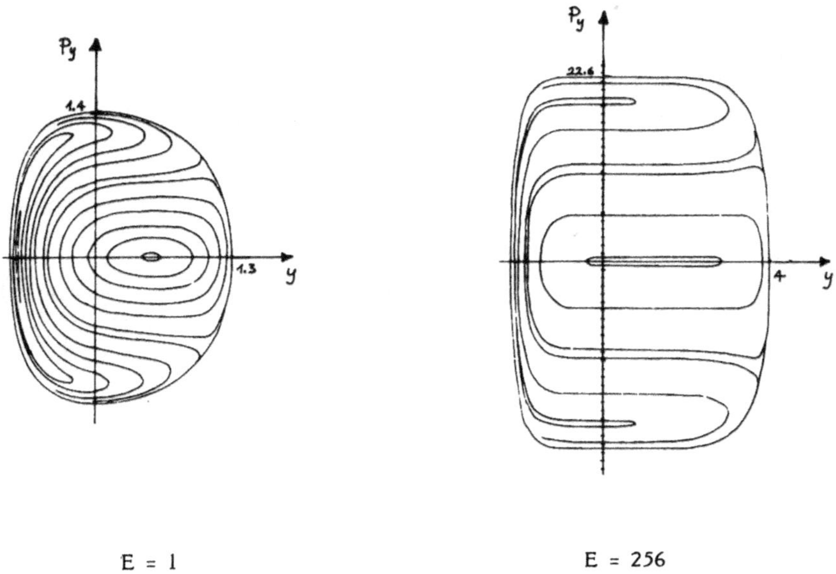

Fig.III.10 - Poincaré maps of the Toda lattice, at energies E = 1 and E = 256 (taken from Ref.[FST-73])

At any energy they studied, they got points of the Poincaré map lying on smooth curves, as if the system had, besides the energy, a further first integral. And indeed, another constant of motion was discovered by Hénon (He-74) and independently by Makanov [Ma-74], whose explicit form is :

$$F = 8p_x\,(p_x^2 - 3p_y^2) + (p_x + \sqrt{3}\,p_y)\,e^{2y - 2\sqrt{3}x} - 2p_x\,e^{-4y} + (p_x - \sqrt{3}\,p_y)\,e^{2y + 2\sqrt{3}x} = ct.$$

Thus the Toda lattice is an integrable system for any set of initial conditions. One can notice that in the limit $(x, y) \to (0, 0)$, the Hamiltonian (III-13) reduces to a two-dimensional isotropic harmonic oscillator Hamiltonian :

$$\bar{H}_0 = \frac{1}{2}\left(p_x^2 + p_y^2 + x^2 + y^2\right), \tag{III-14}$$

and the second integral of motion becomes simply proportional to the orbital momentum :

$$F \xrightarrow[\substack{(x,y)\to(0,0)\\(p_x,p_y)\to(0,0)}]{} 12\,(y p_x - x p_y)\ .$$

Now, let us look at the Hénon-Heiles Hamiltonian [HH-64], which can be viewed as deduced from (III-13) by truncation after third order terms :

$$H = \frac{1}{2}\,(p_x^2 + p_y^2 + x^2 + y^2) + x^2 y - \frac{y^3}{3}\ . \qquad \text{(III-15)}$$

This system is often considered as a model in several physical problems : vibrating three-atomic molecule, effective mean field (produced by other stars) description of the motion of a star in the galaxy. Contrarily to the Toda lattice, the Hénon-Heiles system has not a bounded energy surface S_E for all values of the energy E, but only for $E \leq \frac{1}{6}$. Fig.(III.1) shows the Poincaré map for the surface of section $\Sigma_y = (y, p_y)$ at $E = \frac{1}{12}$ (several trajectories are reproduced on the same map).

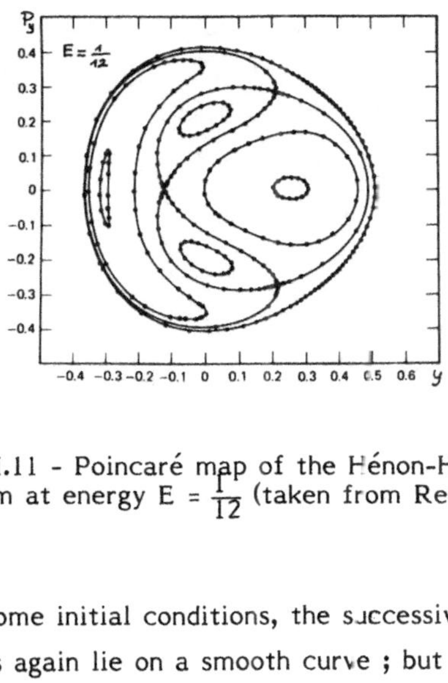

Fig.III.11 - Poincaré map of the Hénon-Heiles system at energy $E = \frac{1}{12}$ (taken from Ref.[HH-64])

For each trajectory, the emerging pattern seems to be an invariant curve, and one may be tempted to conclude that the system is integrable. Therefore, Fig.(III.12) [obtained at $E = \frac{1}{8}$] is a surprise :

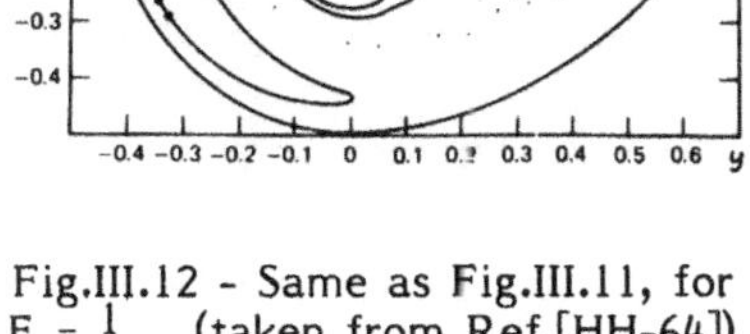

Fig.III.12 - Same as Fig.III.11, for $E = \frac{1}{8}$ (taken from Ref.[HH-64])

For some initial conditions, the successive points again lie on a smooth curve ; but there are also points distributed erratically (which all correspond to the same set of initial conditions). Thus, a chaotic region of the energy surface has appeared by increasing the energy from $\frac{1}{12}$ to $\frac{1}{8}$: the truncation of the Toda lattice Hamiltonian has destroyed integrability ! Looking next at E = $\frac{1}{6}$ (Fig.III.13), one sees that the chaotic region occupies now most of the area of the energy surface, and invariant curves are reduced to small islands.

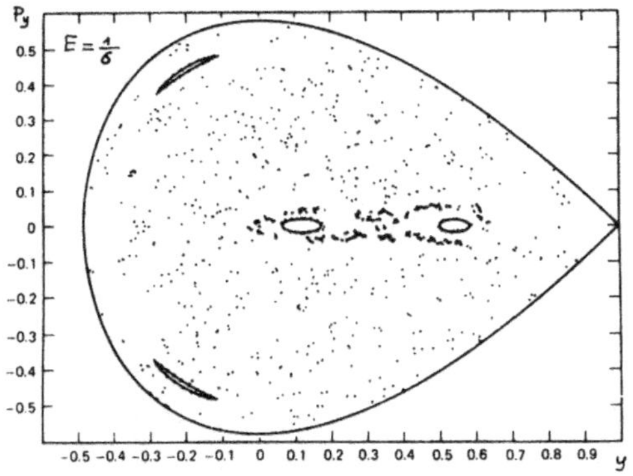

Fig.III.13 - Same as Fig.III.11, for E = $\frac{1}{6}$ (taken from Ref.[HH-64])

The Hénon-Heiles system is a very simple example showing nicely the complicated structure of the phase space, and the sudden changes of this structure by increasing the energy (More comments can be found in Ref.[He-83]). Many other examples have been considered, in particular those illustrating the mechanism of destruction of the invariant tori of a perturbed integrable system ; but these notions require more technical background than proposed here. We would just like to make the following comment about Fig.(III.12). One may interpret the invariant curves as traces of distorted invariant tori which remain present after perturbing the integrable limit (III.14) of (III.15) ; however, the persistance of such tori cannot be predicted by KAM theorem, which requires (only as a sufficient condition) the frequency ratio of the unperturbed oscillator to be "sufficiently irrational", which of course is not true for the Hénon-Heiles system ($\omega_1 = \omega_2 = 1$).

III.3.b) Number-theoretic abstract dynamical systems

1) Rotations of the Circle

Consider a unit circle C (circumference length = 1), take some point on C as origin O, and choose a sense to define the arc joining any point A of C with O (i.e. to measure the abscissa of A). Define now the mapping of C on itself by the rotation

$$\begin{cases} C \xrightarrow{T} C \\ x \in C \longmapsto Tx = x + \omega , \end{cases}$$

where x is the abscissa of any given point of C, and where ω is any real number.

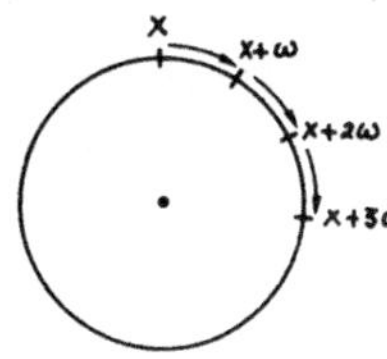

Fig.III.14 - Rotations of the circle

This transformation can also be viewed as the mapping

$$\begin{cases} X=[0,1[\xrightarrow{\ T\ } [0,1[\\ x\in X \longmapsto \mathrm{Frac}(x+\omega)\,, \end{cases} \tag{III-16}$$

where "Frac" denotes the fractional part of a number. The transformation T clearly preserves the Lebesgue measure $d\mu = dx$ on X. Without loss of generality, one can assume that $0 < \omega \leq 1$.

Suppose first that ω is a rational number, which can be written as the irreducible fraction $\omega = \frac{p}{q}$ (p and q are prime to each other). One can easily see that the mapping generates exactly q different points -which therefore are fixed points- $Tx, T^2x, \dots, T^q x = x$, i.e. the initial point x is recovered for the first time at the qth iteration : every orbit is periodic, and the system is <u>not ergodic</u>.

Take next ω irrational ; now, $T^n x$ (n = 1,2,...) takes infinitely many different values. It can be shown that, this time, the mapping is <u>ergodic</u> for the Lebesgue measure (see for instance [AA-67]). It is clear, however, that T is <u>not mixing</u>, since two points initially close to each other remain so for ever : the shape of any volume in X-or length in this case- remains strictly unchanged by the successive iterations. This mapping is therefore an example of a dynamical system which is <u>ergodic</u>, but <u>not chaotic</u> in the sense that it is perfectly stable with respect to initial conditions (it is intuitively evident that the entropy of this system is zero).

The preceding considerations can be extended without difficulty to the translations(*) of the N-dimensional torus T^N. The problem is reformulated by replacing x and ω by vectors $\underline{x}$ and $\underline{\omega}$, the space X_N is now the N-direct product $[0,1[\times [0,1[\times \dots \times [0,1[$, and the measure is the Lebesgue measure $dx_1 \dots dx_N$ on X_N. One gets readily the generalization of the preceding result : if the torus T^N is irrational (see Sect.III.1), the mapping is ergodic on T^N. Notice that this mapping can be viewed as the discrete time evolution of a integrable Hamiltonian system on an irrational torus according to Eq.(III-1). Indeed, $T^n \underline{x} = n\underline{\omega} + \underline{x}$ (n = 1,2,...), which is exactly equivalent to Eq.(III-1) taken at t = 1,2,.... In some sense, the system is an "ergodic clock" (**).

(*) It seems that the term "translations" is preferred in the literature for the torus T^N, whereas "rotations" is more frequently used for the circle, i.e. for the torus T^1 .

(**) As already emphasized, the concept of ergodicity is relative to a given space X. Here, we mention the ergodicity of a class of integrable physical systems on <u>its N-dimensional invariant torus</u>. But it is clear that the physical system itself is not ergodic on the energy shell, except for N = 1.

Let us finally mention a nice application of the preceding results, proposed in [AA-67]. Consider the numbers $\alpha_n = 2^n$ (n = 1,2,...) and ask what is the probability distribution p_k of k (k = 1,2,...,9) for the first decimal of α_n ; the answer is $p_k = \log_{10}(1 + \frac{1}{k})$. What is remarkable in this example is that the result can be derived in a very simple way, using only the ergodicity of (III-16) (see the proof in [AA-67]).

2) Continued fraction expansion

Let us define the one-dimensional automorphism T :

$$\begin{cases} X = [0,1[\xrightarrow{T} [0,1[\\ x \in X \longmapsto Tx = \begin{cases} \mathrm{Frac}(1/x) & \text{if } x \neq 0 \\ 0 & \text{otherwise} \end{cases} \end{cases}$$

This transformation is related to the continued fraction expansion of the starting point :

$$x = \cfrac{1}{a_1(x) + \cfrac{1}{a_2(x) + \cdots}}$$

($a_i(x)$ (i = 1,2,...) are positive integers) by :

$$a_n(x) = \mathrm{Int}\left(\frac{1}{T^{n-1}x}\right)$$

where "Int" denotes the integer part of a number.

The mapping T does not preserve the Lebesgue measure, but preserves the Gauss measure $d\mu = \frac{1}{\ln 2} \frac{dx}{1+x}$. The dynamical system (X , μ ,T) is a K-system, and even a Bernouilli system, whose Kolmogorov-Sinai entropy has been calculated by Rohlin [Ro-61] :

$$h_{K-S}(T) = \frac{\pi^2}{6 \ln 2} \ .$$

Various measure-theoretic results are known for continued fraction expansions. Most of them are due to Khinchin (see [Kh-64] for a very clear account), and also to Lévy [Le-54], who were very successful in proving highly non-trivial results, using old-fashioned methods. Many of their results can now be obtained as straightforward consequences of ergodicity. For instance, let us quote the following properties, whose simple derivation can be found in Ref.[Bi-78].

- The asymptotic relative frequency P_k of k among the partial quotients $a_1(x)$, $a_2(x)$,...., $a_n(x)$,...., is known to be

$$P_k = \frac{1}{\ln 2} \ln \frac{(k+1)^2}{k(k+1)} \qquad \text{(III-17)}$$

for almost all x .

- $$\lim_{n\to\infty} \sqrt[n]{a_1(x)\cdots\cdots a_n(x)} = \prod_{k=1}^{\infty} \left(1+\frac{1}{k^2+2k}\right)^{\ln k/\ln 2} \quad a.e.$$

("a.e." stands for "almost everywhere")

- $$\lim_{n\to\infty} \frac{1}{n} \ln \left| x - \frac{p_n(x)}{q_n(x)} \right| = -\frac{\pi^2}{6\ln 2} = -h_{K\text{-}S}(T) \quad a.e.,$$

where $\frac{p_n(x)}{q_n(x)}$ is the nth-order convergent of x (or rational approximation to x):

$$\frac{p_n(x)}{q_n(x)} = \cfrac{1}{a_1(x) + \cfrac{1}{a_2(x) + \cdots + \cfrac{1}{a_n(x)}}}$$

($p_n(x)$ and $q_n(x)$ are prime to each other)

- $$\lim_{n\to\infty} \frac{1}{n} \ln q_n(x) = \frac{\pi^2}{12\ln 2} \quad a.e.$$

Notice that all the properties derived from ergodicity hold on X = [0,1[, <u>except for a set of numbers of zero measure</u>. In particular, rational numbers are trivially excluded from X for the applicability of these results, since they have only a finite number of non-zero partial quotients a_k . Also excluded are quadratic irrational numbers (i.e. irrational roots of a quadratic equation with integer coefficients), whose sequence of partial quotients $\{a_k(x)\}$ becomes periodic for sufficiently large values of k : for any quadratic irrational, there exist two integers k_0 and p such that, for every $k \geq k_0$, $a_k = a_{k+p}$. (Evidently, Eq.(III-17) cannot hold for such numbers).

3) Pseudo-random number generator

The following mapping is often quoted as a one of the nicest examples of complete randomness generated by a deterministic algorithm. Define T as

$$\begin{cases} X = [0,1] & \xrightarrow{T} [0,1[\\ x \in X & \longmapsto Tx = \mathrm{Frac}\,(rx)\,, \end{cases} \tag{III-18}$$

where r is any positive integer.

One should first note that this application is not one-to-one. Indeed, if we choose $r = 2$, we see that the inverse image of a point x_{n+1} is not uniquely defined, since $x_n = \frac{x_{n+1}}{2}$ and $x'_n = \frac{x_{n+1}+1}{2}$ are both possible solutions of $Tx_n = x_{n+1}$. Anyway, this "time reversal invariance breaking" is not essential.

The transformation T -called the r-adic transformation- is known to leave the Lebesgue measure invariant(*). Let us first give a very simple interpretation of T ; for illustration, let us fix $r = 10$. We write x as :

$$x = 0.\, a_0 a_1 \ldots\ldots a_n \ldots\ldots \,, \tag{III-19}$$

where $\{a_i / a_i = 0,1,2,\ldots,9\}$ is the sequence of digits of the representation of x in decimal basis. Clearly, the effect of T on x is simply to shift the decimals one position to the left :

$$T^n x = 0.\, a_n a_{n+1} \ldots\ldots \,. \tag{III-20}$$

The same result (III-20) holds in the general case where Eq.(III-19) represents the number x, written in base-r (now, $a_i = 0,1,\ldots,r-1$), as can be seen from the general expression)

$$x = \sum_{n=0}^{\infty} \frac{a_n}{r^{n+1}} \,.$$

Now, let us ask for the degree of irregularity of this dynamical system ; simple intuition suggests that it should be strongly chaotic. Indeed, let us consider two initial points x and x' which are close to each other in the sense that their first k digits are the same, and look at the distance between $x_n = T^n x$ and $x'_n = T^n x'$ as n increases from one to infinity. For n less than k, the first $(k-n)$ digits

(*) To "visualize" this property, take for instance $A = [0, \frac{1}{2}]$.
Then, for $r = 2$, $T^{-1}A = [0, \frac{1}{4}] \cup [\frac{1}{2}, \frac{3}{4}]$, so $\mu(A) = \mu(T^{-1}A) = \frac{1}{2}$

of x_n and x'_n remain the same, i.e. the distance $d(x_n, x'_n)$ does not increase in an "uncontrolable" way. But as soon as $n > k$, the two points x_n and x'_n have completely forgotten their original relationship, so that the distance $d(x_n, x'_n)$ may now evolve in a completely erratic manner as n increases (The knowledge of the first p digits of a number x tells nothing about the following digits of x). Therefore, this dynamical system seems to exhibit the strongest possible sensitivity to a perturbation of initial conditions.

This intuition is indeed corroborated by the mathematical property of the dynamical system to be a Bernouilli flow, i.e. to be at the top of the irregularity scale defined in III.1. Moreover, one can show in this case that the successive numbers of the sequence of r-based digits $\{a_0, a_1, \dots, a_n, \dots\}$ for a typical initial condition x (by "typical" we mean for almost all x of the interval X) are completely uncorrelated ; this can be explained in the following way. As mentioned in Sect.III.1, a necessary condition for a system to be Bernouilli is that there exist a partition of the dynamical space having the property that the outcomes of successive measurements of the system are completely uncorrelated. In this case, this means that one can find a partition P_0 of the interval [0,1] in such a way that the successive numbers $x_n = T^n x$ fall into the different atoms of the partition P_0 completely at random ; and this partition P_0 happens to be precisely the set of r segments $[0, \frac{1}{r}, \frac{2}{r}, \dots, \frac{r-1}{r}, 1]$, so that the property of the successive measurements to be uncorrelated simply, means that the sequence $\{a_0, a_1, \dots, a_n, \dots\}$ is completely random, i.e. the successive applications of the mapping T can be viewed as a sequence of games with a roulette wheel having r equiprobable outcomes.

Applying the formula (III-8), we get for the Kolmogorov-Sinai entropy of $T^{(*)}$:

$$h_{K\text{-}S}(T) = -\sum_{i=1}^{r} \frac{1}{r} \ln \frac{1}{r} = \ln r . \qquad \text{(III-21)}$$

Notice that this value of $h_{K\text{-}S}$ is exactly the classical result in information theory for the entropy of a system of Bernouilli trials with r equiprobable possibilities. This entropy, or a priori lack of information on the system, is clearly an increasing function of r, since the degree of uncertainty increases with the number of possibilities (a player who would have the choice of $r > 1$ for a given fixed, r-independent gain, would evidently choose r = 2, the coin toss !) . This property is reflected in (III-21).

Let us now investigate how the mapping T can be used as a pseudo-random number generator(**) and, for this purpose, first look at the consequences of ergodicity. Consider

(*) Here, we admit that the partition P_0 satisfies another necessary condition (that we did not mention) for the system to be Bernouilli.

(**) The process T, applied to a given X having an infinite number of digits, is a truly random number generator. The term "pseudo" refers to the finitness of the number of digits with which the computer works in practice.

the characteristic property of ergodicity (III-5), translated for our discrete time process as :

$$\lim_{n\to\infty} \frac{1}{n} \sum_{p=0}^{n-1} f(T^p x) = \int_0^1 f(x)dx \qquad a.e., \tag{III-22}$$

where $f(x)$ is integrable on [0,1] for the Lebesgue measure dx . Choose now f to be the indicator I_Ω (or characteristic function) of any subset Ω of [0,1] (of non-zero measure) :

$$\lim_{n\to\infty} \frac{1}{n} \sum_{p=0}^{n-1} I_\Omega (T^p x) = \int_\Omega f(x)dx = L(\Omega) \quad a.e., \tag{III-23}$$

where $L(\Omega)$ is the length (the Lebesgue measure) of Ω. This equation simply tells us that the proportion of numbers generated by successive applications of T which belong to any interval $\Omega \subset [0,1[$ asymptotically tends to the length $L(\Omega)$ of Ω . Notice that this property is shared by all dynamical systems which are ergodic on [0,1] for the Lebesgue measure, and in particular holds for the example 1) (rotations of the circle) : all these systems are such that the numbers generated by $\{T^n : n = 1,2,\ldots\}$ asymptotically have a uniform distribution on [0,1]. However, all these systems cannot be used as random number generators. Consider for example the rotations of the circle, which, as we have seen, produce points in an inexorable regular pattern : clearly, in this case, the successive points obtained by the mechanism T have the strongest correlations that can be imagined . On the contrary, the algorithm (III-18) produces completely uncorrelated points, or, in other words, the successive trials associated with successive applications of T are, in this case, completely independent.

Several consequences of the property just discussed are worth to be mentioned.

i) consider the case where the subset Ω is any of the segments $\Omega_k = [\frac{k}{r}, \frac{k+1}{r}]$ ($k = 0,1,2,\ldots,r-1$). Then, $I_\Omega (T^p x)$ is equal to one if the pth digit of x is k , and zero otherwise. Thus, the repeated experiments $\{T^p x \ (x = 01,..n-1)$ in or out of $\Omega_k\}$ can be viewed as independent trials for a random variable which takes the value k ($k = 0,\ldots,r-1$) with probability $\frac{1}{r}$. One can therefore interpret Eq.(III-23) as follows. The l.h.s. represents the relative frequency of the value k , and the r.h.s. the probability $\frac{1}{r}$ for getting k . According to Eq.(III-23), the probability that the l.h.s. is equal to the r.h.s. is one, i.e. the relative frequency is almost surely equal to the probability, which is the strong law of large numbers (i.e. except for a set of starting points of zero measure).

ii) Another illustration of the strong law of large numbers is due to Borel, and can be recovered very nicely directly from the ergodic theorem. A number $x \in [0,1[$ is said to be normal to base r if each digit of its expansion in this basis has the relative frequency $\frac{1}{r}$. A number is said to be normal if it is normal to every basis r. As a consequence of ergodicity of T, one can show that almost all numbers are normal. To prove this statement, it is sufficient to show that almost all numbers are normal to base r, for each fixed r. Indeed, this ensures that all the numbers are normal to base r, except for a set A_r of zero measure ; therefore, the union of all numbers being normal to any base remains of measure one (the measure of $\bigcup_r A_r$ is zero). Let us illustrate this property for r=2. We define

$$f(x) = \begin{cases} 0 \text{ for} & x < \frac{1}{2} \\ 1 \text{ for} & x \geq \frac{1}{2} \end{cases} .$$

Then, the ergodic property (III-22) for $f(x)$ gives

$$\lim_{n \to \infty} \frac{1}{n} \sum_{p=0}^{n-1} f(T^p x) = \frac{1}{2} \qquad \text{a.e.} \tag{III-24}$$

To interpret the l.h.s. of Eq.(III-24), let us write the explicit expression of x in binary basis :

$$x = \sum_{n=0}^{\infty} \frac{a_n}{2^{n+1}} \, , \qquad (a_n = 0, 1)$$

then

$$T^p x = \sum_{n=0}^{\infty} \frac{a_{n+p}}{2^{n+1}} \, .$$

We see that $\quad$ if $T^p x < \frac{1}{2}$ $\quad$ then $\quad a_{n+p} = 0$ $\quad$, whereas

if $T^p x \geq \frac{1}{2}$ $\quad$ then $\quad a_{n+p} = 1$.

Therefore,

$$f(T^p x) = \begin{cases} 0 & \text{if } a_{n+p} = 0 \\ 1 & \text{if } a_{n+p} = 1 , \end{cases}$$

and the l.h.s. of Eq.(III-24) is simply the relative frequency of the digit one in the binary expansion of x , which does converge, for almost all x , to its probability $\frac{1}{2}$.

III.3.c) Two-dimensional abstract dynamical systems

1) Translations on the torus T^2

This mapping has been studied in III.3.c). We noticed that it is ergodic, but not mixing, if the two parameters of the translation are rationally independent.

2) Arnold's cat

The Arnold's cat mapping T is the linear automorphism of the two-dimensional torus $T^2 : \{(x, y) \bmod 1\}$ for the measure $dx\,dy$, defined as

$$T(x, y) = (x+y,\ x+2y) \qquad (\text{mod } 1). \tag{III-25}$$

One can also consider T as a mapping of the unit square or to itself :

$$X = [0,1[\times [0,1[\xrightarrow{\ T\ } [0,1[\times [0,1[.$$

To construct a geometrical realization of this automorphism, one first maps the unit square X into the rhombus MX, where $M = \begin{pmatrix} 1 & 1 \\ 1 & 2 \end{pmatrix}$, and then brings back any point ξ lying outside the square into the square by subtracting unity as many time as necessary (one or two times) to each coordinate of ξ .

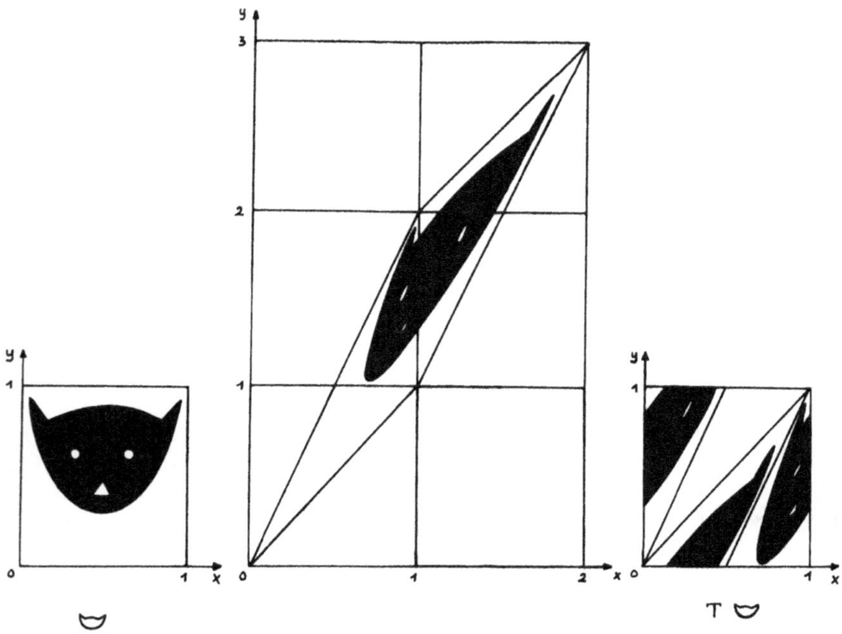

Fig.III.15 - The Arnold's cat mapping : is the starting figure, T represents the first iteration (taken from Ref.[AA-67])

The automorphism T is known to be a K-system ([AA-67]), and the example is nice because this property can nearly be seen (if one does not care about mathematics) "with the eyes". Fig.(III.16) represents the second iteration of T. The cat is already completely unrecognizable, which confirms the first impression of Fig.(III.15) that T undergoes drastic deformations of shapes. It is easy to imagine that the cat will asymptotically be "diluted" into the whole square, which is the mixing property. Now, consider the eigenvalues λ_1 and λ_2 :

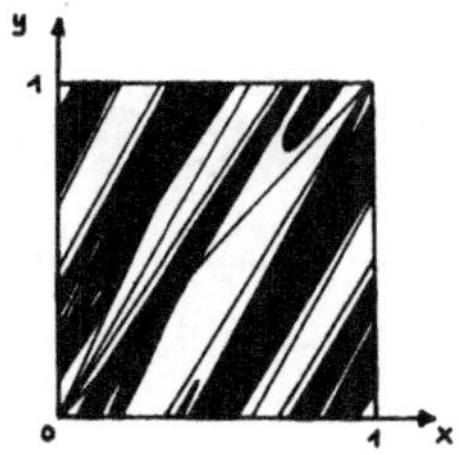

T^2 ☺

Fig.III.16 - The Arnold's cat mapping : second iteration

$$\lambda = \lambda_1 = \frac{1}{2}(3+\sqrt{5}) > 1 > \lambda_2 = \lambda_1^{-1} > 0 .$$

One sees that the direction D_1 defined by the eigenvector associated with λ_1 is expanding under the action of T, whereas the direction D_2 associated to the other eigenvector is contracting : at each step of the process, a rectangle whose sides are parallel to D_1 and D_2 is stretched along D_1 by a factor λ, and compressed along D_2 by the same factor λ. This mechanism is responsible for exponential separation of trajectories, indicating the K-property. After p steps, the expanding factor is simply $\lambda^p = \exp(p \ln \lambda)$, which suggests that the K-S-entropy of the system is simply (this result can be derived rigorously)

$$h_{K-S}(T) = \ln \lambda .$$

Finally, let us notice that the origin O is an invariant point of T, and that the orbits of all points $(x, y) = \left(\frac{p}{q}, \frac{p'}{q}\right)$ are periodic ; one can also show the existence of two invariant curves which are both dense in the square (see again [AA-67]).

3) Baker's transformation

Consider again the unit square, and define the transformation T as

$$\begin{cases} X = [0,1[\times [0,1[\xrightarrow{T} X \\ (x,y) \in X \longmapsto T(x,y) = \begin{cases} \left(2x, \frac{y}{2}\right) \text{ if } 0 \leq x < \frac{1}{2} \\ \left(2x-1, \frac{y+1}{2}\right) \text{ if } \frac{1}{2} \leq x < 1 \end{cases} \end{cases}$$

The name of the transformation is illustrated in Fig.(III.17), in which the unit square transformed under T looks like a piece of dough worked by a baker.

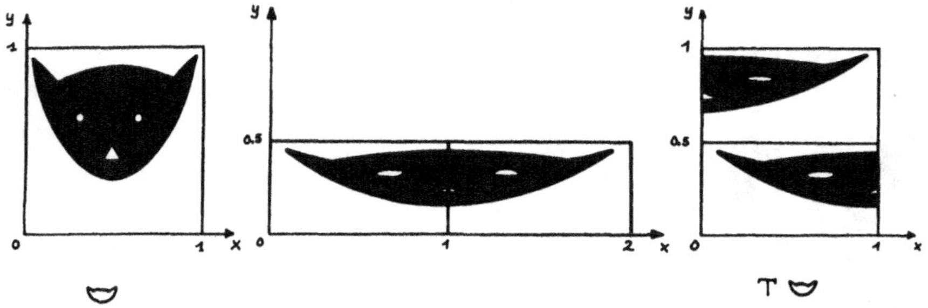

Fig.III.17 - The Baker's transformation : is the starting figure, and T represents the first iteration (taken from Ref.[AA-67]).

The area-preservation of T can be seen on the figure. Also is clear the fact that each iteration of T doubles the number of horizontal "slices" in which the cat is more and more "compressed" and this property suggests the mapping to be mixing. In fact, the dynamical system is much more chaotic than simple mixing, since it is a Bernouilli system. We can easily imagine this strong irregularity by noting that the image of the abscissa x_n of any point of X is, in any case

$$x_{n+1} = \mathrm{Frac}\,(2x_n),$$

which is the dyadic transformation of Eq.(III-18). The results obtained for the one-dimensional mapping Eq.(III-18) ensure that if we consider the partition P_0 of the square into the two atoms $0 \leq x < \frac{1}{2}$ and $\frac{1}{2} \leq x < 1$, then the sequence of measurements of $\{T^n x : n = 0,1,\dots\}$ with respect to the partition P_0 will be completely random, i.e. completely equivalent to a sequence of coin tosses.

Up to now, we have seen some examples of Bernouilli flows, all being abstract dynamical system. Another famous example of Bernouilli mapping is generated by geodesics on surfaces of negative curvature. We shall now investigate a class of physical systems, some of them having also the Bernouilli property.

III.4 EXAMPLES AGAIN : BILLIARDS

A billiard is a dynamical system produced by the free motion of a point particle in a domain D of the plane $\mathbb{R}^2$ bounded by a closed curve Γ -or, more generally, by a set of closed curves Γ_i -, with elastic reflections at the boundary Γ of D (i.e. the tangential component of any trajectory remains constant at Γ, whereas its normal component changes sign)$^{(*)}$. The boundary Γ may contain singular points (vertices) where the tangent is not defined ; we disregard all trajectories coming at some time to a vertex of Γ, which form a set of zero measure. A billiard is therefore a two degrees of freedom system, with at least one constant of motion, the energy $E = \frac{1}{2} m v^2$. Notice that the behaviour of any orbit in phase space depends neither on the mass m of the point particle nor on its velocity v (this peculiar feature of billiards is due to the fact the motion is simply governed by the laws of geometrical optics) ; therefore the properties of this kind of dynamical systems do not depend on the energy E, all the energy surfaces having exactly the same structure.

An orbit in configuration space is an infinite sequence of segments, each one corresponding to a constant velocity vector $\vec{v}$. Therefore, it is natural to search for a Poincaré surface of section Σ such that the successive crossings with Σ correspond to the successive bounces on the "wall" Γ. To do this, let us take as coordinates of any bounce Q_n its curvilinear abscissa s_n along Γ (once an origin O and a sense have been defined on Γ), and $p_n = \sin \alpha_n$, where α_n is the angle of the trajectory leaving Q_n with the normal to Γ at Q_n, pointing towards the interior of D. It is easy to show that the mapping M defined by

$$\begin{pmatrix} s_{n+1} \\ p_{n+1} \end{pmatrix} = M \begin{pmatrix} s_n \\ p_n \end{pmatrix}$$

is area-preserving, i.e. that (see Ref. [Be-81 a] for more details)

$$\det \begin{bmatrix} \dfrac{\partial s_{n+1}}{\partial s_n} & \dfrac{\partial s_{n+1}}{\partial p_n} \\[2ex] \dfrac{\partial p_{n+1}}{\partial s_n} & \dfrac{\partial p_{n+1}}{\partial p_n} \end{bmatrix} = 1 .$$

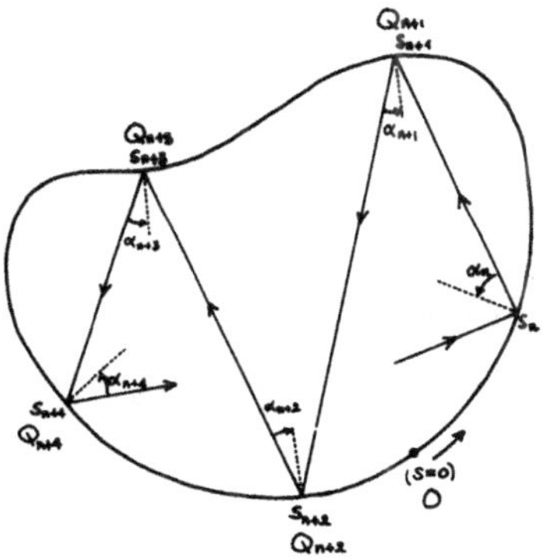

Fig.III.18 - Surface of section coordinates for a billiard

$^{(*)}$ The domain D may also be a non-compact part of $\mathbb{R}^2$: see below the example of the Sinai's billiard.

The study of billiards is of interest in many respects :

i) they are, among the physical Hamiltonian systems, the most studied from a mathematical point of view, and some important analytical properties are known

ii) billiards are very rich in their behaviours : there exist systems lying at the top, as well as at the bottom of the hierarchy of regularity described in Sect.III.1, and also systems which cannot be classified according to this hierarchy, i.e. for which the phase space contains both regular and chaotic regions

iii) the energy surface of billiards is of dimension two, which is the smallest one needed to observe chaotic motion

iv) for our purpose -trying to find signatures of chaos in quantum systems whose classical analogues are chaotic by statistical studies of the spectrum-, drums (quantum billiards) are the most convenient systems : they have an infinite discrete spectrum, and efficient numerical methods are available to compute long series of adjacent levels.

Regular billiards

The rectangle, the circle, and the ellipse are regular billiards, i.e. dynamical systems whose energy surface is a torus. The two independent constants of motion are

i) for the rectangle : the two projections $|v_x|$ and $|v_y|$ of the moduli of the velocity on two axes parallel to the sides of the rectangle

ii) for the circle : the energy, and the angular momentum $L_z = |\vec{L}|$ with respect to the center of the billiard

iii) for the ellipse : the energy, and the product $L_1 L_2$ of the angular momenta with respect to the two foci

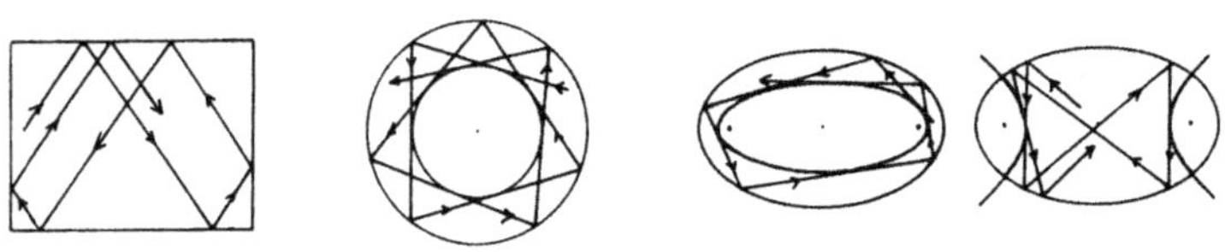

Fig.III.19 - Trajectories in regular billiards. Caustics can be seen for the circular billiard (the caustics are circles of same center as the billiard) and for the elliptic billiard (the caustics are ellipses and hyperbolae confocal with the ellipse defining the boundary of the billiard)

For the circle and the ellipse, the integrability manifests itself by the existence of families of caustics, which reveal the existence of tori (see again Ref. [Be-81 a], and also Ref.[Be-83]).

Chaotic billiards

As a first example, let us consider the Sinai's billiard : the point mass moves in the plane $\mathbb{R}^2$ and bounces on reflecting disks, all of the same radius, forming a periodic lattice as shown in Fig.III.20. By trivial symmetry arguments, one can reduce the study of this system to the study of the motion of a particle in a square, with a circular obstacle around the center. Simple intuition suggests that the dynamical system associated with the Sinai's billiard is strongly chaotic. Indeed, as illustrated in Fig.III.20, a bundle of nearby trajectories reflecting on convex obstacles is drastically defocused, and spreads in a large part of the configuration space after a small number of bounces. The effect of this instability with respect to initial conditions can be estimated : suppose, for instance, that one wants to calculate on a computer an orbit for R = 0.1 ; then, one can show that at each bounce, about one decimal digit is lost for the angle determining the direction of the orbit. This means that, typically, one cannot compute the trajectories in Sinai's billiard for a number of bounces greater than 10-20, which illustrates the dramatic instability of the system ! (see Ref.[Be-78] for more details, and also for other illustrations of the degree of instability, in terms of external perturbations of the physical system).

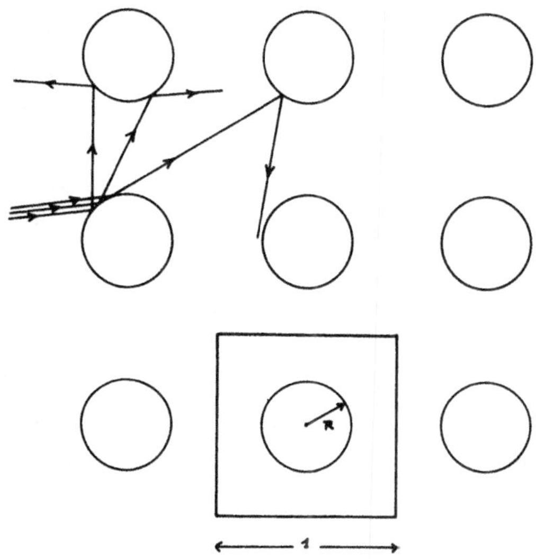

Fig.III.20- The Sinai's billiard

In fact, it has been proved by Sinai [Si-70] that this billiard is a K-system ; it has even been shown more recently to be a B-system. The proof by Sinai -which is very long and very elaborate- has initiated further works on other billiards, conjectured to be also chaotic, and the K-property (and also the B-property) is now known to hold for several billiards, such as, for example the stadium (two half-circles of radius R, joined by two segments parallel to the line passing through their centers), and also the diamond (see Fig.III.21). For the diamond, the chaotic behaviour has the same origin as for the Sinai's billiard, which is the negative curvature of the obstacles. The origin of the instability of the stadium is somewhat less intuitive. A small bundle of trajectories contracts after reflecting on a circular -focusing- part of the boundary, but subsequently expands

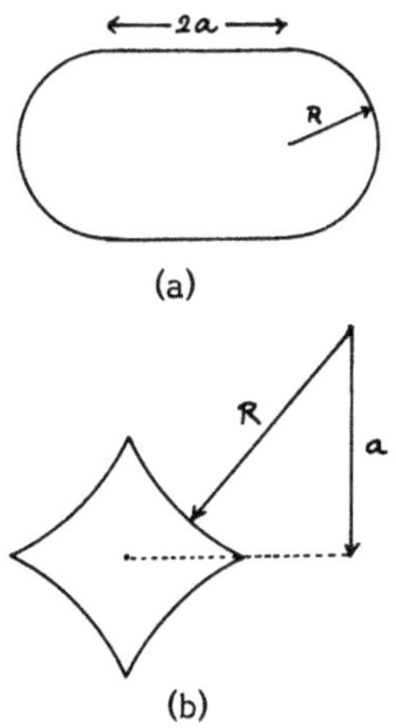

Fig.III.21- (a) the stadium
(b) the diamond

in such a way that the synchronization of successive contractions is made impossible : expansions prevail over contractions, producing the stochastic properties of the billiard.

Notice that in the limits R = 0 (Sinai's billiard : see Fig.III.20) or a = 0 (Stadium : see Fig.III.21-a), the Sinai's billiard and the stadium become integrable systems (a square and a circle respectively). However, it should be emphasized that the K-property (and the B-property) holds for <u>any</u> value of $R > 0$ (Sinai's billiard) or of $a > 0$ (stadium). This means that a regular billiard may be transformed into a strongly chaotic billiard by an infinitely slight perturbation of its boundary ; evidently, the KAM theorem -which would predict the persistance of tori under a small perturbation of the integrable system- does not apply here. This is because the smoothness conditions required for the KAM theorem to be applicable are not satisfied by billiards. So far there exists no equivalent to KAM theorem for billiards. Some results in this direction are known but their field of applicability is still very peculiar ; for instance, Lazutkin [La-73] proved the existence of tori for convex billiards having a very smooth boundary ("very smooth" means that the radius of curvature as a function of the arc length is at least a C^{553} function... which is only a sufficient condition for the tori to exist !).

Finally, let us recall that the ergodicity of these chaotic systems does not exclude the existence of periodic trajectories (these form a set of zero measure of trajectories which do not visit the neighbourhood of all the points of the energy surface). For instance, the trajectory in the stadium passing through the two centers of the circle is periodic ; also are periodic the trajectories hitting the horizontal segments of the stadium at right angle, etc... These periodic orbits can form a family or be isolated. We shall not discuss the questions related to the periodic orbits, though they play a crucial role in the quantization of the classical systems (see [Be-81 b] and [Be-83]).

<u>Pseudo-integrable billiards - Polygons</u>

The chaotic billiards presented above have in common the property that at least one part of the boundary is not a straight line ; this is, in fact, an essential condition to produce the K-property. Indeed, it has be shown that the K-S entropy of billiards in polygons is zero, i.e. trajectories in polygons never show exponential instability ; but can one say more about polygons ?

As a particular kind of polygon, we have already mentioned the rectangle, which is integrable. Also integrable are three kinds of triangles : the equilateral one, and the two right triangles (90°,45°,45°) and (90°,30°,60°). Putting apart the integrable systems, polygons can be divided into two families :

i) irrational polygons (i.e. polygons whose angles are irrational multiples of π). These billiards are conjectured to be ergodic, and perhaps mixing systems (see Ref.[Ho-75]).

ii) rational polygons (all angles are rational multiples of π). For such systems, it is clear that the angles of the segments defining any trajectory with a fixed direction can take only a finite number of different values. Consequently, these systems are certainly not ergodic. In fact, one can show the existence of another first integral F which is independent of H and in involution with H only almost everywhere, i.e. except for a discrete set of singular points (see e.g. Ref.[ZK-75] ; such systems are called pseudo-integrable. Due to the existence of F, the allowed region in phase space is of dimension two, like for an integrable billiard, but the singular points just mentioned prohibit the invariant surfaces from being tori ; these surfaces can be viewed as tori with more than one hole, i.e. two-dimensional manifolds of genus $g > 1$.

Integrable, chaotic and pseudo-integrable billiards will be discussed in Section IV from a quantal point of view.

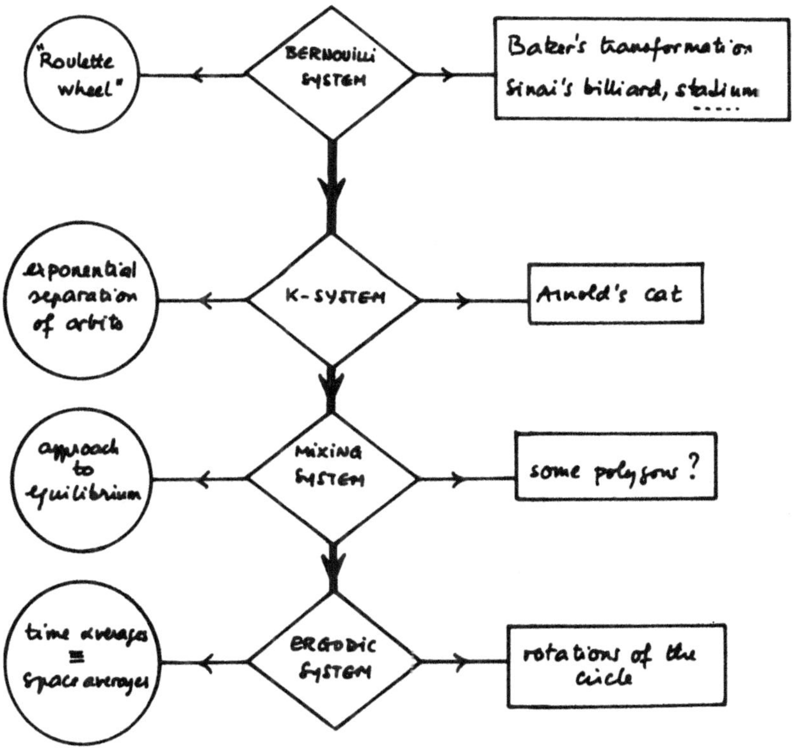

Fig.III.22 - Ergodic theory : the hierarchy of chaos

BIBLIOGRAPHY

Ergodic theory

For an introduction to the concepts of ergodic theory we suggest Refs.[LP-73, Fo-83], in which are stated the main ideas, and [Ba-75, TKS- 83] for a first detailed approach. For a more advanced mathematical study, see [AA-67,Pa-81,CFS- 82] (general ergodic theory), and also [Sh-73,Bi-78,ME-81] (more specialized topics).

Chaotic motion

Among the enormous literature on classical chaotic motion of conservative systems the following references are both elementary enough, and rich in material : [LL-83,Be-78, He-80] (general reports) and [Fo -74,Wh-77] (emphasis on KAM theorem). An extensive bibliography, classified by topics, can be found in [He-80].

IV - QUANTUM CHAOS

As shown in the preceding section, classical chaotic motion is now reasonably understood, at least for the "pure" systems which enter in the classification of the ergodic theory as summarized in Fig.III.22. It is then natural to ask whether there are quantum-mechanical manifestations of classical chaotic motion. We shall use the term "quantum chaotic system" in the precise, and restricted, sense of "quantum system whose classical analogue is chaotic". General discussions on quantum chaotic systems can be found in ref.[Ca-84] ; it appears that there is a large variety, and even disparity, of opinions.

It should be clear from the beginning that the notions which are of most relevance in the classical case may be obvious and useless in the quantum case. Consider for instance the quantum analogue of a classical chaotic billiard, namely a free particle in a box. Remember that its spectrum, determined by the eigenvalue equation (I-22), contains an infinite number of discrete eigenvalues E_n. The time evolution of a quantum state $\Psi(\underset{\sim}{r}, t)$ is given by

$$\Psi(\underset{\sim}{r},t) = \sum_n a_n \psi_n(\underset{\sim}{r}) \exp\left(-\frac{i}{\hbar}E_n t\right) , \qquad \text{(IV-1)}$$

which is quasi-periodic. Therefore there will never be exponential separation of the difference of wave-functions of two close initial states, in contrast to the classical orbits which do show the exponential divergence characteristic of classical chaotic motion (K-systems). We thus expect that the "dictionary" translating classical chaos ⟷ quantum chaos will not be simple.

In what follows we shall restrict ourselves, in the search for manifestations of chaos, to properties of the spectrum and nothing will be said on properties related to wave functions, like structure of nodal lines, properties of the Wigner function, etc.. The tools to characterize fluctuation properties of spectra, described in Section II in connection with random matrix theories, will be thoroughly used.

The first property which has been considered as a possible tool to discriminate between integrable and chaotic systems is the behaviour of the nearest-neighbour spacing distribution for small spacings, i.e., $p(x)$ as $x \to 0$. Berry and Tabor [BT-77], using torus quantization, predict level clustering for integrable systems $p(x) \underset{x\to 0}{\sim} e^{-x}$ in the asymptotic high energy regime(*). This feature can be understood by examining

(*)This prediction excludes harmonic oscillators, for which the arguments of the proof are not applicable. It can be shown, in fact, that the distribution $p(x)$ does not follow the Poisson law for the two-dimensional harmonic oscillator [BGP-84].

the occurrence of degeneracies in a parameter space (see [Be-84]). It can also be viewed as the result of mixing independent spectra(*) . Let us give a heuristic argument illustrating this idea. Consider the circular membrane, which is classically integrable (circular billiard), or better, to avoid degeneracies, the semi-circular drum. The eigenfrequencies of

$$(\Delta + k^2)\Psi = (\Delta + E)\Psi = 0 \qquad \text{(IV-2)}$$

with Dirichlet boundary conditions on the semi-circle are given by the zeros of the Bessel functions $J_\nu(x)$ (ν = 1,2,3,...). Let $j_{\nu,s}$ (s = 1,2,3,..) denote the s-th zero of $J_\nu(x)$. The $j_{\nu,s}$'s (s =1,2,...) extent from $\simeq \nu$ to infinity with a density

$$\frac{1}{\pi}\sqrt{1-\frac{\nu^2}{k^2}} \qquad \text{for } k > \nu \qquad \text{(IV-3)}$$

and are practically fluctuation-free. Consider now an interval at high energy containing N levels k_n (n = 1,2,...,N) ordered with increasing value of k_n . The successive values of k_n , which are the zeros $j_{\nu,s}$, correspond to of the order of N different (and unordered) values of ν . The point now is that zeros of J_ν and $J_{\nu'}$, with ν not too close to ν', are likely to be uncorrelated. Consider, for instance, a stretch of ten eigenfrequencies near the 1400th eigenfrequency. The labelling (ν , s) of successive eigenfrequencies is as follows : (82,4), (70,7), (37,18), (45,15), (5,32), (32,20), (3,33), (1,34), (18,26),(60,10). We are therefore in a similar situation as when randomly superposing picket fences, which leads to a Poisson spectrum(**).

It has been predicted that a drastic change in the spacing distribution $p(x)$ takes place when going from an integrable case (Poisson) to a non-integrable case. Berry [Be-83,Be-84] predicts $p(x) \underset{x\to 0}{\sim} x$, like in GOE, whereas Zaslavsky [Za-81] predicts $p(x) \underset{x\to 0}{\sim} x^\sigma$ where σ is related to the Kolmogorov entropy h_{KS} of the corresponding classical system. On the other hand, Richens and Berry [RB-81] predict that level repulsion will also be present for systems which classically are pseudo-integrable. In what follows we shall discuss some recent results concerning fluctuation-

---

(*) In section II we saw that the result of superposing randomly highly correlated spectra -the argument works even for picket fences- is to produce a Poisson spectrum. We also saw that the effect of superposing shell model spectra corresponding to different values of J^π , each one having level repulsion, is to destroy all kinds of correlations, and in particular level repulsion (see Fig.II.11).

(**) The keypoint of this heuristic argument is the independence of the positions of successive levels in a given interval for large k (or E). Of course, this does not hold for the spectrum $m + \alpha n$ (m, n = 12,3,...; α irrational) of the two-dimensional harmonic oscillator, which leads to strongly correlated levels in any interval of the spectrum.

properties of spectra of systems whose classical analogues are either integrable or chaotic. We shall not confine the discussion to the presence or absence of level repulsion but, for reasons that should be clear to the reader from the discussion of Section II, take as reference patterns the entire Poisson- and GOE- fluctuations.

IV.1 VIBRATIONS OF THE MEMBRANE (QUANTUM BILLIARDS)

Towards the end of Section III, reasons were given for putting special emphasis in the study of billiards. And examples were shown of integrable, chaotic and pseudo-integrable billiards. Let us now study some of their properties in the quantum case, specifically their spectral properties. The cases to be discussed are shown on Fig.IV.1 : circle (integrable), Sinai's billiard and stadium (both strongly chaotic, in fact Bernouilli systems). Mention should be made of the pioneering work in the direction we follow : Refs.[MK-79,CVG-80] for the stadium and ref.[Be-81b] for Sinai's billiard. To determine the eigenvalues of Eq.IV-2 with Dirichlet boundary conditions, an efficient method has been proposed by Berry [Be-81b]. It is inspired on the work in solid state physics, by Korringa, Kohn and Rostoker, to determine the high-energy bands at the centre of the Brillouin zone. Once a sequence of eigenvalues is obtained, there is no ambiguity in separating the average part and the fluctuating part of the spectrum. Indeed, we know that for the systems we are considering the smooth part is given by Eq.(I-24) (see Fig.I.7). The procedure is therefore to first compute a sequence of eigenvalues $\{E_i\}$ and then to unfold the spectrum via Eq.(I-27), where N_{av} is given by Eq.(I-24). One finally has a sequence of points $\{x_i\}$ with mean spacing equal to unity, all over the spectrum.

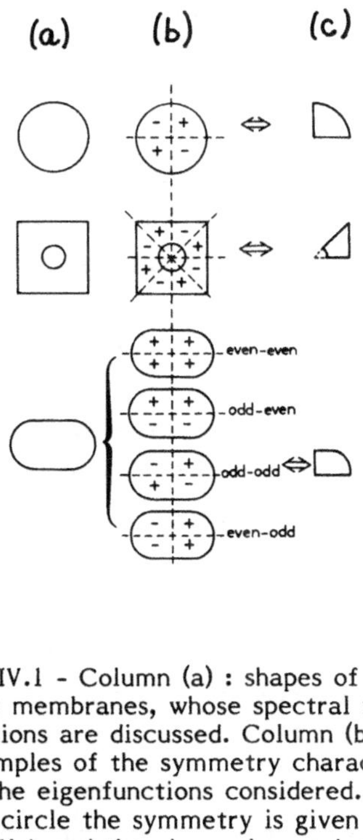

Fig.IV.1 - Column (a) : shapes of different membranes, whose spectral fluctuations are discussed. Column (b) : examples of the symmetry character of the eigenfunctions considered. For the circle the symmetry is given by Eq.IV-4 and the eigenvalues coincide with those of the quarter of a circle. For Sinai's billiard shape, the symmetry is given by Eq.IV-5 and the eigenvalues are those of the desymmetrized Sinai'billiard, as shown on column (c). For the stadium, the four possible symmetry classes are indicated (see Eqs. IV-6). For the odd-odd symmetry, the eigenvalues are those of the quarter of stadium, as indicated on column (c).

Let us now present some results. To illustrate how the analysis is performed, we consider first a "trivial" case, the case of the circular membrane. To avoid two-fold degeneracies, we take a semi-circle or a quarter of a circle. Equivalently, the quarter of a circular membrane corresponds to eigenvalues of the full circle whose associated eigenfunctions have the symmetry property illustrated on column (b) of Fig.IV.1, namely

$$\Psi(x,y) = -\Psi(-x,y) = -\Psi(x,-y) \quad . \qquad \text{(IV-4)}$$

The eigenvalues are given by the squares of the roots $j_{\nu,s}$ of the Bessel functions $J_\nu(x)$, with $\nu = 2,4,6,\ldots$ and $s = 1,2,3,\ldots$ On Fig.I.7a is shown the cumulative density N(E) for the first eigenvalues. On Fig.IV.2 are shown the results for the spacing distribution $p(x)$ and for the average value of Δ_3 as a function of the length L. On Table IV.1 is given the value of the correlation coefficient C between adjacent spacings. As can be seen, the computed values are far from GOE-values and rather close to, although not consistent with, the Poisson-values (except for C, which is consistent with the Poisson-value C=0). We interpret these results as follows. The spectra under study are not translational invariant or stationary. We expect that the characteristic fluctuation patterns (GOE or Poisson) are valid in the asymptotic regime (high energy). The departures of the results presented for the circular membrane from Poisson indicate that the asymptotic regime has not yet been reached. Work is in progress to determine the rate of convergence to the asymptotic regime (Poisson in the present case).

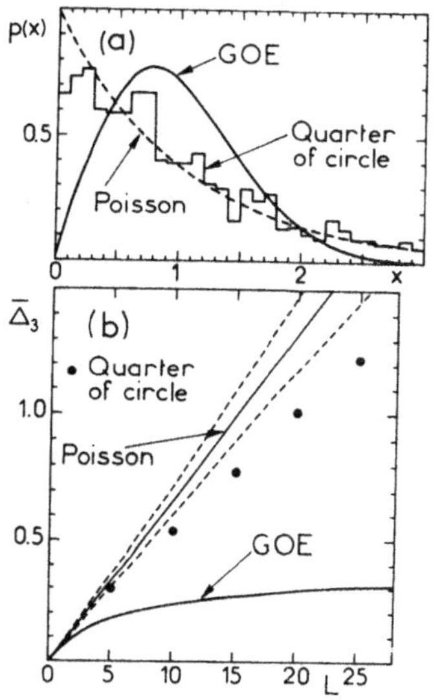

Fig.IV.2 - Results of level fluctuations for the first 675 eigenvalues of a circular membrane corresponding to eigenfunctions with the symmetry indicated on column (b) of Fig.IV.1 : (a) Nearest-neighbour spacing histogram. (b) $\bar{\Delta}_3$ as a function of L ; dashed lines for Poisson and small bars for GOE indicate the effect of the finiteness of the sample as predicted by the theory (one standard deviation). Curves corresponding to the Poisson case (stretch of uncorrelated levels) and to the random matrix theory predictions (GOE) are drawn for comparison.

Let us now consider chaotic systems. We treat the desymmetrized Sinai's billiard as shown on column (c) of Fig.IV.1, or, equivalently, the solutions of the Schrödinger equation IV-2 with Sinai's billiard as the boundary and with the symmetry

$$\Psi(x,y) = -\Psi(-x,y) = -\Psi(x,-y) = -\Psi(y,x) \qquad \text{(IV-5)}$$

	1/4 Circle	Sinai	Stadium (one symmetry)	Stadium (mixed symmetries)	GOE	Poisson
Number of levels	675	740	810	3200		
C	-0.04 ±0.04	-0.30± 0.04	-0.31 ± 0.04	-0.18±0.02	-0.27	0

Table IV.1

Values of the correlation factor C between adjacent spacings (Eq.I-37) for different systems investigated. GOE- and Poisson-values are given for comparison.

To improve the statistical significance of the results, we consider four different cases by changing the value of the parameter R and perform a joint analysis of the results as explained in [BGS-83,BGS-84a]. The results are presented on Fig.IV.3 and Table IV.1. In contrast to the case of the circular membrane, for which the results are close to the Poisson spectrum, we see now that the results are fully consistent with GOE-predictions for $p(x)$ (not only for small values of x), for $\overline{\Delta}_3$ and for C.

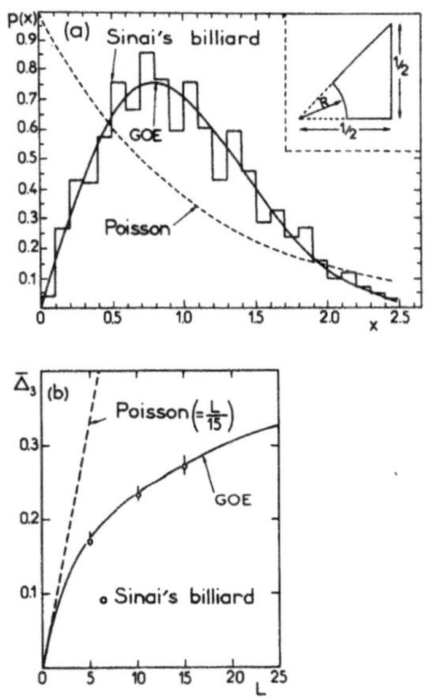

Fig.IV.3 - Results of energy level fluctuations for desymmetrized Sinai's billiards as specified on the upper right corner of the figure. 740 levels have been included in the analysis, corresponding to the 51-th to 268-th level for R=0.1, 21-th to 241-th level for R=0.2, 16-th to 194-th level for R=0.3, 11-th to 132-th level for R=0.4. See caption of Fig.IV.2 for further explanation (taken from [BGS-83,BGS-84a]).

Let us investigate another system whose classical analogue is chaotic, namely one particle in a two-dimensional box whose boundary is the stadium. The solutions of eq.(IV-2) can be classified according to four different symmetry classes :

even-even

$$\Psi(x,y)=\Psi(-x,y)=\Psi(x,-y) \qquad \text{(IV-6)}$$

odd-even

$$\Psi(x,y)=\Psi(-x,y)=-\Psi(x,-y) \qquad \text{(IV-6')}$$

odd-odd

$$\Psi(x,y)=-\Psi(-x,y)=-\Psi(x,-y) \qquad \text{(IV-6'')}$$

even-odd

$$\Psi(x,y)=-\Psi(-x,y)=\Psi(x,-y) \qquad \text{(IV-6''')}$$

and the spectrum, despite the symmetry of the problem, contains no degeneracies. Take now the eigenvalues corresponding to a definite symmetry-class, for instance

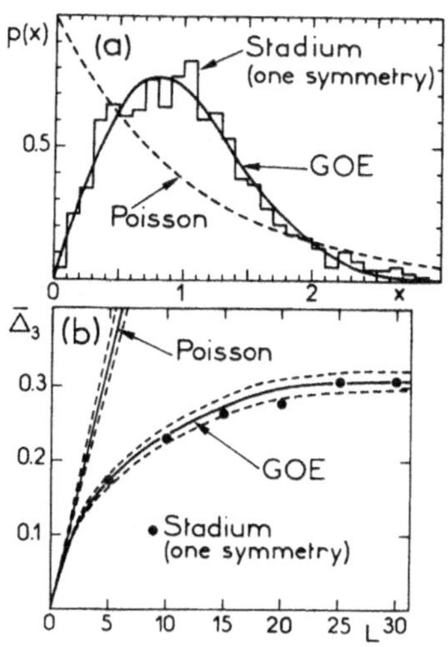

Fig.IV.4 - Results of level fluctuations for the first 810 eigenvalues of a membrane whose boundary is a stadium. They correspond to eigenfunctions with odd-odd symmetry (see Fig.IV.1). The ratio 2a/R of the straight line segment to the radius is 1 (see Fig.III.21a). See caption of Fig.IV.2 for further explanations (taken from [Sc-84,BGS-84b]).

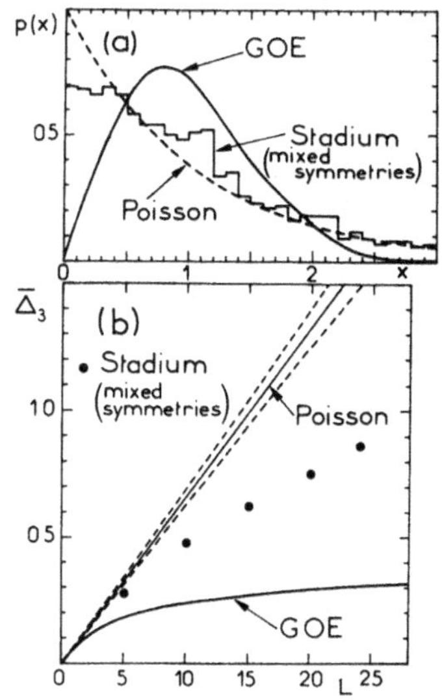

Fig.IV.5 - Same as in Fig.IV.4 but with the four different symmetry classes, as specified in Fig.IV.1. The spectrum analyzed contains the first 3200 eigenvalues. See caption of Fig.IV.2 for further explanations (taken from [Sc-84, BGS-84b]).

the odd-odd case. Results are presented on Fig.IV.4 and Table IV.1. Again we have a remarkable agreement with GOE-predictions. A similar agreement is obtained when analyzing the eigenvalues belonging to the other three symmetry classes. Consider finally the spectrum which contains all levels corresponding to the four symmetries of the stadium (IV-6,6',6",6'"). The results change drastically. They are shown on Figs.I.7b, IV.5 and Table IV.1. The spectrum fluctuations are intermediate between GOE and Poisson. The results would be closer to Poisson-fluctuations if more than four different families characterized by different quantum numbers would be present. This is in exact analogy with what happens when superposing several different GOE spectra. Or with compound nucleus resonances, when no attention is payed to quantum numbers and the spectrum results from mixing several pure series.

IV.2 OTHER EXAMPLES

The systems treated in the preceding subsection were classically either integrable or strongly chaotic (highly-unstable ergodic systems) and correspondingly we saw that the spectra show Poisson- or GOE- fluctuation patterns. What happens then for systems displaying a more complicated structure in phase-space ? We have seen in Section III that there exist systems whose Poincaré sections look as belonging to an integrable system at some particular energies but that, by changing the energy, islands of chaoticity appear and increase until they cover the full Poincaré sections, showing that in fact the system is non-integrable and chaotic. Does this behaviour have some manifestation in the corresponding quantum systems ? And is there a transition parameter characterizing the relative importance of chaotic/non-chaotic regions which manifests in the spectral fluctuations ? The problem is very interesting but difficult and, although several efforts in this direction are worth mentioning -the study of a two-dimensional harmonic oscillator plus quartic terms in ref.[HYK-84], of a system of kinetically coupled Morse oscillators in ref.[MT-84], of the Hénon-Heiles Hamiltonian (Eq.III-15) in ref.[PS-83]- it is still premature to draw general conclusions from them. Mention should also be made of the study of a one-parameter family of billiards which classically goes continuously from the integrable to the chaotic regime ; in the quantum case the spacing distribution $p(x)$ seems to continuously go from a Poisson to a Wigner distribution [Ro-84].

In this direction, a very recent work by Seligman, Verbaarschot and Zirnbauer deserves special mention [SVZ-84]. These authors consider a two-dimensional system consisting of two interacting particles moving in one-dimensional potential wells. The Hamiltonian is

$$H = \frac{1}{2}\left(p_1^2 + p_2^2\right) + V_1(x_1) + V_2(x_2) + V_{int}(x_1 - x_2) , \qquad \text{(IV-7)}$$

where V_1 , V_2 and V_{int} have the same functional form

$$V_j(x) = \lambda_j \left(x^2 + \mu_j x^4 + \nu_j x^6\right) \qquad j = 1, 2, int . \qquad \text{(IV-8)}$$

In Eq.IV-8 λ_j , μ_j and ν_j are parameters. With an adequate choice of the numerical values of μ_j , ν_j , λ_1 and λ_2 , the properties of the system are studied as a function of a single parameter λ_{int} . For $\lambda_{int} = 0$ the system is separable and therefore integrable. The study of classical trajectories and Poincaré sections indicate that the system is probably, for large values of λ_{int} , classically chaotic. By varying the strength of the interaction λ_{int} , the fraction f of phase space filled by chaotic trajectories, in the energy region occupied by the first 400 levels of each parity, can be varied from 0 to ~ 1. Some results for level fluctuations are reproduced on Fig.IV.6. Figs.IV.6(a) to (e) correspond to the fraction of phase space taking the estimated values $\sim$ 1.0, 0.7, 0.4, 0.1, and 0.0, respectively. (a) corresponds to the chaotic regime and (e) to the integrable case. As can be seen, (a) agrees beautifully with GOE and (e) with Poisson. In situations characterized by $1 > f > 0$, the fluctuation patterns are intermediate between GOE and Poisson. For the integrable or quasi-integrable case ((e) and (d)), there is a flattening of $\bar{\Delta}_3$ (L) for L $\gtrsim$ 9 which is not fully understood.

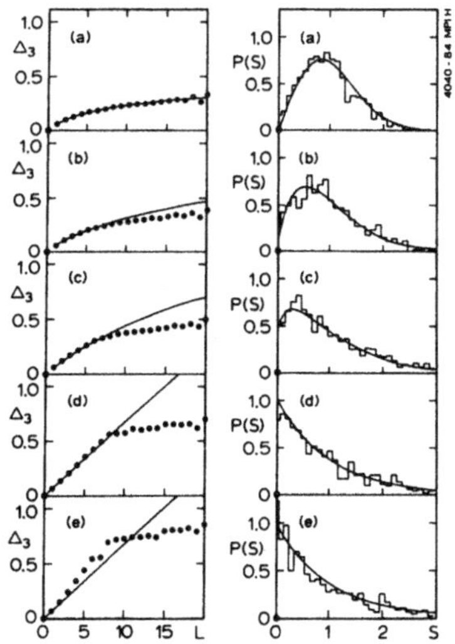

Fig.IV.6 - Results of level fluctuations ($\bar{\Delta}_3$ (L) and $P(x)$) for the Hamiltonian IV.7. Dots and histograms represent the results obtained from the 40-th to the 400-th level of each parity. The continuous lines correspond : in (a) and (e) to GOE and Poisson respectively ; in (b), (c) and (d) to a continuous interpolation between GOE and Poisson derived from a one-parameter random matrix model (taken from [SVZ-84]).

IV.3 DISCUSSION

We close this Section with some remarks and a short discussion. A more general one is postponed until the next Section.

Two main conclusions can be drawn from the results discussed : Integrable systems show Poisson fluctuation patterns whereas strongly chaotic systems show GOE fluctuations. An example has been given of a strongly chaotic system whose states belong to four different symmetry classes and the fluctuations are, separately, of GOE type, in perfect analogy with what happens, for instance, in real physical systems like nuclei, with pure sequences. We have also seen a nice example of a continuous transition from integrable to chaotic regime with a corresponding transition in the spectral fluctuations from Poisson to GOE. However, many questions remain open. We have already mentioned that the spectral fluctuations of the systems discussed are not translational invariant and that the characteristic fluctuation patterns are expected to be asymptotic. It is important to determine the rate of convergence to the asymptotic regime and, at a given energy E, the range on which Poisson or GOE fluctuations are expected to be valid. Clearly, the range does not extend from zero to E and in this connection an important role is played by the different sorts of classical closed orbits (see [Be-81b ,84]). Another question of current interest is to determine the parameter that governs the spectral fluctuations when the regime is not fully chaotic. The authors of ref.[SVZ-84] tentatively propose an average of the fraction of phase space filled by chaotic trajectories in the energy range considered. This seems very plausible but more experience is needed before a final conclusion can be drawn.

Let us also mention some problems created when heavily relying on numerical results (see the introductory Section for extremistic views and warnings). Obviously, the longer the spectrum span studied, the more difficult and time consuming is the work to be furnished. But in order to have a chance to guess asymptotic properties, it is of crucial importance to test the approximate translational invariance of the results and one thus needs many levels. It may therefore be misleading to rely exclusively on results obtained from the first few dozens of lowest eigenvalues as is often done. Furthermore, it is also clear that if one is comparing,for instance, two billiards differing only by small irregularities, in order to see differences one needs small wavelengths, i.e., high energies. Also worth mentioning are some other types of numerical difficulties. At the end of Section IV some properties of rational and irrational polygon billiards were mentioned. It is our understanding that, without a solid theoretical tool, it is hardly possible to investigate on a computer questions where the rationality or irrationality plays an essential role. (In this respect, the study of the harmonic two-dimensional oscillator is illuminating [BGP-84]).

We have, for instance, studied a billiard in a rational polygon (i.e. a pseudo-integrable system) discussed in ref.[RB-81]. Richens and Berry predict level repulsion. We have computed several spectra containing each of the order of 300 levels but no sign of attaining an asymptotic regime has been found, for instance for the average value of Δ_3. Presently we cannot draw conclusions from these numerical experiments.

BIBLIOGRAPHY

- Ref.[Be-83], entitled "Semiclassical Mechanics of Regular and Irregular Motion", is an excellent review of the subject.

- In ref.[Ca-84] the point has been made on the present state of the art ; it includes the more recent developments.

V- SUMMARY, REMARKS, CONCLUSIONS

Our main concern in theses lectures has been the study of fluctuation properties of spectra. We have seen that as a prerequisite, one needs a clean identification of the smooth behaviour of the spectrum. What is left out from the average behaviour are precisely the level fluctuations. Ensembles of random matrices have been introduced to describe the Hamiltonian and special emphasis has been put in the description of the Gaussian Orthogonal Ensemble (GOE), the prototype model valid for time-reversal and rotational invariant systems. The derivation of the GOE using information theory is most enlightening; it appears as the model obtained when the only ingredient of the theory is the real symmetric nature of the Hamiltonian. It is important to remember that GOE is a parameter-free theory. Reasons have been given to consider the failure of GOE in describing global properties (experimental level densities) as unimportant. Indeed, one expects that global and local properties are disconnected. On the other hand, it has been shown that GOE-fluctuations are not specific of GOE. They are shared presumably by a large class of ensembles of random matrices. We have examined some of them.

We have discussed how to characterize level fluctuations. In full generality, one needs the set of k-level cluster functions Y_k. Some of the most relevant qualitative features of GOE fluctuations have been emphasized : level repulsion (small probability of occurrence of small spacings) and spectral rigidity (for instance, logarithmic increase with L of the variance of the number of levels to be found in an interval of length L). This is in contrast with what happens for a spectrum obtained by adding spacings coming from random independent trials distributed like e^{-x}, viz.a Poisson spectrum. In this case there is by construction no level repulsion but level clustering (the variance of the number of levels increases linearly with L). The effect of level repulsion is that levels appear rather evenly distributed, and when spectral rigidity is present the spectrum looks incompressible. It is important to notice that the spacing distribution p(x) contains no information about spacing correlations, one of the main characteristics of GOE-fluctuation patterns. The role of exact symmetries is prominent and GOE-predictions apply to levels having the same set of exact quantum number (J^{π}).

The comparison between GOE-fluctuation predictions and experimental data has been reviewed. Due mainly to a thorough effort in high resolution measurements of compound nucleus resonances, a very stringent comparison between theory and nuclear data can be performed. GOE-predictions are fully consistent with experiment, not only for 2-point measures, where the comparison may become significant at as low as a few percent level, but even for measures containing up to 4-point cluster functions.

The present quality of atomic data allow a comparison which, although quite significant, is yet far from what can be achieved for the nuclear case. There are a few very recent and promising attempts to compare GOE-fluctuations with molecular spectra. An extensive experimental effort in the atomic and molecular case is called for. It is worth mentioning that the close agreement between GOE-predictions and data in the nuclear case can be used to impose restrictions on mechanisms that would change the fluctuations. In particular if time reversal invariance is not an exact symmetry, the appropriate model would be the Gaussian Unitary Ensemble (GUE, see Fig.II.7b). But even a small perturbation of a GOE-matrix by a GUE-matrix induces major changes in the fluctuation properties. This notion is being pursued to derive an upper bound on the time-reversal symmetry breaking part of the nuclear Hamiltonian [FK-82]. It is not yet clear whether with the amount of presently available nuclear data such an upper bound may be competitive with the ones derived by other means (sensitivities down to 10^{-3}). In passing, we mention a conjecture supported by extensive numerical data [Od-82] : the fluctuation properties of the imaginary part of the zeros of the Riemann zeta function are those of the eigenvalues of GUE matrices (or of unitary matrices belonging to the equivalent circular unitary ensemble).

In summary, we are facing a remarkable generality of fluctuations which is two-fold, experimental (nuclei, atoms, probably molecules, covering five or six orders of magnitude in scale, which is fixed by the average level spacing) and theoretical (it is shared by a large variety of matrix ensembles). It applies to very different systems, governed as well by strong short range interactions than by electromagnetic long range interactions. What is then the origin of this universality of level fluctuation laws ? They apply to complex systems, but complex in which sense ? What is the origin of the randomness ? To get insight in these questions we turn our attention to classical conservative Hamiltonian systems, were the notions of simple and complicated have been thoroughly investigated. In the classical case, we follow the path that leads from extreme regularity (integrable systems) to strongly chaotic motion. The most important tools are shortly reviewed -structure of phase space, Poincaré sections, how the phase space is filled when the system evolves, stability of orbits, etc.- and the hierarchy of irregular systems is defined : ergodic, mixing, K- and Bernouilli systems. It is discussed in which sense a system governed by causal equations of motion may be unpredictable. Examples of very different nature are given, illustrating the different categories. In particular, the properties of several two- dimensional systems are described using the tools introduced. It is seen that in general the structure of the Poincaré sections depends on the energy and that for some systems, by varying the energy, one undergoes a transition from almost regular motion to chaotic motion. Special emphasis is put on billiards, which have the simplifying feature that all their energy surfaces have the same structure. It appears that there is no need of many degrees of freedom (in fact, two are sufficient) in order to attain strongly chaotic motion and several two-dimensional strongly defocusing billiards are discussed (Bernouilli systems).

We then proceed to search for quantum-mechanical manifestations of classical chaotic motion and restrict ourselves to spectral properties. To simplify the discussion and with the aim of finding clean signatures of chaoticity, we treat as pure as possible limiting systems, namely the Schrödinger equation for one particle moving in a box in two dimensions (equivalently the transverse vibrations of a membrane) corresponding to integrable and to strongly chaotic billiards (Bernouilli systems). It appears that the spectrum-fluctuations are of the Poisson type for the integrable case and of GOE-type for the chaotic billiards. We are thus led to make the following Conjecture 1 : The spectrum of the Laplacian with Dirichlet (or Neumann) boundary conditions on an irregular boundary has asymptotically (high energy) GOE-fluctuations(*). By irregular we mean such that the corresponding classical billiard is a Bernouilli system (possibly that ergodic is sufficient). These results, of course, apply also to transverse vibrations of a membrane or, in three dimensions, to the electromagnetic oscillations of a cavity. Porter has foreseen this possibility very clearly. He wrote : "That matrix ensembles will most likely be relevant to the fields of acoustics and elasticity is rather evident" [Po-65a]. But experiments should be performed, for instance by observing the resonances of a microwave cavity of irregular shape [Dy-83]. Therefore, the following picture emerges : At a "macroscopic scale", we have universality properties for these systems. The number of eigenvalues up to a given energy depends only on macroscopic features of the boundary, such as surface, perimeter. At the other extreme, at a "microscopic scale", fluctuations also show universality patterns : Poisson-pattern for integrable systems, GOE-pattern for strongly chaotic systems.

It is unfortunate that in the literature one uses the terms "regular" and "irregular" spectra to indicate spectra originating from a regular and irregular system respectively. With this denomination, a spectrum like the one shown in Fig.I.8a would be "regular" whereas the ones shown in Fig.I.8c,d would be "irregular". But, as we have seen, levels from an "irregular" spectrum are much more regularly distributed than from a "regular" spectrum. The interplay between the symmetry or regularity of the shape of a box or a potential and the regularity of the associated eigenfrequencies or eigenvalues is an important and familiar topic in acoustics and nuclear physics. In acoustics, for instance, one major practical problem consists in designing auditoriums such that the response is as uniform as possible and it is well known from acousticians that for that purpose irregular walls are desirable, whereas when the room is more symmetrical the response of the room is more irregular [Mo-81]. In nuclear physics

---

(*)One can imagine variations to this Conjecture 1. To attack the problem mathematically, it may be convenient, instead of putting the complication in the shape of the boundary, to put it in the metric of the space and consider free motion without walls. We remind, for instance, that the geodesic flow on a surface of negative curvature is a Bernouilli system [Or-74].

it is known that there are many facts well explained by the properties of the mean field. In particular, shell effects belong to everyday practice of nuclear physicists and are responsible for dramatic effects in a multitude of phenomena like the abundances of elements, the occurrence of stable deformations, the heights of fission barriers, the existence of shape isomerism, etc.. And large shell effects are nothing but extreme cases of the clustering or bunching of eigenvalues that we have been discussing. Would the mean field be completely irregular, shell effects would not exist.

However, the origin of the success of GOE in describing the fluctuations of the compound nucleus resonances has not to be searched in properties of the mean field because we very well know that a single-particle theory is absolutely unable to correctly predict the positions and widths of these resonances. The origin of this success is to be found in more general properties which are not specific of shapes of boxes or potentials. Remember that spectra of other systems (like atoms) show GOE-fluctuations. Remember also that some particular systems, not necessarily billiards, which have been studied, when they undergo a transition from the integrable to the ergodic strongly chaotic regime in the classical case, the corresponding spectrum fluctuations in the quantum case undergo a transition from Poisson-patterns to GOE-patterns. We thus make Conjecture 2, which is more general than Conjecture 1 : Spectrum fluctuations of quantal time-reversal invariant systems whose classical analogues are strongly chaotic have GOE fluctuation patterns [BGS-84a]. If the conjecture happens to be true, it will then have been established the universality of the laws of level fluctuations in spectra already found in nuclei, to a lesser extent in atoms and to a much lesser extent in molecules. They should be tested systematically and should also be found in other systems, such as hadrons, etc..

One of the main themes in these lectures has been the translation of a scheme like the one shown in Fig.III.22 into the quantum case. We think that although some significant steps have been performed, most remains to be done. For instance, one should look for formal proofs of the equivalence of GOE- and spectral fluctuations of chaotic billiards (we remember that most of the arguments rely only on a numerical basis, although some attempts to give formal proofs should be noticed [Pe-83,Be-84]). One should discover "simple" systems showing spectrum fluctuations corresponding to the other canonical ensembles of random matrices, namely GUE and GSE. For instance, we expect that chaotic systems which are not time-reversal invariant will show GUE fluctuations ; work in this direction is in progress.

To close these lectures let us finally remark that we have been dealing with objects which are, at first sight, disconnected : quantal objects like compound nucleus resonances, classical objects like eigenmodes of vibrating membranes, frequencies in a cavity, mathematical objects like eigenvalues of random matrices, the structure in phase space of irregular motions, the zeros of the Riemann zeta function, etc.. The unifying ability is one of the great privileges of theoretical physics, which moves between two distinct worlds, the world of physical phenomena and the world of mathematics. Wigner, the "physicist", who has been at the origin of the theory of level fluctuations in terms of random matrices, says, in his article suggestively entitled "The Unreasonable Effectiveness of Mathematics in the Natural Sciences" : "The first point is that mathematical concepts turn up in entirely unexpected connections. Moreover, they often permit an unexpectedly close and accurate description of the phenomena in these connections. Secondly, just because of this circumstance, and because we do not understand the reasons of their usefulness, we cannot know whether a theory formulated in terms of mathematical concepts is uniquely appropriate" [Wi-67b]. Whereas Poincaré, the "mathematician", who has been at the origin of the study of the stability of dynamical systems, says : "La Physique ne nous donne pas seulement l'occasion de résoudre des problèmes..., elle nous fait pressentir la solution"... Almost with perfect reflexion antisymmetry !

Acknowledgements

We are indebted to Rizwan U. Haq, Akhilesh Pandey, Charles Schmit and Thomas Seligman for numerous discussions and for permission to present results prior to publication. We are grateful to Freeman J. Dyson for correspondence providing criticism and suggestions.

REFERENCES

[AA-67] V.I. ARNOLD and A. AVEZ, *Problèmes ergodiques de la mécanique classique*, Gauthier-Villars, Paris, 1967.

[Ar-76] V.I. ARNOLD, *Méthodes Mathématiques de la Mécaniaue Classique*, Mir, Moscou, 1976.

[Ba-68] R. BALIAN, Nuov. Cim. **57** (1968) 183.

[Ba-75] R. BALESCU, *Equilibrium and Nonequilibrium Statistical Mechanics*, Appendix A, Wiley-Interscience Publ., N.Y., 1975.

[Be-1873] J. BERTRAND, Comptes Rendus Académie des Sciences LXXVII (1873) 849.

[Be-78] M.V. BERRY, in *Topics in Nonlinear Dynamics* (ed.S. Jorna) Am.Inst.Conf. Proc. 46 (1978) 16 .

[Be-81a] M.V. BERRY, Eur.J.Phys. **2** (1981) 91.

[Be-81b] M.V. BERRY, Ann.Phys., N.Y. **131** (1981) 163.

[Be-83] M.V. BERRY, in *Chaotic Behaviour of Deterministic Systems*, Les Houches, Summer School Lectures XXXVI, R.H.G. Helleman and G. Jocs (eds.), North-Holland, Amsterdam, 1983, p.171.

[Be-84] M.V. BERRY in [Ca-84].

[BF-71] O. BOHIGAS and J. FLORES, Phys. Lett. **34B** (1971) 261.

[BFF-81] T.A. BRODY, J. FLORES, J.B. FRENCH, P.A. MELLO, A. PANDEY and S.S.M. WONG, Rev.Mod.Phys. **53** (1981) 385.

[BG-75] O. BOHIGAS and M.J. GIANNONI, Ann.Phys., N.Y. **89** (1975) 393.

[BGP-84] O. BOHIGAS, M.J. GIANNONI and A. PANDEY, to be published.

[BGS-83] O. BOHIGAS, M.J. GIANNONI and C. SCHMIT, in [Ca-84].

[BGS-84a] O. BOHIGAS, M.J. GIANNONI and C. SCHMIT, Phys.Rev.Lett. **52** (1984) 1.

[BGS-84b] O. BOHIGAS, M.J. GIANNONI and C. SCHMIT, preprint 1984.

[BH-76] H.P. BALTES and E.R. HILF : *Spectra of Finite Systems*, Bibliographisches Institut, Mannheim 1976.

[BHP-83] O. BOHIGAS, R.U. HAQ and A. PANDEY, in *Nuclear Data for Science and Technology*, K.H. Böchhoff (ed.), Reidel, Dordrecht, 1983, p. 809.

[BHP-84] O. BOHIGAS, R.U. HAQ and A. PANDEY, preprint.

[Bi-78] P. BILLINGSEY, *Ergodic Theory and Information*, Robert E. Krieger Publishing Company, Huntington, New York, 1978.

[BLM-76] E.G. BILPUCH, A.M. LANE, G.E. MITCHELL and J.D. MOSES, Phys.Rep. **C28** (1976) 145.

[BT-77] M.V. BERRY and M. TABOR, Proc.Roy.Soc.Lond. **A356** (1977) 375.

[Ca-84] G. CASATI (ed.), Proceedings of *Conference on Quantum Chaos*, Como 1983, Pergamon, to appear.

[CFS-82] I.P. CORNFELD, S.V. FOMIN and Ya.G. SINAI, *Ergodic Theory*, Springer-Verlag, N.Y., 1982.

[CG-83] H.S. CAMARDA and P.D. GEORGOPULOS, Phys.Rev.Lett. **50** (1983) 492.

[CVG-80] G. CASATI, F. VALZ-GRIS and I. GUARNERI, Lett.Nuov.Cim. **28** (1980) 279.

[DH-80] P.J. DAVIS and R. HERSH, *The Mathematical Experience*, Birkhäuser, Boston, 1980.

[DM-63] F.J. DYSON and M.L. MEHTA, J.Math.Phys. **4** (1963) 701, reprinted in [Po-65].

[Dy-62a] F.J . DYSON, J.Math.Phys. **3** (1962) 140, reprinted in [Po-65].

[Dy-62b] F.J. DYSON, J.Math.Phys. **3** (1962) 1199, reprinted in [Po-65].

[Dy-70] F.J. DYSON, Comm.Math.Phys. **19** (1970) 235.

[Dy-71] F.J. DYSON, Rev.Mex.Fis. **20** (1971) 231.

[Dy-72] F.J. DYSON, J. Math.Phys. **13** (1972) 90.

[Dy-83] F.J. DYSON, private communication 1983.

[Ed-74] H.M. EDWARDS, *Riemann's Zeta Function*, Academic Press, 1974.

[FK-82] J.B. FRENCH and V.K.B. KOTA, Ann.Rev.Nucl.Part.Sci. **32** (1982) 35.

[Fo-74] J. FORD in Lecture Notes in Physics n°28 : *Lectures in Statistical Physics*, Springer-Verlag, Berlin-Heidelberg, 1974, p.204.

[Fo-83] J. FORD, in Physics Today (April 1983) - See also J. FORD, in *Long-Time Prediction in Dynamics*, C.W. Horton, Jr., L.E. Reichl, V.G. Szebehely, (eds.) J. Wiley, 1983.

[FST-73] J. FORD, S.D. STODDARD and J.S. TURNER, Prog.Theor.Phys. **50** (1973) 1547.
[FW-70] J.B. FRENCH and S.S.M. WONG, Phys.Lett. **33B** (1970) 449.
[Ga-61] M. GAUDIN, Nucl.Phys. **25** (1961)447, reprinted in [Po-65].
[GC-68] I.J. GOOD and R.F. CHURCHHOUSE, Math. of Computation **22** (1968) 857.
[Ge-72] A. GERVOIS, Nucl.Phys. **A184** (1972) 507.
[Go-51] H. GOLDSTEIN, *Classical Mechanics*, Addison-Wesley, Reading, Massachusetts, 1951.
[He-74] M. HENON, Phys.Rev. **B9** (1974) 1921.
[He-80] R.H.G. HELLEMAN in *Fundamental Problems in Statistical Mechanics*, Vol.5, E.G.D. Cohen (ed.), North-Holland, Amsterdam,1980, p.165.
[He-83] M. HENON in *Chaotic Behaviour of Deterministic Systems*, Les Houches, Summer School Lectures XXXVI, R.H.G. Helleman and G. Joos (eds.), North-Holland, Amsterdam 1983, p.53.
[HH-64] M. HENON and C. HEILES, Astron.J. **69** (1964) 73.
[HKC-83] E. HALLER, H. KOPPEL and L.S. CEDERBAUM, Chem.Phys.Lett. **101** (1983)215.
[HM-63] C.B. HASELGROVE and J.C.P. MILLER, *Tables of the Riemann Zeta Function*, Royal Society Mathematical Tables 6, Cambridge University Press 1963.
[Ho-75] A. HOBSON, Journ. of Math.Phys. **16** (1975) 2210.
[HPB-82] R.U. HAQ, A. PANDEY and O. BOHIGAS, Phys.Rev.Lett. **48** (1982) 1086.
[HWR-78] G. HACKEN, R. WERBIN and J. RAINWATER, Phys.Rev. **C17** (1978) 43. This paper contains a complete set of references to earlier papers by the Columbia group.
[HYK-84] H. HIROOKA, Y. YOTSUYA, Y. KOBAYASHI and N. SAITO, Phys.Lett. **101A** (1984) 115.
[Ka-59] M. KAC, *Statistical Independence in Probability, Analysis and Number Theory*, Math.Ass.America, 1959.
[Ka-66] M. KAC, Am.Math.Monthly **73** (1966) 1.
[Kh-64] A.Y. KHINCHINE, *Continued Fractions*, University of Chicago Press, Chicago, 1964.
[La-70] C. LANCZOS, *The Variational Principles of Mechanics*, fourth edition, University of Toronto Press, Toronto, 1970.
[La-73] V.F. LAZUTKIN, Izv.A.N.SSSR, ser matem **37** (1973) ; english translation in Math. USSR Izv. Vol. **7** (1973) 185.
[Le-54] P. LEVY, *Théorie de l'addition des variables aléatoires*, Gauthier-Villars, Paris, 1954.
[LL-83] A.J. LICHTENBERG andM.A. LIEBERMAN, *Regular and Stochastic Motion*, Springer-Verlag, New-York, 1983.
[LP-73] J.L. LEBOWITZ and O. PENROSE, in Physics Today **26** (February 1973) 23.
[Ly-68] J.E. LYNN, *The Theory of Neutron Resonance Reactions*, Clarendon Press, Oxford 1968.
[Ma-74] S.V. MANAKOV,Zh.Eksp. Teor.Fiz. **67** (1974) 543 ; english translation in Sov.Phys. JETP **40** (1974) 269.
[Ma-83] V.R. MANFREDI, private communication, 1983.
[MdC-72] M.L. MEHTA and J. des CLOIZEAUX, Ind.J.Pure Appl.Math. **3** (1972) 329.
[Me-60] M.L. MEHTA, Nucl.Phys. **18** (1960) 395, reprinted in [Po-65].
[Me-67] M.L. MEHTA, *Random Matrices and the Statistical Theory of Energy Levels*, Academic Press, 1967.
[Me-71] M.L. MEHTA, Comm.Math.Phys. **20** (1971) 245.
[ME-81] N.F.G. MARTIN and J.W. ENGLAND, *Mathematical Theory of Entropy*, Encyclopedia of Mathematics and its Applications, vol.12, Gian-Carlo Rota(ed.), Addison-Wesley, 1981.
[MF-75] K.K. MON and J.B. FRENCH, Ann.Phys., N.Y. **95** (1975) 90.
[MK-79] S.W. Mc DONALD and A.N. KAUFMAN, Phys.Rev.Lett. **42** (1979) 1189.
[Mo-81] P.H. MORSE, *Vibrations and Sound*, 5th edition,American Inst. of Physics, 1981.
[Mo-73a] H.L. MONTGOMERY, Proc.Symp.Pure Maths. **24** (1973) 181.
[Mo-73b] J. MOSER, *Stable and Random Motions in Dynamical Systems*, Princeton University Press, Princeton, 1973.
[Mo-74] H.L. MONTGOMERY, Proc. of Int. Congr. of Mathematicians, Vancouver, 1974.
[Mo-76] H.L. MONTGOMERY, Proc. Symp. Pure Maths. **28** (1976) 307.
[MP-83] M.L. MEHTA and A. PANDEY, J.Phys. **A16** (1983) L601.

[MSP-84] S. MUKAMEL, J. SUE and A. PANDEY, Chem.Phys.Lett. **105** (1984) 134.
[MT-84] T. MATSUSHITA and T. TERASAKA, Chem.Phys.Lett., in press.
[MW-79] C. MAHAUX and H. WEINDEMÜLLER, Ann.Rev.Nucl.Part.Sci. **29** (1979) 1.
[Od-82] A. ODLYZKO, private communication, 1982.
[Or-74] D.S. ORNSTEIN, *Ergodic Theory, Randomness, and Dynamical Systems*, New Haven and London, Yale University Press, 1974.
[Pa-78] A. PANDEY, Doctoral dissertation, University of Rochester, 1978, unpublished.
[Pa-79] A. PANDEY, Ann.Phys., N.Y. **119** (1979) 170.
[Pa-81] W. PARRY, *Topics in Ergodic Theory*, Cambridge University Press, 1981.
[Pa-81a] A. PANDEY, Ann.Phys. **134** (1981) 110.
[Pe-83] P. PECHUKAS, Phys. Rev.Lett. **51** (1983) 943.
[Po-65] C.E. PORTER (ed.), *Statistical theories of spectra : fluctuations*, Academic Press, 1965.
[Po-65a] C.E. PORTER, Introductory article in [Po-65].
[PS-83] J.C. PARIKH and V.B. SHEOREY, preprint 1983.
[RB-81] P.J. RICHENS and M.V. BERRY, Physica **2D** (1981) 495.
[Ro-61] V.A. ROHLIN, Izv.Akad.Nauk.SSSR Ser.Mat., **25** (1961) 499 ; english translation, Amer.Math.Soc.Transl.Ser. 2, **39** (1964) 1.
[Ro-84] A. ROBNIK, J.Phys. **A17** (1984) 1049.
[RP-60] N. ROSENZWEIG and C.E. PORTER, Phys.Rev. **120** (1960) 1698, reprinted in [Po-65].
[Sc-84] C. SCHMIT, Contribution to *Workshop on Semi-Classical Methods*, Grenoble 1984.
[Sh-73] P. SHIELDS, *The Theory of Bernouilli Shifts*, University of Chicago Press, 1973.
[Si-70] Ya.G. SINAI, Usp.Mat.Nauk **25** (1970) 141 ; english translation in Russ.Math.Surv. **25** (1970) 137.
[Si-79] C. SIMO, But.Soc.Cat. C.F.Q.iM. **I** (1979) 145.
[SVZ-84] T.H. SELIGMAN, J.J.M. VERBAARSCHOT and M.R. ZIRNBAUER, preprint Heidelberg MPI H-V13,1984.
[SW-49] C.E. SHANNON and W. WEAVER, *The Mathematical Theory of Communication*, University of Illinois, Urbana, Illinois, 1949.
[SZ-72] M. SOYEUR and A. ZUKER, Phys.Lett. **41B** (1972) 135.
[TKS-83] M. TODA, R. KUBO and N. SAITO, *Statistical Physics I*, Springer Series in Solid-State Sciences, Peter Fulde (ed.), vol.30, Springer-Verlag, Berlin-Heidelberg, 1983.
[To-67] M. TODA, J.Phys.Soc. Japan **22** (1967) 431.
[To-70] M. TODA, Prog.Theor.Phys.Suppl. **45** (1970) 174.
[VZ-84] J.J.M. VERBAARSCHOT and M.R. ZIRNBAUER, submitted to Ann.Phys. (N.Y.).
[We-84] H. WEIDENMÜLLER, preprint Heidelberg MPI-H-V10, 1984.
[Wh-77] K.J. WHITEMAN, Rep.Prog.Phys. **40** (1977) 1033.
[Wi-56] E.P. WIGNER, Contribution to *Conference on neutron physics by time-of-flight*, Gatlinburg, Tennessee 1956, reprinted in [Po-65].
[Wi-67a] E.P. WIGNER, SIAM Rev. **9** (1967) 1.
[Wi-67b] E.P. WIGNER, *Symmetries and Reflections*, Indiana University Press, Bloomington, 1967.
[Wi-72] E.P. WIGNER, in *Statistical Properties of Nuclei*, J.B. Garg (ed.), Plenum, New York, 1972.
[Wi-79] K.G. WILSON, Scientific American **241**, n°2 p. 158.
[Wo-70] S.S.M. WONG, Nucl.Phys. **A159** (1970) 235.
[Za-77] D. ZAGIER, The Math.Intelligencer **0** (1977) 7.
[Za-81] G.M. ZASLAVSKY, Phys.Rep. **80** (1981) 157.
[ZK-75] A.N. ZEMLYAKOV and A.B. KATOK, Math.Zametki **18** (1975) 291 ; english translation in Math. Notes **18** (1976) 760.
[ZVW-83] M.R. ZIRNBAUER, J.J.M. VERBAARSCHOT and H.A. WEIDENMÜLLER, Nucl.Phys. **A411** (1983) 161.

SPECTRAL AVERAGING AND PARTITION FUNCTIONS*

J. B. French
Department of Physics and Astronomy
University of Rochester
Rochester, N.Y. 14627

1. Introduction

The aim of statistical spectroscopy, of which spectral averaging is one part and spectral fluctuations the other, is to produce and apply a kind of statistical mechanics which takes account of small particle number, the importance of shell structure, and the role of symmetries and special features of the Hamiltonian. It should be valid of course at high excitations, the obvious domain of statistical behaviour. But it should extend also into the ground-state region, giving results there which are directly useful (we shall see an example), or it should join on to theories which are valid in that domain.

By *spectral averaging* we shall mean local averaging over states of a system which has a specified microscopically defined Hamiltonian. Then by a *spectral distribution* we mean the locally averaged (smoothed) distribution in energy of the quantity of interest; for example the level density will be the spectral distribution which corresponds to the spectrum itself. In some cases multivariate distributions (for example with respect to E,J), or distributions with different variables, may also be considered. In any case such distributions, expressed as explicit functions of the microscopic matrix elements (of H and other operators involved) will be the subject of our discussion.

We shall make some use also of *ensemble averaging*, i.e. averaging at fixed energy over the values given, by an ensemble of H's, for quantities of interest. The major use of ensemble averaging is in dealing with fluctuations, there being no other method for doing this, except numerical calculations in the case of small systems. Fluctuations of course define the limits of accuracy of spectral averaging. Whether or not the fluctuations carry information tells us whether these errors represent a weakness of spectral averaging (if they do) or a strength (if they do not); we shall see an example which relates to this. Apart from that ensemble averaging can be used in determining general laws both about averages and fluctuations. The technical difference of course is that averages concern one-point functions while fluctuations derive from two-point functions, and higher.

*Supported in part by the U. S. Department of Energy.

A general review of our subject would supplement a discussion of its purposes with a full account of the methods of calculation. It would also describe the spectral-averaging results and laws which have followed from theoretical considerations or numerical calculations. No such review now exists though one is in progress. There are on the other hand several more or less readable[1-4] accounts which describe its aims and say something about its other aspects. It seems pointless however to spend our time reproducing things which are alread in these accounts, and it is quite out of the question to attempt here the general review. Therefore we shall proceed differently, starting with a very brief discussion, drawn largely from those published, then turning, as our principle topic,to a self-contained solution of a long-standing problem, that of the nuclear level density[5] taking account of interactions. By doing this we shall come to terms with a non-trivial problem and solve it by making use of connections between statistical behavior, symmetries and model-space geometry. In the last stages of the problem we shall have recourse to ensemble rather than spectral averaging.

The solution of the level-density problem, which has only very recently been carried out (with V. K. B. Kota[6]) should give a good feeling for the kinds of ideas and procedures which are effective in statistical spectroscopy. Moreover because the level density and (smoothed) partition function are related by Laplace transformation, and therefore carry the same information, the level density is the central construct in a large part of statistical nuclear physics. We shall sometimes refer to it as the *spectrum function*, this more impressive term being introduced precisely to stress that it carries enormously more information than the numerical level density. Indeed the major goal of spectral averaging is reached by constructing spectrum functions and various symmetry decompositions of them.

The Hamiltonian is of course representable as a point in the matrix-element space. For given particle number this point defines the entire spectrum, just as a point in phase space defines the entire "history", backwards and forwards, of a classical system. Motion of the point leads to a continuous deformation of the spectrum. Infinitesimal motion in a direction fixed by an operator K leads to a theory for the expectation values of K in the eigenstates of H. Little or nothing however has been done[7] with more general motions of H though this could well have interesting consequences both abstract and practical, perhaps including among the latter a theory of strongly deformed or fissioning nuclei.

2. The Rules of the Game

We distinguish five general laws whose brief descriptions we take, with some extensions, from Reference 4.

2.1 CLT's in Spectroscopic Model Spaces

Central limit theorems which act in many-particle model spaces and symmetry-defined subspaces determine the smoothed Hamiltonian spectrum (the level density), and, thereby, smoothed expectation-value and strength distributions, and much about the goodness of symmetries and other matters of interest. The eigenvalue densities, usually close to Gaussian, are defined by very few trace parameters (distribution moments). Thus the effect of CLT action is to filter away most of the input information which defines H. Similarly for other distributions.

The most elementary CLT is valid in spectroscopic systems[8] with non-interacting particles when blocking effects are negligible. It arises from the convolution of eigenvalue densities under particle addition, $\rho_m = \rho_{m-1} \otimes \rho_1$. This ensures that the logarithm of the characteristic function ($\phi(t)$, the Fourier transform of ρ) is additive, as then also are the *cumulants* K_ν defined by $\ell n\, \phi(t) = \Sigma (it)^\nu K_\nu / \nu!$ The centroid $K_1 = E$ fixes the *location*, the variance $K_2 = \sigma^2$ the *scale*, and for $\nu \gtrsim 3$ the translation- and scale-invariant *reduced cumulants* $k_\nu = K_\nu / \sigma^\nu$ determine the *shape* of the distribution. It follows that $k_\nu(m) = m^{1-\nu/2} k_\nu(1)$ so that the asymptotic density is characterized by $E(m) = mE(1)$, $\sigma^2(m) = m\sigma^2(1)$, $k_{\nu \gtrsim 3}(m) \to 0$, which defines a Gaussian and fixes its parameters. For interacting particles there is no law of this generality but the Gaussian form follows on the ensemble average. Moreover the ensemble used, the embedded GOE (EGOE,which is the standard Gaussian orthogonal ensemble defined in the two-particle but acting in the m-particle space),is known to be ergodic (see 2.5 below). Hence the individual spectra are asymptotically Gaussian "for almost all Hamiltonians", the measure being defined by the ensemble. This is an example of the use of ensemble averaging to produce a general result. On the other hand given a specific H we can find whether or not it is exceptional, and can construct its non-asymptotic spectrum if it is not, by evaluating low-order cumulants, say up to $\nu = 4$; if these indicate a density close to Gaussian (as they usually will for realistic interactions with a half-dozen particles or more) we can use them to give corrections to the Gaussian form. Thus

$$\rho(E) = \rho_G(E, \sigma : E) \left\{ 1 + \sum_{\nu \gtrsim 3} \left\langle P_\nu \left(\frac{H - E}{\sigma} \right) \right\rangle^m P_\nu \left(\frac{E - E}{\sigma} \right) \right\} \tag{1}$$

where ρ_G is Gaussian with indicated centroid and width and the $P_\nu(z) = \nu!^{-\frac{1}{2}} He_\nu(z)$ are the orthonormal polynomials defined by ρ_G as weight functions. Our notation is that $\langle\langle G \rangle\rangle^m = d(m) \langle G \rangle^m$ = the trace of G in the m-particle space. It follows then that $\langle H^p \rangle^m$ is the p'th moment of $\rho(E)$ while $\langle\langle H^p \rangle\rangle^m$ is the p'th moment of the true level density, $I(E) = d \times \rho(E)$.

We can think of (1) as representing $\rho(E)$ in terms of a sequence of excitations of its asymptotic form with coefficients R_ν which for low order are essentially the cumulants. Truncating the series at ν_0 then gives a smoothed density in which, if $\nu_0 \ll d$, no trace of the individual levels will be found. On the other hand we can recover a smoothed spectrum from any given approximate density by observing that the exact distribution function $F(E) = \int_{\infty}^{E} \rho(z) dz$ has a staircase form with jumps d^{-1} (for a non-degenerate spectrum) at the positions of the levels, which are then defined by $F(E_r) = (r-\frac{1}{2})/d$ (taking the midpoint of the jump). Applying this prescription[9] to the approximate F(E) produces the smoothed spectrum. Because of the CLT action the coefficients R_ν, unless $\nu \sim d$, go to zero with increasing particle number at a rate which increases very rapidly with ν. In typical cases, say 10 particles with $d \sim 1000$, only the first 5 or 6 values of R_ν, and the very high ones, are appreciably different from zero. This is summarized by the following statement:

2.2 Separation of Spectral Averages and Fluctuations About the Average

Fluctuations (deviations from uniformity) in the spectra escape the CLT action and may be said to have an existence and character independent of the smoothed density. Formally, in many-particle spaces the power spectrum shows only long-wave-length excitations (λ~spectrum span), which may be incorporated into the shape of the smoothed spectrum, and excitations of very short wave-lengths (λ~level spacing), all intermediate-λ excitations being eliminated by the CLT action. This gives us a permit for very different treatments of averages and fluctuations. For averages we construct the smoothed eigenvalue density for the given H from its low-order moments. For fluctuations we average the quantities of interest over states generated by an ensemble of Hamiltonians.

We shall unfortunately not be able to say much about the next general law, though we shall see its operation in rather simple cases. It indicates rather deep relationships between symmetries and statistical behavior and gives methods for calculating which define the information which survives in many-particle spaces.

2.3 Information Propagation by Symmetries

Physically interesting quantities ("pieces of information"), being invariant under basis transformations, are representable as traces of operators in the model space or its subspaces. Then the information is defined in a certain set of simple subspaces (two-particle ones for a two-body operator and so forth). If the subspaces are defined by symmetries, the information which survives the CLT action "propagates" from the defining subspaces throughout the entire subspace lattice. In the common case of "Casimir propagation" the complicated traces are just linear combinations of the simpler ones; in other cases further features of the model subspaces come into play. This propagation, which occurs only when the subspace decomposition is defined by symmetries, both defines the information which is interesting and gives methods for calculating it.

The next law, related to 2.2, shows that spectra are highly rigid structures and that level repulsion has far more consequences than is generally assumed. The last law describes ensemble ergodicity.

2.4 Spectral Rigidity

Spectral fluctuations in a many-particle system are small. Specifically in a large space of dimensionality d, the individual levels "move" by only a small amount, $\sim \pi^{-1}(\ell n\ d)^{\frac{1}{2}}D$, (with D the local level spacing) from their expected positions as determined, for example, by ensemble averaging. The relative motion of two levels which are on the average r×D apart is even smaller, $\sim \pi^{-1}(\ell n\ r)^{\frac{1}{2}}$, so that the motions are highly correlated. This result of Dyson and Mehta[10] constitutes a sophisticated extension of the von-Neumann-Wigner level repulsion,[11] whose effects are then seen to extend over the entire spectrum. The level fluctuations carry little or no information except about symmetries; similarly for fluctuations in other quantities.

2.4 Ergodic Behavior

In order that an ensemble value for a given quantity be relevant for a single Hamiltonian it must be valid for "almost all" ensemble members. Equivalently its variance over the ensemble should be small, preferably vanishing in some natural limit. The really essential difference between an ensemble average and the corresponding spectral average is that correlations present in the latter are missing in the former. It is then not surprising that ergodic behavior (equality of the two averages in the limit) will result if the correlations, i.e. the autocorrelation function, decrease rapidly enough with the separation distance in the spectrum. The first ergodic results for the classical ensembles are due to Grenander[12]; complete ergodicity for these ensembles was proved by Pandey[13]. There are also partial results for the embedded ensembles defined above.

3. The Role of N

N counts the number of single-particle states used in a calculation. If we restrict N we are essentially making shell-model calculations, though not by the usual methods and not at all restricted to small spaces. But why should we not consider the complete s.p. spectrum, as Bethe[14] did in the original level-density calculation, and thereby go beyond the shell-model domain into more general many-body problems? A simple derivation[15] of the essential Bethe result, that $I(E) \propto E^{-1}\exp(bE^{\frac{1}{2}})$, shows that it follows directly from the famous Hardy-Ramanujan partition formula. But, while Bethe's result is for non-interacting particles (NIP), we must deal with interactions in order to produce the spectrum function, to extend into the ground-state region, and so forth. But with interactions the energies are not additive, the problem is not essentially combinatorial, and there is no longer a parallel with the H-R problem.

Therefore authors have in the past proceeded reluctantly to finite N, basically in order to make use of the CLT's which Bethe did not need (except for the J-decomposition of the density) and which in any case are inoperative with unbounded spectra. Thus in a sense we replace the H-R formula in unrestricted spaces by using CLT's in restricted ones. As we have stressed above, CLT's work in bounded spaces even with interacting particles. Closely related to that is that N introduces the group U(N) of unitary transformations in the single-particle space, which brings with it classifications of operators and a model-space geometry which gives measures for the magnitudes of operators and correlations between them. Beyond that it brings in a whole host of U(N) subgroups, some of which are of immediate interest (the symmetry being more or less good); others, even if the symmetry is badly broken, as it is with configuration groups, may be useful in the partitioning of model spaces and in the analysis of experimental data. We shall see that configuration groups are important on both counts.

We shall give in the next section a preliminary account of a method which combines the advantages of infinite N and finite N. But first we give some simple results[16] about U(N) which we shall need for that purpose.

U(N) enters because states generated by distributing m fermions over N single-particle (s.p.) states form the artisymmetric irreducible representation[16] [m] of U(N), while their adjoints belong to [N-m]. This has the consequence that a k-body operator G(k) belongs to (k+1) irreps, $[N-\nu,\nu]$ with $\nu = 0,1,\dots k$. It is then easy to see that $G^0(k) = \binom{n}{k}G^0(0)$ (this because $\binom{n}{k}$ is k-body and invariant under U(N), whose irreps occur with unit multiplicity); G^0 then fixes the average energy for a given m. Moreover $G^1(k) = (n-1)G^1(1)$ which simply renormalizes the s.p. spectrum. For k=1 with $G = \Sigma\, \varepsilon_i n_i$ we have $G^0 = n\bar{\varepsilon}$ and $G^1 = \Sigma\, (\varepsilon_i-\bar{\varepsilon})n_i$, where $\bar{\varepsilon} = N^{-1}\, \Sigma\, \varepsilon_i$, the n_i of course are the single-state number operators, and $n = \Sigma\, n_i$; when k = 1 (NIP) the unitary reduction involves simply separating out the trace operator. The general decomposition for a number-conserving operator is

$$G(k) = \sum_{\nu=0}^{k} G^{\nu}(k) = \sum_{k=0}^{k} \binom{n-\nu}{k-\nu} G^{\nu} \tag{2}$$

where $G^{\nu} = G^{\nu}(\nu)$ is ν-body so that each term in the expansion is k-body. We shall not need the explicit construction[16] of the G^{ν} in the general case.

Only the scalar irrep ν=0 has a non-vanishing trace in the m-particle system, so that we have the basic result that $\langle G(k)\rangle^m = \binom{m}{k}\langle G(k)\rangle^k$. Moreover since the product of two irreps contains the scalar irrep only if the two are equivalent, we have $\langle G^{\nu}(k)F^{\nu'}(k')\rangle^m = 0 = \langle G^{\nu}F^{\nu'}\rangle^m$ if $\nu\neq\nu'$. This orthogonality indicates that there is a unitary geometry defined by associating with every traceless Hermitian operator G its norm in the m-particle space,

$$|G|^m = \{\langle (G)^2\rangle^m\}^{1/2} \tag{3}$$

For non-traceless Hermitian G we would remove the ν=0 part, thus defining the norm as the square root of the spectral variance. The norm so defined is a proper norm in the mathematical sense, so that for example the sum and product of small operators are themselves small. Moreover the norm is compatible with the unitary norm for the states so that, for example, operating on a normalized state vector with a small operator will not produce a large vector. Of course the simple introduction of a geometry is not necessarily a matter of physical interest. But our geometry is made effective by the operations of the CLT which, in the asymptotic limit, ensure that only the centroids and variances are of consequence.

The scalar product of two vectors, along with their norms, defines the cosine of the angle between them. Similarly we have for traceless operators F, G acting in the m-particle space

$$\zeta_{F-G}(m) = \langle FG\rangle^m / |F|^m \times |G|^m \tag{4}$$

which then defines $\mathrm{Cos}\theta_{F-G}$ or, in statistical terms, the correlation coefficient between the operators. It is easy to see that ζ is in fact the correlation coefficient between the similarly ordered sequences of matrix elements in any representation or, if the operators are simultaneously diagonalizable, between their eigenvalues. It follows of course that $-1 \leq \zeta \leq 1$.

Finally we indicate how these unitary norms and correlation coefficients are to be calculated. It is good to make use of a hole $\leftrightarrow$ particle symmetry deriving from the invariance of the fermion commutation rules under $a_i \leftrightarrow a_i^+$. This tells us that holes are also fermions but defined with respect to the filled state [N], the hole vacuum, rather than the particle vacuum [0]. There is then a one-to-one correspondence between the irreps [m] and [N-m]. The essential technical point is that the non-scalar contracted unitary tensor G^ν, written in normal form, can take on no commutator (contraction) terms when rewritten in antinormal form for such terms would belong to an irrep with smaller ν. The operation obviously gives rise to a sign change if ν is odd, so that we have $G^\nu \xrightarrow{h\leftrightarrow p} (-1)^\nu G^\nu$. It now follows immediately that[16]

$$\langle G^\nu(k)F^{\nu'}(k')\rangle^m = \delta_{\nu\nu'}\binom{m-\nu}{k-\nu}\binom{m'-\nu}{k'-\nu}\binom{m}{\nu}\binom{N-m}{\nu}\binom{N-\nu}{\nu}^{-1}\langle G^\nu F^\nu\rangle^\nu \,. \tag{5}$$

The first factor $\delta_{\nu\nu'}$ expresses our orthogonality, the next two derive from the contraction shown in (2), the next two define a polynomial of maximum order 2ν (called for by the basic result following (2)), which is symmetrical in m,(N-m) (the h$\leftrightarrow$p result), and the last binomial coefficient renormalizes things so that we have an identity when $k=k'=\nu$. The ν-trace follows directly from the matrix elements which define the irreducible operators.

We are now ready to proceed to our general problem.

4. The Level Density and Spectrum Function

4.1 Three Propositions

Writing $H = H(0) + h + V$, respectively 0,1,2-body operators, we begin with three "propositions" concerning the role of the two-body interaction V. These propositions, whose justification will emerge as we proceed, are stated in terms of configurations (e.g. $d_{5/2}^{m_1}\, d_{3/2}^{m_2}\, s_{1/2}^{m_3}$) built on spherical orbits which we label by $\alpha,\beta\ldots$, and sets of these configurations with given harmonic-oscillator excitation. We label this excitation as $\underline{s}$; thus, if the 1s and 1p shells can be regarded as filled, $(1d,2s)^m$ with $0 < m \leq 24$ has s=0 while $(1d,2s)^{m-r}(1f)^r$ has s=r. We make a corresponding decomposition of V, $V = \sum_t V_t$ with t=0,2,4,... (only even values occurring because of parity conservation) where V_t generates $\pm t\hbar\omega$ excitations. Thus V_0 connects an s-configuration only with s while V_2 connects s with s $\pm$ 2. It is important to note that configurations of different s will on the average be separated by $\hbar\omega\Delta s$; but because the configuration centroids have a wide energy span ($\sim$50 MeV for $(ds)^{12}$) there will be many cases, at higher excitation, where different—s configurations are close-lying. But since these must necessarily involve very different partitionings among the common orbits they cannot be directly connected by the (two-body) V.

Our procedure is to start with the NIP case (V=0). The density is then a set of degenerate "spikes" which, when locally smoothed, would have a characteristic NIP form. The calculation of this form can be done by spectral averaging or by other methods; we regard that part of the problem as solved and waste no time over it. Let the NIP density be $I_0(E)$. We now turn on V, one piece at a time, making use of the propositions which are:

(1) The part of V which only shifts the configuration centroids (i.e. the positions of the spikes) simple renormalizes the s.p. energies and therefore preserves the NIP structure of the density. Note that this part of V necessarily has t=0.

(2) The part of V which spreads the states of a configuration without admixing between different major shells (i.e. the other part of V_0) spreads equally all configurations of given oscillator excitation s. Its effects are represented by convolutions.

(2) $V_{t>0}$, which admixes between different major shells, has a very large unitary norm. But these large $t \gtrsim 2$ variances do not describe large configuration spreadings, for the corresponding distributions are not close to Gaussian but are instead multimodal (this because of the large separation between the admixing states). The quantities of interest are then not the configuration variances but rather the modal variances. These are little affected by $V_{t>0}$, which can then be ignored, or treated in perturbation theory.

The essential content of these propositions is that the level density is given by a convolution of a Gaussian "interaction distribution" with a renormalized NIP density, or by a sum of such convolutions (one for each s). Since everything is calculable in terms of the matrix elements of H, we have then a simple form for our spectrum function which takes account of the interactions and yet displays the NIP density as a convolution factor.

4.2 The Centroid Generator

Using the first proposition we should now extract that part of V which only moves the spikes while preserving their degeneracies. Such a separation is of course only defined uniquely with respect to an orthogonalization procedure, the separation of V being then into two orthogonal parts. The group which does this, and defines the part which we are seeking as $V^{[0]}$, the scalar part of V, is the group $\Pi U(N_\alpha)$ of simultaneous unitary transformations in every spherical orbit (Π here indicating the direct product). Then $[0] = [0,0,0\ldots]$, being scalar in every separate orbit, and $V = V^{[0]} + V^{[\Lambda]}$ where $V^{[\Lambda]}$ is non-scalar. It follows then that $V^{[0]}$ is a quadratic polynomial in the number operators n_α for each orbit; more specifically it must be linear in the two-body operators,

$$\begin{aligned} n_{\alpha\beta} &= n_\alpha n_\beta & \alpha \neq \beta \\ &= \binom{n_\alpha}{2} & \alpha = \beta \end{aligned} \tag{6}$$

These are projection operators for the two-particle states $(1_\alpha, 1_\beta)$ whose dimensionality is $N_{\alpha\beta}$ defined similarly Moreover it should be clear that (if we ignore the possibility of radial degeneracy, or otherwise deal with it) the non-scalar $V^{[\Lambda]}$ is non-scalar in every orbit and thus has no effect on any centroid. $V^{[0]}$ is obviously defined by the quantities $V_{\alpha\beta}$, the average interaction between two particles, one in orbit α and one in β (which may coincide), thus

$$V^{[0]} = \sum_{\alpha \leq \beta} V_{\alpha\beta} n_{\alpha\beta}; \qquad V_{\alpha\beta} = N_{\alpha\beta}^{-1} \sum_{i\epsilon\alpha, j\epsilon\beta} V_{ijij} \tag{7}$$

In a given region of excitation, and given particle number, only some finite set N=N(E) of s.p. states will be "active" (neither filled nor empty), these of course defining the Fermi surface. The N-set changes slowly with excitation energy and particle number. Its exact choice is not important; the states of one or two major shells should be satisfactory. To see now whether $V^{[0]}$ shifts the centroids in the region of interest in a manner corresponding to renormalizing the s.p.e. we simply decompose it with respect to U(N),

$$V^{[0]} = \sum_{\nu=0}^{2} V^{\nu,[0]} \tag{7'}$$

As described above the $\nu=0$ part generates an overall shift, not of present interest, while $\nu=1$ renormalizes the s.p.e. The part then which disturbs the effective NIP centroid structure is $V^{2,[0]}$. By simple intuitive arguments, or more formally e.g. by (3.20) of Reference 16, we find (the sums being taken over the active orbits)

$$V^{0,[0]} = \bar{V}\binom{n}{2} \quad ; \quad \bar{V} = \binom{N}{2}^{-1} \sum_{\alpha\lessgtr\beta} N_{\alpha\beta} V_{\alpha\beta}$$

$$V^{1,[0]} = (n-1) \sum \varepsilon_\alpha(2) n_\alpha$$

$$\varepsilon_\alpha(2) = (N-2)^{-1} \sum (N_\gamma - \delta_{\alpha\gamma})(V_{\alpha\gamma} - \bar{V}) \tag{8}$$

$$V^{2,[0]} = V^{[0]} - V^{0,[0]} - V^{1,[0]}$$

$V^{1,[0]}$ displays explicitly the s.p.e. renormalization of $h = \Sigma\, \varepsilon_\alpha(1) n_\alpha$; thus

$$h \to \hat{h} = h + V^{1,[0]} \quad ; \quad \varepsilon_\alpha(1) \to \varepsilon_\alpha = \varepsilon_\alpha(1) + (m-1)\varepsilon_\alpha(2) \tag{9}$$

To see now whether or not $V^{2,[0]}$ can be ignored we consider the norms of, and correlation coefficients between, the three significant operators h^1, $V^{1,[0]}$, $V^{2,[0]}$ the norms being taken over the space defined by (N,m). The necessary formal results are given above, and we take various "standard" interactions used in light and intermediate nuclei as testing examples. It would be good to answer the question in terms of the range, spin-dependence, and so forth, of the interaction but that has not yet been done. Observe first that h^1 and $V^{1,[0]}$ are correlated with each other; the correlation coefficient in fact is large and positive ($\gtrsim 0.9$ in the ds shell, 0.99 in the space generated by the first 15 orbits), an interesting result for which we shall later seek an understanding. Moreover, $V^{2,[0]}$ is correlated with neither. It follows easily from Section 3 that $|F^\nu(k)|^m \propto m^{k-\nu/2}$ for large m, so that $|h^1|^m \propto m^{\frac{1}{2}}$, $|V^{1,0}|^m \propto m^{3/2}$ and $|V^{2,0}|^m \propto m$. The single cross term also grows as m. We find on examining interactions that for small m, $V^{1,[0]}$ and $V^{2,[0]}$ are of comparable magnitude but much smaller than h^1, and much smaller than the cross-term. The consequence is that where $V^{2,[0]}$ is comparable with $V^{1,[0]}$ they are both negligible; as m increases $V^{2,[0]}$ grows more slowly than $V^{1,[0]}$ and at the same rate as the much larger cross term. It is in fact dominated everywhere, giving norm corrections $\sim$2% in ds and 0.2% in the larger case. We take for granted that this result extends to heavier nuclei (we shall of course explore that) and conclude then that the configuration centroids are determined by renormalized s.p.e. which give an NIP density, say $I_1(E)$, with a parameter very slowly varying with energy.

4.3 Zero-$\hbar\omega$ Spreading and Admixing. The Variance Generator

The only things that $V_0^{[\Lambda]} = V_0 - V_0^{[0]}$ does are to spread configurations and to admix states within a single major shell. The question is whether there is some simple pattern to the spreadings, which, because of the CLT action, we can reasonably take as Gaussian.[17] Since the centroids have been fixed we therefore need consider only variances. Numerical calculations have indicated that, for m-particle configurations of the same oscillator excitation s, the variances vary only very little from σ_A^2, the properly weighted average, even though the dimensionalities of the configurations vary enormously (from 1 to 4×10^5 in $(ds)^{12}$, for example). Therefore we should do three things:

a) Calculate the average variance. Labeling the configurations by c, their dimensionalities, centroids and variances being d_c, E_c and σ_c^2, and the overall centroid and variance as E, σ^2, we have

$$\begin{aligned}\sigma_A^2 &= \sum_c (d_c/d)\sigma_c^2 = \sigma^2 - \sum_c (d_c/d)\{E_c - E\}^2 \\ &= \langle(V-E)^2\rangle^m - \langle(V^{[0]}-E)^2\rangle^m = \langle(V-V^{[0]})^2\rangle^m = \langle(V^{[\Lambda]})^2\rangle^m\end{aligned} \tag{10}$$

which would follow more directly from the orthogonality of $V^{[\Lambda]}$ and $V^{[0]}$. Since $V^{[\Lambda]} = V^{2,[\Lambda]}$ (i.e. is of fixed U(N) rank $\nu=2$), with defining matrix elements given by subtracting those of $V^{[0]}$ from those of V, σ_A^2 follows by using (5).

b) Explain in physical terms why the fluctuations are small and in fact remain so even if we combine orbits to form new ones or make unitary transformations to a new s.p. basis and thereby new configurations. This is not yet fully understood, though thinking about the interactions in real space removes most of the surprise. It implies a sum rule, namely that the sum of squares of V_0 matrix elements connecting the many-particle state ψ with all others is essentially the same[18] for all ψ.

c) Calculate the variance of the variances σ_c^2 and thereby a quantitative measure of the σ_c fluctuations. Since the variance of the variance of V involves operators of particle ranks 2-8 this might seem complicated. It is in fact quite simple. We start by constructing a "variance generator" K (analogous to the centroid generator $V^{[0]}$). The variance of a two-particle configuration $(1_\alpha, 1_\beta)$, one particle in orbit α and one in β (which may coincide with α), is the sum of squares of the matrix elements which connect $(1_\alpha, 1_\beta)$ with every $(1_\gamma, 1_\delta)$, averaged over the starting states. Let $W^\Gamma_{\alpha\beta\gamma\delta}$, with $\Gamma \equiv J$ or J,T as appropriate, be the matrix elements of $V^{[\Lambda]}$. Then with $[\Gamma] = (2J+1)$ or $(2J+1)(2T+1)$, the $(1_\alpha, 1_\beta)$ variance is

$$\sigma^2(1_\alpha, 1_\beta) = N_{\alpha\beta}^{-1} \sum_{\substack{\Gamma \\ \gamma\leq\delta}} [\Gamma](W^\Gamma_{\alpha\beta\gamma\delta})^2 \equiv N_{\alpha\beta}^{-1} \sum_{\gamma<\delta} K_{rstu} \tag{11}$$

where we have adopted an orbit ordering to take account of the $\alpha \leftrightarrow \beta$ and $\gamma \leftrightarrow \delta$ symmetries of $K_{\alpha\beta\gamma\delta}$. To write the operator form for K, i.e. the variance generator we observe that operators with maximal unitary rank ($\nu=k$) are symmetric (k even) or antisymmetric (k odd) under the hole-particle transformation $F \rightarrow F^x$, defined by $\{a_i \leftrightarrow a_i^+;\ i=1,\ldots N\}$. Under this transformation $n_\alpha \leftrightarrow N_\alpha - n_\alpha = n_\alpha^x$ and thus we have a symmetry of K which fixes its form as

$$K = \sum_{\substack{\alpha\leq\beta \\ \gamma\leq\delta}} \left\{ \langle n_{\alpha\beta} n^{x}_{\gamma\delta} \rangle^{2} \; K_{\alpha\beta\gamma\delta} n_{\alpha\beta} n^{x}_{\gamma\delta} \right\} \tag{12}$$

a result given originally as (4.19) of Reference 16 and since then rederived in several more or less complicated ways by other authors. Observe now that every state of a many-particle configuration $\underline{c}$ is an eigenfunction of K with eigenvalue σ_c^2. Our interest of course is in $\langle K^2\rangle^m$ whose evaluation in terms of the interaction matrix elements is now simple. Then writing, for the RMS fluctuation in σ_c, $\delta(m) = \langle K^2\rangle^m/2\sigma_A^4(m)$ we have schematically

$$\sigma_c(m) = \sigma_A(m)\{1 \pm \delta(m)\} \tag{13}$$

For ds-shell interactions (N=24) we find $\sigma_c(m) \sim 8(1\pm.03)$ MeV. Spaces whose s.p. spectra span several major shells are immense, and then simple averaging may be misleading. The appropriate averages, also easily calculable,[19] are over the subspaces with fixed s-$\hbar\omega$ excitation, which so far are non-interacting since t=0. For example a 10-orbit interaction (N=80) with m=24 has s_{max}=44 and gives σ_c = (30±1.5) MeV on the gross average. But for s=0,2,4 we find σ_c=7.5(1±.03), 11.9(1±.11) and 15.2(1±.10) MeV, so that we still have essential constancy in the configuration variances for fixed (m,s) but an increase with s, and thus with the excitation energy, which seems not at all surprising. The fluctuations, still small, also increase with s (from 3 to 10% in our example) but the larger fluctuations, occurring, as they do, at high excitations are of no interest at all.

Thus there is indeed a simple pattern to the t=0 configuration spreadings; the variances are constant for fixed s and increase slowly with s. But it is easy to see that the sum of Gaussians of fixed variance σ_A but with various centroids (and dimensionalities) is the convolution ⊗ of that Gaussian with the centroid distribution.

Thus, ignoring the small variance fluctuations we see now that for $E \lesssim \hbar\omega$, for which only s=0 is relevant, the level density generated by the t=0 Hamiltonian is the convolution of the renormalized NIP density $I_1(E)$ with an "interaction Gaussian" of unit integral and variance $\sigma_A^2(s=0)$. Thus $I_2(E) = I_1 \otimes \rho_G^{(s=0)}[E]$. For the general case we have a sum of convolutions

$$I_3(E) = \sum I_3^{(s)}(E) = \sum I_1^{(s)} \otimes \rho_G^{(s)}[E] \tag{14}$$

which displays the simple connection between the interacting and non-interacting-particle densities. The only extra thing called for in its evaluation is the s-decomposition of the NIP density, methods for which are well known.

4.4 $V_{t>0}$ Excitations

As we have said, these excitations give rise to large configuration variances, so large indeed that if they belonged to Gaussian distributions they would be incompatible with shell-model stability even for low-lying configurations. For example with $(ds)^{10}$ in a 10-orbit "no-core" space we find $\sigma_c^2(t)$ = 64,280,550 MeV^2 for t=0,2,4 giving widths 8,19,30 MeV. But of course the distributions are not Gaussian but multimodal, describing the weak admixing of well separated configurations instead of the strong admixing required for Gaussian. Formally the intensity-admixing parameter, $\tau(t) = \sigma^2(t)/(t\hbar\omega)^2$, is small (0.18, 0.09, 0.03 for t=2,4,6 in our example) so that the large variances simply represent the motion over a large energy range $(t>0)$-$\hbar\omega$, of small admixing amplitudes. Involved here is an old question, about the distinction between true variances and the effective variances, which for example comes up in sum-rule problems. The latter variances, which are the interesting ones, are in our case the much smaller modal variances. The $V_{t>0}$ contributions may be treated perturbatively, or even ignored (in which case the modes are the unperturbed (Gaussian) configuration densities and the states have good configuration symmetry). We have here a reminder that one must not proceed blindly with Gaussian assumptions which are inappropriate when the generating interaction is weak; incidentally the fact that a distribution is multimodal would show up immediately on calculation of the third and fourth cumulants, though the explicit form of the distribution may be hard to arrive at.

To show how these things happen consider first the problem of the non-diagonal interaction, Z, between two spikes of dimensionalities d_1, d_2, separated by a spacing Δ. Agree that the first spike has s=0 and the second s=2 so that $\Delta \sim 2\hbar\omega$. Let us not deal directly with the eigenvalue density I(E) but with the partial densities $I^1(E)$ and $I^2(E) = I(E) - I^1(E)$ in the two subspaces. Recall that $I^1 = d_1 \times \rho^1$ has moments $T_r^{(1)}(H^p) \equiv \langle\langle H^p \rangle\rangle^{(1)} = d_1 \langle H^p \rangle^{(1)} = d_1 M_p^{(1)}$ so that M_p is the moment of ρ^1 the normalized partial density. I^1 and I^2 are discrete distributions taking on values in the range (0,1) at the eigenvalues of H. A zero in I^1 indicates that the H eigenvector has no component in the first space, and a unit value that it belongs entirely in it. Now, where Δ = g-f, we have

$$H = Z + P = \begin{pmatrix} 0 & \vdots & \tilde{X} \\ \cdots & \cdot & \cdots \\ X & \vdots & 0 \end{pmatrix} + \begin{pmatrix} f & \vdots & 0 \\ \cdots & \cdot & \cdots \\ 0 & \vdots & g \end{pmatrix} \tag{15}$$

where X_{ij} is real and $\tilde{X}$ is the transpose of X. Thus X generates transformations $(1 \rightarrow 2)$, i.e. from $[m,0] \rightarrow [m-2,2]$, and $\tilde{X}$ the inverse ones.

The Degenerate Case: P=0=Δ

It will be seen that the densities for arbitrary Δ follow from those with Δ=0. Hence we start with that case. Because $X, \tilde{X}$ operate between the two spaces we have $\langle\langle H^p \rangle\rangle^{m_1} = \langle\langle H^p \rangle\rangle^{m_2}$ which would, if valid for all p, imply that $I^1(E) = I^2(E)$.

But of course it is not valid for p=0, these traces being d_1, d_2. A moment's thought will show that instead of $I^1 - I^2 = 0$ we have

$$I^1(E) - I^2(E) + (d_1 - d_2)\delta(E) = 0 \tag{16}$$

It follows from this, since a δ-function density cannot occur with a negative sign, that the larger space (#2) contains (d_2-d_1) H-eigenvectors with E=0 and that, apart from that, the two partial densities are equal. Moreover, since $M^1_{2\nu+1} = \langle X(\tilde{X}X)^\nu \rangle^{(1)} = 0$, $I^1(E)$ (and therefore $I^2(E)$) is an even function.

If we take for Z an "off-diagonal GOE" (Gaussian orthogonal ensemble), in which distinct matrix elements are independent with $\bar{X}_{ij}=0$, $\bar{X}^2_{ij} = v^2$ we find (Kota, 1983, unpublished) that, with $d=d_1+d_2$,

$$I^1(E) = \frac{d}{2\pi v^2|E|}\left\{(R_+^2 - E^2)(E^2 - R_-^2)\right\}^{1/2} \; ; \quad R_- \lesssim |E| \lesssim R_-$$

$$R_\pm = d^{-1/2} v \, \{d_2 \pm d_1\} \tag{17}$$

which gives a semicircle when $d_2=d_1$ and two separated "deformed semicircles" when $d_2 > d_1$ (in which case I_2 has of course the extra E=0 density spike).

Because the independence of the matrix elements implies multibody interactions GOE models are unrealistic for densities. Let us instead specify that the $[m,0] \leftrightarrow [m-2,2]$ excitations are generated by a 2-body interaction, and that Z represents the ensemble of such interactions which reduces to the off-diagonal GOE when m=2. The ordinary counting which gives Gaussian moments is based on the fact that the H's in $M_{2\nu} = \langle H^{2\nu} \rangle$, which must be paired for non-vanishing contributions, can be paired in all possible ways so that $M_{2\nu} \to (2\nu-1)!!(M_2)^\nu$. In the present case however we have $M^{(1)}_{2\nu} = \langle (\tilde{X}X)^\nu \rangle^{(1)}$, involving a sequence of alternate $1 \to 2$ and $2 \to 1$ excitations. Then we have only $(\tilde{X}X)$ correlations and, easily, $M^{(1)}_{2\nu} \to \nu!(M^{(1)}_2)^\nu$ where $M^{(1)}_2 = \sigma^2_{12}$, the partial $1 \to 2$ variance; $d_1\sigma^2_{12} = d_2\sigma^2_{21}$ = sum of squares of matrix elements connecting the two spaces. The corresponding density is than

$$I^1(E) = d_1|E|\sigma^{-2}_{12} \exp\{-E^2/\sigma^2_{12}\} \; ; \quad -\infty \lesssim E \lesssim \infty \tag{18}$$

$I_2(E)$ is not found by $d_1 \to d_2$, $\sigma^2_{12} \to \sigma^2_{21}$, as one might have guessed, but instead by (16), σ^2_{12} (the partial variance from the smaller to the larger subspace) occurring in the exponential factor in both densities.

The Non-degenerate Case

The relationship between the partial densities with different Δ follows from the fact that P in (15) behaves as a constant in each subspace. We then expect some sort of (generalized) convolution connecting densities with different Δ, which we could recognize from the moments since $M_p(\rho_A \otimes \rho_B) = \sum \binom{p}{s} M_s(A)M_{p-s}(B)$. We have

$$M_p^1 = \langle (Z + P)^p \rangle^1 = \sum_{s=0}^{p} \langle \{Z^s P^{p-s}\} \rangle^1 \tag{19}$$

where {} stands for the sum of all different arrangements of Z's and P's, and only even s contributes. Start with f=0,g=1 and think of the Z's as "barriers" which form (s+1) slots, #1,2,...s+1, into which the (p-s) P's may be inserted. The number of ways of doing this is of course $\binom{p}{s}$; but since P vanishes in subspace #1 we must count only terms in $\{Z^s P^{p-s}\}$ with all P's in even-numbered slots. We can accomodate this restriction by reducing the number of barriers to s/2-1 and then, with [p/2] the integer part of p/2, P effectively 1, and $s=2\nu$

$$M_p^{(1)} = \sum_{\nu=0}^{[p/2]} \binom{p-\nu-1}{\nu} \langle Z^{2\nu} \rangle^{(1)} \tag{20}$$

To come closer to the convolution form it is good to return to the original form (15) of P. A simple extension of the counting gives

$$M_p^{(1)} = \sum_{\nu=0}^{[p/2]} \sum_{t=0}^{p-2\nu} \binom{t+\nu-1}{\nu-1} \binom{p-\nu-t}{\nu} f^{p-2\nu-t} g^t \langle Z^{2\nu} \rangle^{(1)} \tag{21}$$

and now, taking f = -1, g = +1 (so that $\Delta=2$) and evaluating the t sum, we find

$$M_p^{(1)} = (-1)^p \sum_{\nu=0}^{[p/2]} \binom{[p/2]}{\nu} \langle Z^{2\nu} \rangle^{(1)} \tag{22}$$

The fact that $M_{2\nu+1}^{(1)} = -M_{2\nu}^{(1)}$ indicates that $\rho^1(E)$ is given in terms of the even distribution[22] $\rho^{1,+}(E) = \frac{1}{2}\{\rho^1(E) + \rho^1(-E)\} \Longleftrightarrow \{M_{2\nu}^{(1)}\}$ by

$$\rho^1(E) = (1 - E^{-1})\rho^{1,+}(E) \tag{23}$$

and consequently must vanish for $|E| < 1$. For $\rho^{1,+}$ we have from (22) that

$$M_{2\mu}^{1,+} = \sum_{\nu=0}^{\mu} \binom{\mu}{\nu} \langle Z^{2\nu} \rangle^{(1)} \tag{24}$$

Observe[20] that $\{\langle Z^{2\nu} \rangle^{(1)}\} \Longleftrightarrow \rho_{\Delta=0}^{1,+}(E) = \rho_{\Delta=0}^{1}(E)$, the latter equality arising because the degenerate densities are even. But if an even density $\rho \Longleftrightarrow \{M_{2\mu}\}$ then, with $M_{2\mu} = \hat{M}_\mu$ we have $\{\hat{M}_\mu\} \Longleftrightarrow q(x) = x^{-\frac{1}{2}}\rho(x^{\frac{1}{2}})$ for $x \geq 0$; $q(x) = 0$ for $x < 0$. Then (24) becomes

$$\hat{M}_\mu^{1,+}(\Delta) = \sum_{\nu=0}^{\mu} \binom{\mu}{\nu} \hat{M}_\nu^{1+}(0) \times 1^{\mu-\nu} \tag{25}$$

where we have introduced $1^{\mu-\nu}$ to complete the convolution form given above. Since $1 = M_{\mu-\nu}(g(x) = \delta(x-1))$ we have, using the notation q_Δ, ρ_Δ, etc., where as above Δ

is the spike separation so that $\Delta=0$ is the degenerate case,

$$q_2 = q_0 \otimes g \; ; \; q_2(x) = q_0(x-1) \tag{26}$$

$$\rho_2^{1,+}(E) = |E|(E^2-1)^{-\frac{1}{2}}\rho_0^{1,+}((E^2-1)^{\frac{1}{2}}) \; ; \; |E| \gtrsim 1 \tag{27}$$

$$\rho_2^1(E) = \left|\frac{E-1}{E+1}\right|^{\frac{1}{2}} \rho_0^1(E^2-1)^{\frac{1}{2}}) \tag{28}$$

$$\Rightarrow |E-1|\sigma_{12}^{-2} \exp\{-(E^2-1)/\sigma_{12}^2\} \quad |E| \gtrsim 1 \tag{29}$$

in which the first line describes the convolution, the second follows from the definition of q(x), the third by using (23) and the evenness of ρ_0^1. The last line follows from the special form (18) for the density in the degenerate case; other forms, such as GOE or those defined not by ensembles but by single matrices, could be used here instead, for the essential convolution result (28) follows from the special binary form (15) of the Hamiltonian. Taking f+g=0 in (15) gives an H with a zero "binary trace" (though the standard trace is not zero unless $d_1=d_2$) and it is this which converts the moment (20) into a convolution form (25)).

The results (28,29) are for special values (-1,1) of the spike energies, which give -1 for the ρ^1 centroid, and $\Delta=2$. For the general case observe that $E^2-1 = (E+1)^2 - 2(E+1) \Rightarrow (E-E_c)^2 - \Delta(E-E_c)$ by translational invariance and homogeneity, while $E-1 \Rightarrow (E-E_c-\Delta)$. Then with $W = E-E_c$, which measures the energy from the lower centroid we have as the extension of (29) the bimodal form,

$$\rho_\Delta^1(W) \Rightarrow |W-\Delta|\sigma_{12}^{-2} \exp\{-W(W-\Delta)/\sigma_{12}^2\} \; ; \quad W \lesssim 0, \; W \gtrsim \Delta \tag{30}$$

while $\rho_\Delta^2(W)$ then follows from (16), the sum of the unnormalized densities, $d_1\rho^1 + d_2\rho^2$, giving the complete eigenvalue density.

In the weak interaction limit $\tau = \sigma_{12}^2/\Delta^2 \ll 1$, we recover the perturbation-theory result. In the example referred to at the beginning of this section $\tau(t) = \sigma^2(t)/(t\hbar\omega)^2 \simeq 0.18, 0.09, 0.03$ for $t = 2,4,6$. Considering only the $s=0\to2$ excitations we then find that the variance of the lower mode is $0.08\,\sigma_{12}^2 \simeq 20(\mathrm{MeV})^2$. This will add to the 64 MeV^2 internal variance of the s=0 configurations to give a width ∿9 MeV, scarcely larger than the 8 MeV found by taking only t=0 excitations but very much less than the 19 MeV which would come from a naive treatment considering the configuration rather than modal variance.

The methods described in sections 4.1-4.4 seem quite adequate for dealing with the theory of the smoothed eigenvalue densities and partition functions for interacting nucleons as long as a satisfactory effective interaction is available. When that is the case it is probably so that multi-$\hbar\omega$ excitations have little effect on the smoothed spectrum function, as has been demonstrated in our example. Many extensions of the theory are called for, including symmetry decompositions and non-scalar densities, a self-contained accurate theory of the NIP spectrum function, a more complete treatment of the interactions with distant configurations, new forms for expectation values and strength distributions. It seems possible also that ensemble averaging methods may be of value in the theory of effective interactions, and probable that the entire theory can form the basis for a statistical theory of reactions.

5. Final Remarks

We began with a compact discussion of general notions, emphasizing those important for the calculation of the spectrum function or, equivalently, the (smoothed) partition function. We end with remarks about general features of spectra and strength distributions which emerge from applying the theory outlined. More detailed treatments are given in the references. We note however that the new methods of the last section should make possible considerably more general applications.

When the spectrum shape is close to an asymptotic form, as we expect when the number of active particles is large (half a dozen or more), it should be stable under small "deformations" of the Hamiltonian, say $H \to H + \alpha K$. Here K is an arbitrary operator of low-particle rank and we are interested in effects of first order in α. The stability arises because the same CLT which applies to H is valid for (almost all) H_α. This gives rise to a remarkable result:

Theorem: $\langle E|K|E\rangle_{sm} \equiv K(E)$, the smoothed expectation value of K in the Hamiltonian eigenstates near E, is basically linear in E. With E measured in units of σ_H, the Hamiltonian spectral width, and $\langle E|K|E\rangle$ in units of σ_K, the slope is given by ζ_{K-H}, the H-K correlation coefficient.

For the origin of the linearity observe that the effect of adding the traceless operator $\alpha(K-\mathcal{E}_K)$ must, for small enough α, correspond simply to a scale change if the CLT spectrum is to survive; thus the level shifts induced must be proportional to $(E-\mathcal{E})$. The constant of proportionality follows by considering the second moment. The result is then

$$K(E) \xrightarrow{CLT} \langle K\rangle^m + \langle K(H-\mathcal{E})\rangle^m(E-\mathcal{E})/\sigma^2 \qquad (31)$$

which, as indicated is valid in the CLT limit. There must of course be enough particles for the CLT to be operative, and the larger the number the better the result. Even so it is valid only to within fluctuations; for, as we add particles, the density is smoothed (2.1) on a scale determined by the spectrum span but the number of levels increases much more rapidly than the span. We could say that, at the same time as we smooth the spectrum, we insist (when our interest is with level-to-level fluctuations) in examining it on an even finer scale. It is then not immediately obvious what we should see; what happens in fact is that the short-wave-length fluctuations (measured in units of the local spacing, and described by low-order correlation functions) are invariant under the smoothing so that they take on their own existence independent of the density and its long-wave-length fluctuations. The intermediate-wave-length fluctuations are eliminated. This is what we have described in (2.2). We remark also that if, by increasing the number of s.p. states considered, we enlarge the space, while keeping the particle number fixed, the linearity must eventually fail, really because (as in 4.4) the interaction between far-separated states becomes too weak to generate the "chaos" which is characteristic of strong interactions. Small deviations as $H \to H_\alpha$ can be accomodated by considering low-order shape changes in the density, say up to $\nu = 4$ in (1) or in (32) below. But partitioning of the space, as we have done with the level density, is easier and usually better. Sometimes one does both[17].

The formally exact equation, of which (31) is the CLT fluctuation-free limit, follows by recognizing that $\langle G\rangle^m = \int G(x)\rho(x)dx$ so that $\rho(x) = \langle\delta(H-x)\rangle^m$. Then

$$K(E) = \rho^{-1}(E)\langle K\delta(H-E)\rangle^m = \sum_{\nu\geq 0} \langle KP_\nu(H)\rangle^m \, P_\nu(E) \tag{32}$$

where we have expanded $\delta(H-E)$ in terms of orthonormal polynomials defined by $\rho(x)$ as weight function. The number of such polynomials is strictly d, the dimensionality, but we can just as well take it infinite and use continuous instead of lattice polynomials. Since $P_0(x) = 1$ and $P_1(x) = (x-E)/\sigma$ the first two terms of (32) reproduce the CLT result (31), the next few the shape corrections which, by their size, enable us to assess the rate of convergence. Note that expectation values have a polynomial rather than exponential form so that high accuracy can be expected in the smoothed K(E).

Among the important expectation values are those deriving from sum rules. For example the non-energy-weighted sum for single-nucleon transfer in a single orbit (which is measurable by stripping and pickup reactions) determines the orbital occupancy in the target. The question which we can now answer is what this tells us about the Hamiltonian. The essential thing is that, in the correlation trace $\langle n_s(H-E)\rangle^m$ where s is the s.p. transfer state, (H-E) has no unitary symmetry $\nu = 0$ part (since it is traceless), and n_s has no $\nu = 2$ part (being one-body); thus the trace depends only on $H^{\nu=1} = \sum\{\xi_i(1) + (n-1)\,\xi_i(2)\}n_i = \sum \xi_i\, n_i$ where

$$\xi_i(2) = (N-2)^{-1}\{\sum_j W_{ijij} - N^{-1} \sum_{jk} W_{jkjk}\} \tag{33}$$

Then for the occupancy, in CLT approximation, we have from (31)

$$\begin{aligned} n_s(E) &= \frac{m}{N}\left\{1 + \frac{N-m}{N-1}\,\xi_s\right\}\frac{(E-E)}{\sigma^2(m)} \\ \sigma^2(m) &= \langle (H^1)^2\rangle^m + \langle (H^2)^2\rangle^m \end{aligned} \tag{34}$$

Thus, except for a damping effect in $\sigma^2(m)$, the occupancy measures directly an induced s.p. energy which, as defined in (33), is a Hartree-Fock-like parameter of the interaction H. It corresponds also to the ε_α of (8), the difference being that we are dealing here with the entire H, not just $V^{[0]}$. The results, which appear explicitly as a function of particle number (so that the data on a whole range of targets can be combined), have been used[21] in deciding between different ds-shell interactions. The most striking aspect of the analysis is the use, in the ground-state domain, of statistical methods and symmetries for the direct determination of an important interaction parameter.

As we have said deviations from linearity in expectation values can be dealt with by taking higher terms in (32). But partitioning, already discussed in (4.4), is simpler and, when done according to irreps of a group (really the only feasible way), gives methods for studying the corresponding symmetry. If symbolically we write the decomposition as $(m) \to \sum_\Gamma (m,\Gamma)$ then traces and densities decompose similarly, $\langle\langle G\rangle\rangle^m \to \sum_\Gamma \langle\langle G\rangle\rangle^{m,\Gamma}$ and $I(E) \to \sum I_\Gamma(E)$. We find for the expectation values, as the extension of (32) that, with P_ν^Γ the polynomials defined by $\rho_\Gamma(E)$ as weight,

$$K(E) = \sum_\Gamma (I_\Gamma(E)/I(E))\, K_\Gamma(E)$$
$$K_\Gamma(E) = \sum_\nu \langle KP_\nu^\Gamma(H)\rangle^{m,\Gamma} \times P_\nu^\Gamma(E) \qquad (35)$$

When Γ defines a good symmetry then I_Γ is an eigenvalue distribution, and, since I_Γ/I is the natural branching ratio, (35) is the obvious result. But the decompositions are valid also when the symmetry is not good. In that case, as we have said, I_Γ is not an eigenvalue distribution though it is discrete, taking on values only at the H eigenvalues. These values are the summed intensities of the projections of the Γ states onto the H eigenstate; they lie in the range (0,1), in contrast to an eigenvalue distribution which gives the extreme values only. Even so the expectation-value expansion (35) maintains its validity once we recognize that in evaluating the traces we must take account of the excitations to and between the other subspaces. Thus by a formally simple device the equation, while maintaining its form and its linearity in the I_Γ accomodates the whole hierarchy of interactions between subspaces. This linear decomposition into interacting structures has been often misunderstood.

We turn briefly now to a surprising relationship between expectation values, fluctuations and collectivities[22]. We start with a well known result[23]:

> <u>Porter-Thomas Theorem</u>: In complicated spaces transition amplitudes $E \xrightarrow{T} E'$ when locally renormalized behave like independent zero-centered Gaussian random variables.

With T the transition operator the amplitude is simply $(E'|T|E)$, its absolute square being the transition strength $R(E',E)$ which renormalized is a χ_1^2 variable. The renormalization gives $(E'|T|E) \to (E'|T|E)/\{\overline{R(E',E)}\}^{1/2}$, the bar indicating a local average ranging over neighborhoods of E, E′ (or one of them if the other is fixed). "Independent" really means uncorrelated in low order. The modern proof of the theorem follows from a GOE averaging coupled with a specific proof of ergodic behavior of the strengths and a general argument that the more realistic embedded GOE (EGOE, as described following (17)) would, because of the renormalization and the local nature of the fluctuations, give the same result.

Consider now, for a given transition operator T, the summed strength $M(E)$ originating at a single state at E, and $\Sigma^2(E)$, its variance. If there were altogether d strengths of the same magnitude we would have $M(E) \sim \chi^2_d$ and, for the "relative variance", $\Sigma^2(E)/\{M(E)\}^2 = 2/d$, a standard result for a χ^2 distribution in d variables. Thus the relative variance would determine the number of final states connected with E (analogous to "open channels"). On the other hand high collectivity defined with respect to T^+T (the Hermitian square of T, whose expectation value defines $M(E)$) gives small d (the strength originating with E going in the extreme case to a single final state) and large relative fluctuations in $M(E)$; conversely low collectivity is associated with small fluctuations.

However the strengths originating with E will by no means have equal values and thus we must use a properly weighted sum. A little consideration[22] will show that $\hat{d}$, the effective number of available states is given by

$$\frac{1}{\hat{d}} = \frac{\Sigma^2(E)}{2\{M(E)\}^2} = \frac{\int I(z)\ \{\overline{R(z,E)}\}^2 dz}{\{\int I(z)\ \overline{R(z,E)} dz\}^2} \tag{36}$$

We have shown how to calculate the level density I(z). We have for the smoothed strength a double polynomial expansion, analogous to (32) to which it reduces on integrating over one variable, and deriving from $\{\rho(z)\rho(E)\}^{-1}\langle T^+\delta(H-z)T\delta(H-E)\rangle^m = R(z,E)$. The utility of (36) would appear to be not in the analysis of data but rather in the prediction of collectivities, given a model space, Hamiltonian and transition operator. In the few model cases[22] for which $\hat{d}$ has been calculated the theory has been very successful in its predictions. It seems probable also that (36) could be used in other ways in the study of collective behavior.

Acknowledgement

The work reported here on level density and the spectrum function has been done in collaboration with Dr. V. K. B. Kota.

References

1. *Theory and Applications of Moment Methods in Many-Fermion Spaces*, ed. by B. J. Dalton *et al.* (Plenum, N.Y. 1980).
2. J. B. French: In *Proceedings of Summer School on Nuclear Spectroscopy*, ed. by G. F. Bertsch and D. Kurath (Springer, Berlin, 1980).
3. J. B. French and V. K. B. Kota, Ann. Rev. Nucl. Part. Sci. 32, 35, (1982).

4. J. B. French, Nucl. Phys. A396, 87c (1983).
5. There are probably hundreds of papers about level density; see Reference 1 for good reviews. All but a few of the theoretical papers ignore the residual interactions, most of these doing so without apology or explanation. This was essential and appropriate in the earliest papers (see (14) for the two fundamental ones) but by no means so far most of the others.
6. J. B. French and V. K. B. Kota, Phys. Rev. Lett. 51, 2183 (1983).
7. But see T. R. Halemane, J. Math. Phys. 12, 2961 (1981).
8. The model space of a "spectroscopic system" is generated by distributing m particles over N s.p. states, with m, N finite. It is convenient to use two functions, I(E) and $\rho(E) = d^{-1}I(E)$ where I(E) dE = no of states in (E,E+dE); then $\int I(E)\, dE = d$, $\int \rho(E)\, dE = 1$. These densities are uniquely defined by their moments, and similarly for other operators. Not necessarily so in infinite spaces, nor would $\rho(E)$ exist.
9. K. F. Ratcliff, Phys. Rev. C3, 117 (1971).
10. F. J. Dyson and M. L. Mehta, J. Math. Phys. 4, 701 (1963).
11. J. von Neumann and E. P. Wigner, Phys. Z. 30, 467 (1929).
12. U. Grenander, *Probabilities on Algebraic Structures* (Wiley, New York, 1963).
13. A. Pandey, Ann. Phys. (N.Y.) 119, 170 (1979).
14. H. A. Bethe, Phys. Rev. 50, 332 (1936). See also C. van Lier and G. E. Uhlenbeck, Physica 4, 531 (1937).
15. See Reference 1, p.7. Here a uniform s.p. spectrum in the neighborhood of the Fermi surface is assumed. Other assumptions give different pre-exponential factors but the exponential dependence remains.
16. F. S. Chang, J. B. French, and T. H. Thio, Ann. Phys. (N.Y.) 66, 137 (1971). The irreps are described in terms of the column structure, rather than the row structure (which is more usual), of the corresponding Young shapes. Thus $[N-\nu,\nu]$ has one column with $N-\nu$ blocks and one with ν. The more common labeling would be $[1^{N-2\nu},2^{\nu}]$.
17. For the accurate location of the ground state (a problem which does not arise in the NIP case) it is good to do better, by including one or two correction terms in (1) for those configurations which contribute significantly in the ground-state domain.
18. If the states are classified by good symmetries (e.g. J. or J, T) the result is true for all states ψ of the same symmetry, but the variances do depend on the symmetry.
19. By the same methods[16] used for calculating configuration averages.
20. By $\{M_p\} \Longleftrightarrow \rho$ we mean that the indicated moments define, and are defined by, the indicated density.
21. V. Potbhare and S. P. Pandya, Nucl. Phys. A256, 253 (1976).

22. J. P. Draayer, J. B. French and S. S. M. Wong, Ann. Phys. (N.Y.) 106, 472, 503 (1977).

23. C. E. Porter and R. G. Thomas, Phys. Rev. 104, 483 (1956); see also J. M. C. Scott, Philos. Mag. 45, 1322 (1954).

24. We take advantage of the space available on this last page to indicate how the things discussed above relate to other aspects of statistical nuclear physics. It could reasonably be said that the whole subject is the natural development of ideas introduced with Bohr's compound nucleus, Bethe's level density and Wigner's ensemble treatment of spectral fluctuations, but taking account, *inter alia*, of things learned in the meantime about: continuum states (including those generated in heavy-ion reactions); departures from statistical equilibrium in compound states; nuclear symmetries; the detailed properties of the free-space nucleon-nucleon interaction, and the effective interaction which follows from it and is used in shell-model analysis; many advances in statistical mechanics. Some of these things refer specifically to the energy region of high excitation, others mainly to the ground-state domain.

The ideal theory which would encompass all these and other phenomena, at high energies and low, does not yet exist. Nobody however believes that the microscopic details are individually of consequence at high excitations; either one should measure a lot of them, and subject the set to statistical analysis (as one does with energy-level fluctuations), or one should measure an average over them, as one does, often indirectly, by the statistical theory of reactions. This being agreed to, the high-energy region is left to those interested in the development and application of methods of statistical mechanics; two examples are the combinatorial "exciton" theory of Griffin, used for the treatment of non-equilibrium processes, and the transport theories of Weidenmüller, Agassi, and others for heavy-ion reactions. It seems altogether likely that the procedures described above could be usefully applied to many high-excitation processes, just as it has been to the level density; but eventually one would need its extension to a proper treatment of continuum states.

For low excitations the evidence is good to excellent that expectation values (31, 32), and even strength fluctuations (36), extrapolate to the ground state. It is not known whether the same extrapolation extends to the spectrum itself, i.e. whether the fluctuation-free spectrum which derives[9] from the smoothed spectrum function would properly capture all the significant information even if the exact symmetries were correctly dealt with. Looked at otherwise the question is whether the ground states (for exact symmetries) are in some way "special", resisting statistical treatment, as the reviewer at the 1975 Tucson Conference has declared; the question seems open. Probably the way to study this problem is to produce a proper method for statistically extending exact matrix results to larger spaces. Ensemble averaging for these parts of H which do not generate strong collectivities would seem to be of value here.

MEAN FIELD METHODS IN LARGE AMPLITUDE NUCLEAR COLLECTIVE MOTION

K. Goeke[+] and F. Grümmer
Institut für Kernphysik
Kernforschungsanlage Jülich
D-5170 Jülich, West Germany

P.-G. Reinhard[++]
Institut für Kernphysik
Universität Mainz
D-6500 Mainz, West Germany

Abstract

The time dependent Hartree-Fock method (TDHF) is reviewed and its success and failure are discussed. It is demonstrated that TDHF is a semiclassical theory which is basically able to describe the time evolution of one-body operators, the energy loss in inclusive deep inelastic collisions, and fusion reactions above the Coulomb barrier. For genuine quantum mechanical processes as e.g. spontaneous fission, subbarrier fusion, phase shifts and the description of bound vibrations, the quantized adiabatic time dependent Hartree-Fock theory (quantized ATDHF) is suggested and reviewed. Realistic three-dimensional calculations for heavy ion systems of $A_1+A_2 \leq 32$ are presented. Applications to various nuclear observables as e.g. astrophysical S-factors in subbarrier fusion, ground state rotational bands, elastic HI-scattering, etc., are shown. The decoupling properties of ATDHF-paths are discussed in detail. Recent theories based on path integral approaches (PIA) are reviewed, which aim at quantizing TDHF in order to describe stationary vibrational states and subbarrier fusion. The relationship of quantized ATDHF to PIA is explored in the case of bound vibrations. A derivation of the quantization prescription by means of the generator coordinate method is presented. The above methods to go beyond TDHF concern the quantum mechanical content of TDHF. For the description of width of one-body observables in inclusive reactions the time dependent generator coordinate method (TDGCM) is reviewed. It appears in one-dimensional calculations as able to enlarge the fluctuations of one-body observables, obtained in TDHF, by an order of magnitude.

1. Introduction

Over the last ten years there has been a constant activity among nuclear theorists towards a consistent description of large amplitude collective nuclear motion as

[+] also: Institut für Theoretische Kernphysik, Universitat Bonn, D-5300 Bonn, West Germany

[++] Present address: Institut für Theoretische Physik, Universitat Erlangen, D-8520 Erlangen, West Germany

e.g. fission and fusion processes, nonharmonic vibrations of soft nuclei, heavy ion reactions, etc.

The new impulse in this rather old field was the desire not to introduce the collective dynamics by an educated guess of collective coordinates but rather to describe the system fully by means of a systematic many body theory using as only input an effective two body interaction. This new emphasis has resulted from the suspicion that particularly for large amplitude collective motions one introduces uncontrollable limitations on the dynamics of the system by imposing certain geometrical shapes for the system rather than allowing its free evolution. The first important step in this direction was the formulation and application of the time dependent Hartree-Fock theory[1-5] (TDHF).

In spite of the great success of TDHF and the enormous stimulation it has caused, its limitations have become rather apparent and consist for instance in a lack of a quantization procedure for the description of discrete bound states, the impossibility of applying TDHF to barrier penetration phenomena and the inadequate implementation of boundary conditions in scattering processes. In addition only one-body operators can be evaluated without providing a good description of their fluctuation, the most important example being the particle number width in deep inelastic collisions. These difficulties gave rise to basically four different theoretical developments to be discussed in this lecture, which all incorporate to some extent elements of TDHF but in the end go conceptually and in the results beyond TDHF.

One group of approaches includes theories like adiabatic TDHF combined with generator coordinate methods[6-14]. The second group of theories[15-18] comprises path integral approaches (PIA). The third group of theories are approaches to consider not the time evolution of a Slater determinant but of a more general state incorporating certain RPA correlations[19,20]. The fourth sort of theories, not discussed in this lecture, includes approaches to go beyond TDHF by the inclusion of some sort of Boltzmann's collision term[21-27]. For these the reader is referred to the articles in Ref. 5. In the present lecture we do not discuss the formal details of the theories but will rather give a general qualitative review with numerical results.

2. TDHF: Success and Failure

The TDHF approach assumes the initial condition of a HI reaction, i.e. two undisturbed ions of A_1 and A_2 particles with a certain relative velocity and a certain impact parameter, to be represented by a Slater determinant of $A=A_1+A_2$ single particle states. If the well separated ions are assumed in their HF ground states this is certainly a reasonable approximation. The basic assumption of TDHF is now that the total wave function can for all times be assumed as a Slater determinant $\phi(\vec{r}_1,\ldots,\vec{r}_A,t)$. With this assumption the equation of motion for the total system of $A_1+A_2=A$ nucleons is precisely given by the TDHF-equation

$$i\hbar \frac{\partial}{\partial t} \phi(\vec{r}_1,\ldots,\vec{r}_A,t) = \{\sum_{i=1}^{A} h_i[\rho]\}\phi(\vec{r}_1,\ldots,\vec{r}_A,t) \tag{2.1}$$

Here $h_i[\rho]$ is the TDHF single particle Hamiltonian depending on the density matrix $\rho(t)$ of the total system:

$$h_i[\rho] = -\frac{\hbar^2}{2m}\nabla_i^2 + U[\rho](\vec{r}_i) \tag{2.2}$$

where $U[\rho](r_i)$ is the mean field including all exchange terms and ρ is given by the single particle wave functions $\varphi_\alpha(\vec{r},t)$ as

$$\rho(\vec{r},\vec{r}\,',t) = \sum_{\alpha=1}^{A} \varphi_\alpha^*(\vec{r},t)\varphi_\alpha(\vec{r}\,',t) \tag{2.3}$$

Today there is a huge number of numerical solutions of the TDHF equations. They all are done representing the single particle wave functions by their values on a two- or three-dimensional grid in the coordinate space and solving eq. (2.1) by finite differences in time. The effective two body interaction, whose average over the time dependent density gives $U[\rho(t)]$, is usually taken to be of Skyrme type. This has the advantage that the exchange terms are local and can therefore be easily handled. The Bonche-Koonin-Negele interaction[1)] is also frequently used, it corresponds to the t_0-t_3-terms in the Skyrme force plus a direct Yukawa term, sometimes also a direct Coulomb term is added.

The first success of TDHF actually consisted in providing a deep insight into the geometry of nuclear shapes relevant for HI reactions. This was due to the fact that one was able to perform snapshots of the mass density distribution revealing e.g. details on neck formation in the entrance and exit channel. An example of this is given in Fig. 1, where the density distribution in the deep inelastic collision of $^{16}O+^{40}Ca$ is drawn. (The figures of this TDHF-section are taken from Ref. 4.) However the first experimentally significant data was the fusion cross section of light heavy ions. Here fusion is defined by the sticking together of the ions for more than one or two revolutions, see Fig. 2 as example. The result for the fusion is given in Fig. 3. One actually realizes a fair agreement of the TDHF with experiment, in particular if one takes into account that there are no free parameters besides those of the force which itself is determined by reproducing rms-radii and energies all over the periodic table. An interesting feature here is the occurrence of so-called fusion ℓ-windows (see Fig. 4). However, as will become clear in the last section of this lecture, it is not at all clear if those fea-

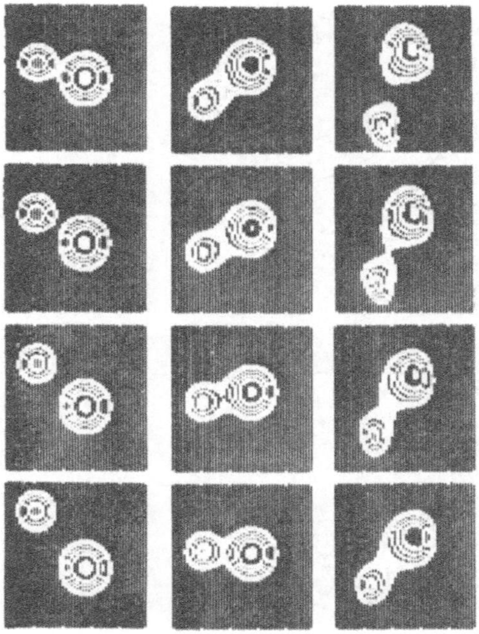

Figure 1

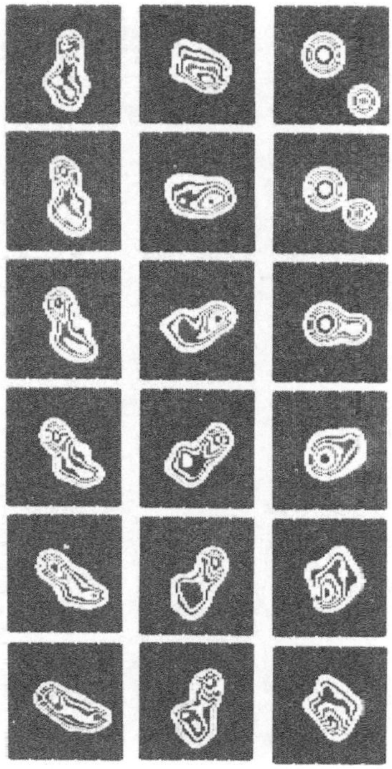

Figure 2

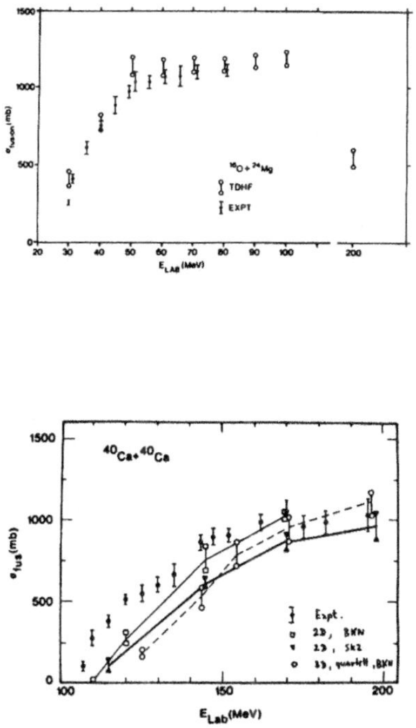

Figure 3

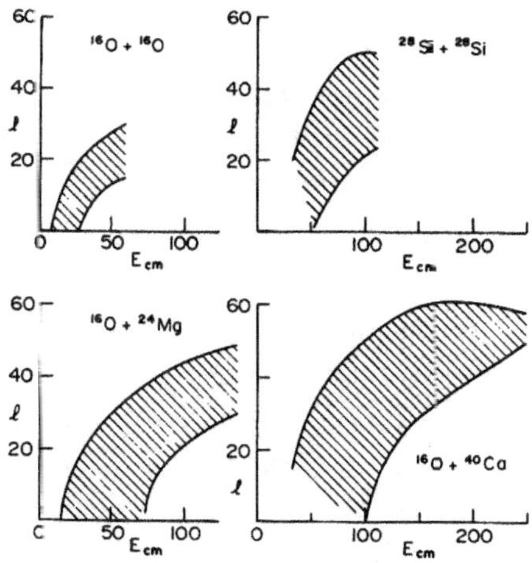

Figure 4

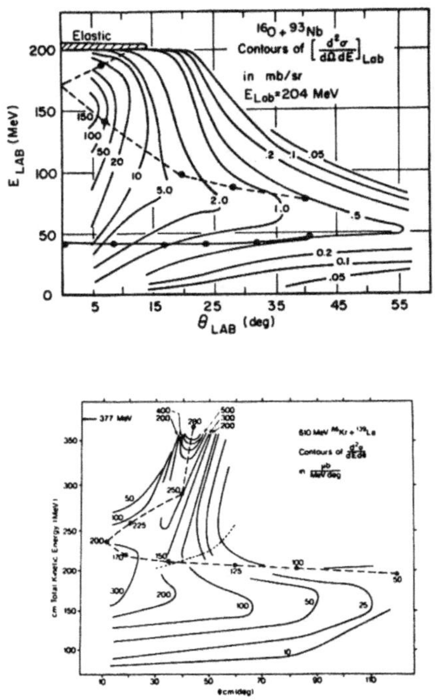

Figure 5

tures survive if one takes into account correlations by means of e.g. the TDGCM-method. The second experimental data reproduced by TDHF are the Wilczynski-plots of various HI reactions, see e.g. Fig. 5. Here the ridges of the rather complicated mountains of the plot are reasonably well reproduced. This also should be considered as a great success of TDHF, since by now it is the only fully microscopic theory which does this.

From these two points one can learn that TDHF seems to be a suitable theory for the description of the energy flow from the relative motion in the entrance channel to the internal excitations in the outgoing fusion or deep inelastic channel. Furthermore, mean values of one body operators are also well described as e.g. the mean particle number in the projectile-like fragment in deep inelastic collisions. The reason for both features lies in the fact that TDHF is just a theory for the propagation of the one-body density matrix and that one uses effective interactions, fitted to experimental data, in particular rms-radii and energies in an HF framework.

From these considerations one realizes immediately that TDHF will not be able to reproduce the widths of one-body operators being themselves two-body operators, since the two-body density matrix of TDHF is just a trivial product of the one-body one. Indeed, the width of the particle number distribution of either fragment in a deep inelastic HI event is underestimated in TDHF by an order of magnitude.

Precisely this motivates theories which try to incorporate collision terms or which try to propagate RPA correlated states rather than considering only pure Slater determinants.

A second sort of shortcomings concerns the failure of TDHF to describe penetration processes. If the energy of an HI reaction process is less than the height of the Coulomb barrier the TDHF describes merely a nearly elastic rebouncing of the two ions without giving appreciable tunneling. This is not principally forbidden, but the restriction to a Slater determinant and one common potential to move in simply presents the tunneling to take place. Similar things happen if one likes to describe fission under the barrier. In such a case the TDHF solution vibrates inside the potential barrier and does not get out.

A third shortcoming of TDHF lies in the fact that it cannot describe discrete vibrational levels, corresponding to stationary states. If the TDHF solution is moving inside some potential well describing e.g. the potential energy surface associated to a vibration then, if at all, it provides periodic solutions for a full continuum of energies. It does not, however, give a prescription which of the energies are to be selected in order to account for the stationary 0-phonon, 1-phonon, etc., states.

A further drawback of TDHF is that one can only give the classical cross section, given by the deflection function, of an HI reaction and not S-matrix elements and phase shifts. The reason for this and for some of the other shortcomings lies in the fact that TDHF is a semiclassical or even classical theory of coupled scalar fields, each of them associated to a single nucleon. Hence real quantum mechanical and many body aspects cannot be described. One can even map TDHF on an equivalent set of classical canonical equations[12] in suitable parameter spaces.

3. The Quantized ATDHF-Theory

3.1 Theory

In this section we discuss the possibility to extract from the many body Hamiltonian of the total $A_1+A_2=A$ system a collective sub-Hamiltonian which describes the degree of freedom associated to the relative motion of the ions. This will then enable us to describe subbarrier fusion, spontaneous fission and bound vibrations and also to some extent phase shifts. The sub-Hilbert-space is assumed to be spanned by a manifold of Slater determinants $\{|\phi_{R\dot{R}}\rangle\}$ or $\{|\phi_{qp}\rangle\}$ in general. The set of $\{|\phi_{qp}\rangle\}$ forms the so-called collective path. The collective path is supposed to be constructed such that any state $|\psi_n\rangle$ of the collectively moving system without intrinsic excitations is to a good approximation contained in the Hilbert subspace spanned by the path $\{|\phi_{qp}\rangle\}$:

$$|\psi_n\rangle = \int dq\ dp\ f_n(q,p)|\phi_{qp}\rangle \tag{3.1}$$

with superposition functions $f_n(q,p)$. Actually eq. (3.1) is some sort of generalized separation ansatz. It decomposes the problem of solving the exact Schrödinger equation in the case of collective motion into three simpler ones: (1) Finding an optimal collective path $\{|\phi_{qp}\rangle\}$ which decouples maximally the collective motion from the internal excitations. (2) Invoking, for the now given path, a quantization procedure which in the end allows to determine the $f_n(q,p)$ and hence the collective wave functions with all their properties. (3) Estimating the coupling of the $|\psi_n\rangle$, thus obtained, to the residual Hilbert-space, thereby giving a measure for the validity of the whole approach in the particular case considered. Actually there are various groups who have suggested solutions to some of the above points[6-14]. In this paper we follow the approach of Goeke and Reinhard since it is the only one which includes a validity condition and has been applied in a realistic way, see Refs. 12, 13, 14, 36, 37 for details. This approach we call quantized ATDHF.

We start from the variational principle

$$\delta \int dq\ dp\ dq'\ dp'\ f^*(q,p)\langle\phi_{qp}|H-E|\phi_{q'p'}\rangle f(q',p') = 0 \tag{3.2}$$

and have to vary with respect to the path $|\phi_{qp}\rangle$ and with respect to the superposition $f^*(q,p)$.

To this end we consider the variation with respect to the path using the Gaussian overlap approximation. The second approximation is an adiabatic expansion about p=0 in powers of p, where we consider an adiabatic path up to order p^1 which can be parametrized as

$$|\phi_{qp}\rangle = \exp ipQ(q)\}|\phi_q\rangle \tag{3.3}$$

i.e. it is fixed completely by the knowledge of the first order expansion $|\phi_{qp}\rangle \simeq |\phi_{qp=0}\rangle + ip\ Q(q)|\phi_{qp=0}\rangle$ where $Q(q)|\phi_q\rangle = -i\ \partial_p|\phi_{qp}\rangle_{p=0}$. As a third point (not really a new approximation) we have to recall that the path is an expansion basis for a series of states in the collective spectrum; thus the path is demanded to be independent of the actual superposition function $f_n(q,p)$. This postulate is consistent with demanding a local and energy independent quantized collective Hamiltonian. Altogether we obtain in order p^0

$$\langle\phi_q|[a^+a,H - \frac{\partial V}{\partial q}\ Q]|\phi_q\rangle = 0 \tag{3.4a}$$

and in order p^1

$$\langle\phi_q|[a^+a,[H,Q] + \frac{i}{M} P]|\phi_q\rangle = 0 \tag{3.4b}$$

where a^+a can be any 1-ph operator and $P|\phi_q\rangle = i\partial_q|\phi_q\rangle$. The classical potential is

$$V(q) = \langle\phi_q|H|\phi_q\rangle \tag{3.5}$$

and the mass parameter is given by

$$\frac{1}{M(q)} = \langle\phi_q|[Q,[H,Q]]|\phi_q\rangle \tag{3.6}$$

with (here the index ph indicates the 1p-1h and 1h-1p part)

$$Q(q) = (\frac{\partial V}{\partial q})^{-1} H_{ph}(q) \tag{3.7}$$

The eqs. (3.4) are a coupled system determirg (not yet uniquely) $|\phi_q\rangle$ and Q(q). They have independently been derived by Villars[9] and Goeke and Reinhard[10]. For a numerical treatment is is essential to realize that they can be recombined to one differential equation for the path[12,14]

$$\frac{\partial}{\partial q}|\phi_q\rangle = \frac{M}{(\frac{\partial V}{\partial q})} [H,H_{ph}]_{ph}|\phi_q\rangle \tag{3.8}$$

The above formulation is independent of the way in which the parametrization is chosen. For convenience one may introduce a measuring operator D which defines the parameter via $q = \langle\phi_q|D|\phi_q\rangle$.

The second point in quantized ATDHF, viz. the quantization, is now straightforward although technically involved. The variation (3.2) with respect to $f_n^*(q,p)$ yields a generalized Griffin-Hill-Wheeler integral equation for $f_n(q,p)$. With the same two approximations as before, i.e. GOA and adiabatic expansion, one can derive[35] from it a differential equation of the type of a Schrödinger equation. The fact that one deals with a set of two conjugate parameters, q and p, corresponds to a Peierls-Thouless double projection method and ensures an accurate asymptotic behaviour of the collective mass. One ends up with a collective Hamiltonian[35]

$$H_c(q,\frac{d}{dq}) = -\frac{d}{dq}\frac{1}{2M(q)}\frac{d}{dq} + V(q)-Z(q) \tag{3.9}$$

where M(q) and V(q) are consistently those of eqs. (3.6) and (3.5). The Z(q) include quantum corrections which remove the spurious zero-point energies contained in V(q) originating from the zero-point motion of the wave packet $|\phi_q\rangle$ along the collective path. They are given by[35]

$$Z(q) = \frac{1}{2} \langle \phi_q | Q^2 | \phi_q \rangle \frac{\partial^2 V}{\partial q^2} + \frac{1}{2} \langle \phi_q | P^2 | \phi_q \rangle \frac{1}{M} \tag{3.10}$$

Actually the two basic equations of quantized ATDHF are eq. (3.8) and eq. (3.9). As has been shown in Refs. 14, 28 the eq. (3.8) has not a unique solution but depends, as any first order differential equation, one some initial choice $|\phi_q^{(o)}\rangle$. After that eq. (3.8) describes simply in the many dimensional space of Slater determinants a fall line with respect to the metric tensor[28,29]. One particular fall line is distinguished from the others by the fact that it fulfills best the validity condition

$$(p(q))^2 \left\| \frac{\partial Q}{\partial p} \right\|_{p=0} \ll 1 \tag{3.11}$$

with $\frac{\hbar^2}{2\mu} p^2(q) \cong E(\text{saddle}) - V(q)$. This particular ATDHF solution is then really the wanted collective path which is decoupled to second order in p^2. See below for a detailed discussion of the decoupling.

In practice, one proceeds as follows. One solves eq. (3.8) by the Euler step method, i.e. a neighboured point $|\phi_{n+1}\rangle$ is calculated outgoing from $|\phi_n\rangle$ by

$$|\phi_{n+1}\rangle = \{1 - \varepsilon \, [H, H_{ph}^{(n)}]_{ph}^{(n)}\} |\phi_n\rangle \tag{3.12}$$

Successive application gives a set of Slater determinants which are afterwards labeled by the distance R of the ions extracted from the total quadrupole moments $\langle \phi_n | r^2 Y_2 | \phi_n \rangle$. Fig. 6 illustrates at the mass parameter how one finds the saddle point and how the various fall lines converge to the collective path (i.e. the saddle-HF-fall line). The calculations are performed using a BKN-force plus Coulomb interaction representing the single particle wave functions by a 16x16x24 mesh of 1 fm gridsize. The must be smaller than 10^{-4} MeV^{-2} in order to obtain stable solutions of eq. (3.12). Each dot in Fig. 6 is the result of ten steps

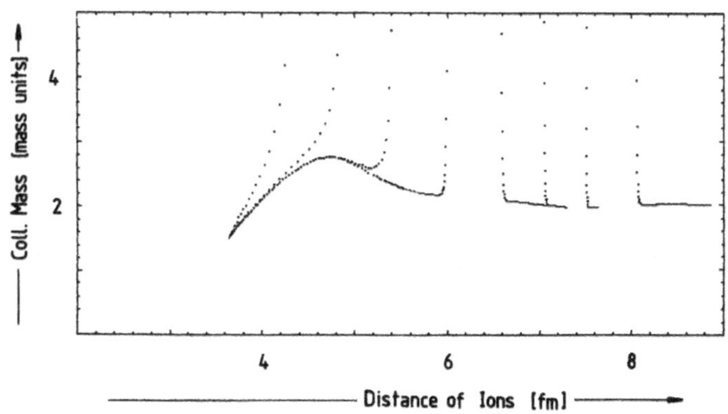

Figure 6

(3.12). The initial solutions have been choser to be α-α harmonic oscillator configurations with an R value ranging from R=4 fm to R=8 fm. One notices the saddle point at R=6.2 fm and the HF ^{8}Be point at R=3.4 fm. One realizes that the single dots accumulate to a clear line representing the potential M(R) for all R values greater than the R-value at the ^{8}Be HF-point.

3.2 Applications

Results of the $^{16}O+^{16}O\leftrightarrow^{32}S$ system are given in Fig. 7, where the classical potential, V(R), the quantum corrected, V(R)-Z(R), and the mass parameter are plotted. One realizes that the mass parameter approaches the correct asymptotic value. It has apparently two bumps where its value deviates noticeably from the reduced mass value μ=8. The zero point energies Z(R) are plotted in Fig. 8. They originate from spurious rotation of the total system, spurious center-off-mass motion of the total system in x- and z-direction and spurious relative motion of the fragments. The total Z(R) is a rather smooth curve. Fig. 9 shows some density distributions occurring in the $^{16}O+^{16}O\leftrightarrow^{32}S$ process. One realizes that all shape degrees of freedom are involved in the reaction. The above calculations have been performed with the Bonche-Koonin-Negele interaction[1]. However this force is rather schematic and does not describe ground state energies and surface properties of nuclei too well. One can see this at Fig. 10, where the relative deviations from the experimental data are given for the binding energy E (positive values mean underbinding), the diffraction radius R_d, the rms-radius and the surface width. All these quantities are rather important for the accurate description of low energy heavy ion data. Hence we generalized the BKN-interaction by not fixing the power α of the density dependent term to unity. The resulting Hartree-Fock Hamiltonian $H_{1\text{-body}} \equiv W_0$ is given by

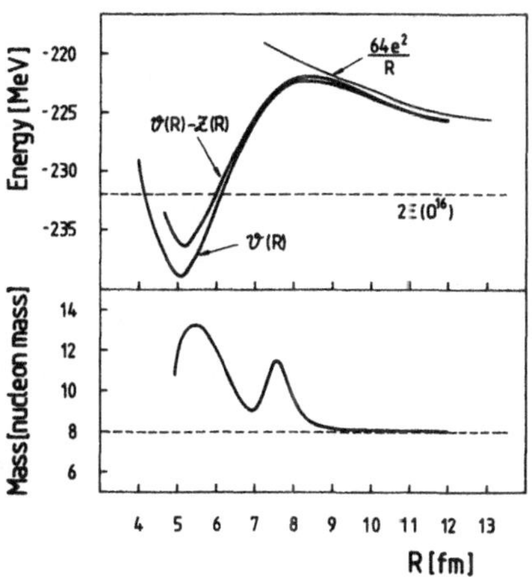

Figure 7

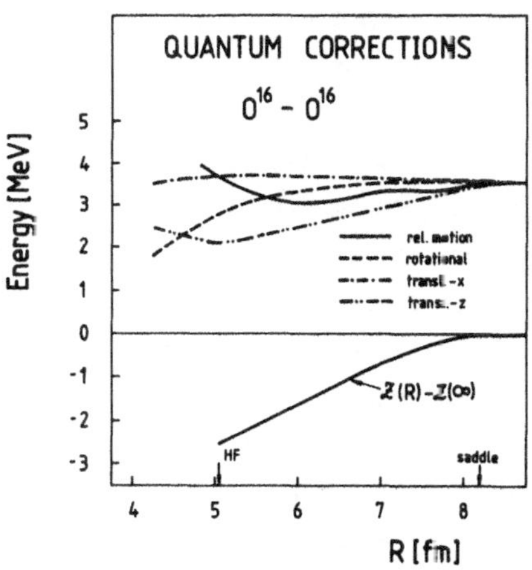

Figure 8

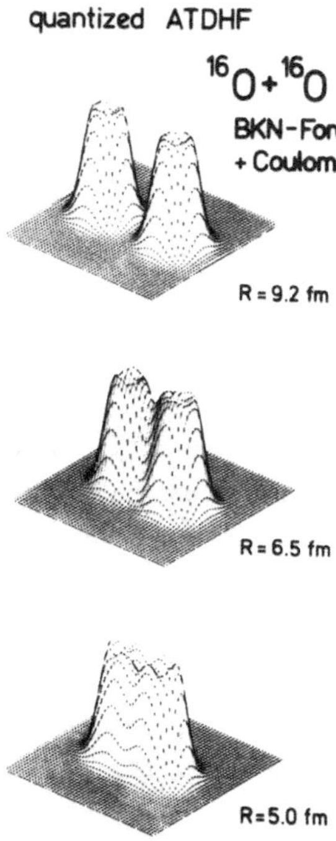

Figure 9

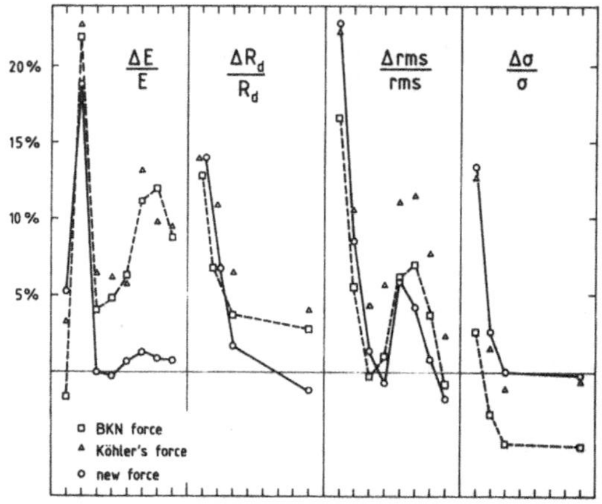

Figure 10

$$W_o = -\frac{\hbar^2}{2m}\nabla^2 + \frac{3}{4}t_o\,\rho_o(r) + \frac{1}{16}(\alpha+2)t_3\,\rho_o^{\alpha+1}$$

$$+ V_y \int \frac{\exp\{|r-r'|/a\}}{|r-r'|/a}\,\rho_o(r')d^3r' + \frac{1}{4}e^2\int \frac{\rho_o(r')}{|r-r'|}\,d^3r'$$

The parameters t_o, t_3, α, V_y, a are fitted to the binding energies of ^{16}O and ^{40}Ca and to the elastic electron scattering data. The resulting agreement of the HF calculations with the data is indicated by the solid lines in Fig. 10 in contrast to BKN (dashed line) and Köhler's force. The new force has a drastic effect on the collective potential and the collective mass of the $^{16}O+^{16}O \leftrightarrow ^{32}S$ system, see Fig. 11. It reduces the height of the barrier and changes the slope. One realizes that this is bound to improve fusion cross sections which are known[36,37] not to be too well reproduced with the BKN force in quantized ATDHF. Actually Fig. 12 shows the fusion cross section above the barrier taken from Ref. 38. The quantities are calculated by classical trajectory calculations involving a friction force fitted by Gross and Kalinowski. The potential used is the quantum corrected V-Z evaluated once with BKN and once with the new force, which gives a noticeably better agreement. The friction force used in the calculations is phenomenological and fitted to many fusion data by Gross and Kalinowski. Actually various approximations to the full calculation, where e.g. M(R) is set to be equal μ (dashed line) and Z(R)

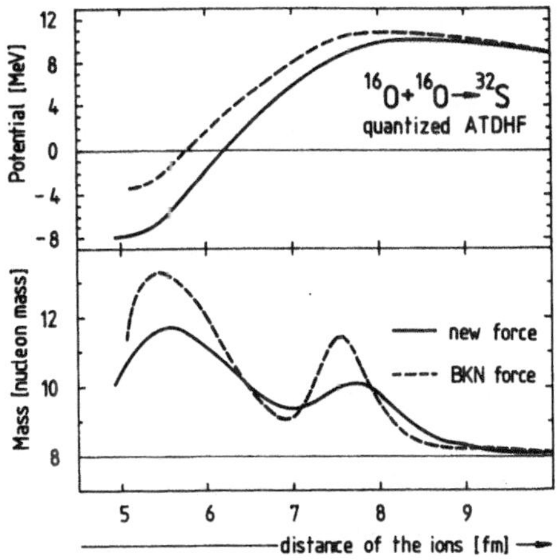

Figure 11

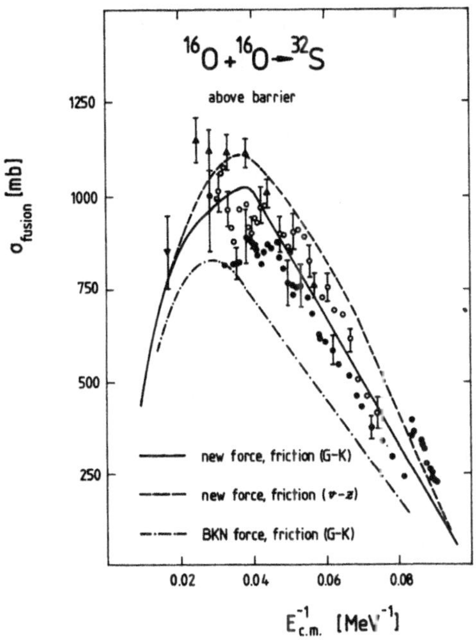

Figure 12

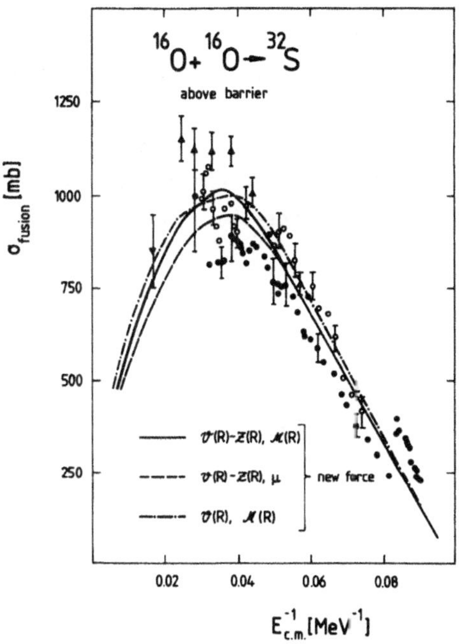

Figure 13

is set equal to zero (dash-dotted line) are displayed in Fig. 13. It turns out that their effect is not very big for fusion above the barrier.

The fusion cross section below the barrier can be evaluated by generalized WKB-techniques. It is given by

$$\sigma_{fus}(E_{CM}) = \lambda\!\!\!^{-2} \sum_{L} (2L+1) T_L(E_{CM}) \tag{3.13}$$

and the transmission coefficients are

$$T_L = \exp[-2\int_a^b \{\frac{2M(R)}{\hbar^2}[V(R)-Z(R)+\frac{\hbar^2}{2\theta(R)}L(L+1)-E_{CM}]\}^{1/2}dR] \tag{3.14}$$

where a and b are the classical turning points. For subbarrier fusion it is customary to consider the astrophysical S-factor rather than the fusion cross section itself. The S-factor $S(E_{CM})$ is defined as

$$S(E_{CM}) = \sigma_{fus}(E_{CM})E_{CM}\exp\{\frac{2\pi Z_1Z_2e^2}{\hbar v}\} \tag{3.15}$$

It is given in Fig. 14 for the $^{16}O+^{16}O\rightarrow^{32}S$ fusion. One realizes an excellent agreement with experiment for the new force and a rather mediocre one for the BKN interaction. This is to be considered as a great success since the force is only fitted to ground state data and the rest is done by the quantized ATDHF theory.

Actually the evaluation of subbarrier fusion cross sections requires great care. In Fig. 15 for instance there are discussed various approximations which correspond to those made in Fig. 13 above the barrier. At low energies one obtains changes of half an order of magnitude for the fusion cross section. Thus all details like zero point energies, CM-corrections, rotational corrections, q-dependent mass parameters have to be taken into account.

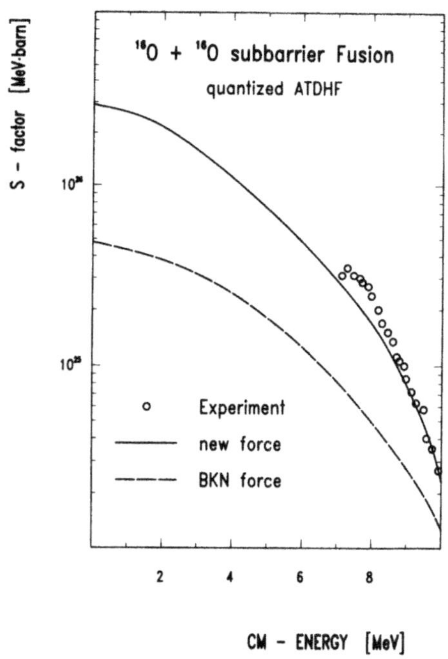

Figure 14

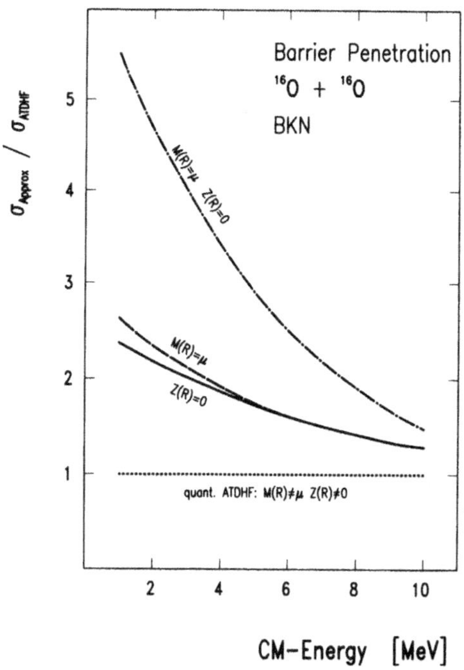

Figure 15

Further results on the $^{12}C+^{12}C$-system can be found in Ref. 37. The $\alpha+^{16}O\leftrightarrow^{20}Ne$-system is reviewed in a different article by D. Provoost et al. in these proceedings.

3.3 The Validity Condition

It is of great importance to investigate the decoupling properties of the collective path evaluated by ATDHF since this tells how relevant it is to prefer ATDHF to simpler methods as e.g. constrained HF. To this end it is convenient to rename the q,p as q_o, p_o and to consider them, without loss of generality, as members of a complete labelling of the full manifold of Slater determinants:

$$|\phi(\underline{q},\underline{p})\rangle = |\phi(q_o q_1,\dots,p_o p_1\dots)\rangle \tag{3.16}$$

The labels $\underline{q}$ and $\underline{p}$ can be constructed such that

$$H(\underline{q},\underline{p}) = \langle\phi(\underline{q},\underline{p})|H|\phi(\underline{q},\underline{p})\rangle \tag{3.17}$$

has the properties of a classical Hamiltonian and that the generators

$$\begin{aligned} Q_\alpha(\underline{qp})\ \phi(\underline{q},\underline{p}) &= -i\ \left(\frac{\partial}{\partial p_\alpha}\right)_{ph} |\phi(\underline{q},\underline{p})\rangle \\ P_\alpha(\underline{qp})\ \phi(\underline{q},\underline{p}) &= i\ \left(\frac{\partial}{\partial q_\alpha}\right)_{ph} |\phi(\underline{q},\underline{p})\rangle \end{aligned} \tag{3.18}$$

fulfil orthogonality and completeness relations[12,39]. Since $|\phi(q_o,0\dots0\dots)\rangle$ is assumed to be time even one can easily show[40] that the following equation holds up to second order in δp_o:

$$\begin{aligned} H(q_o,\delta q_1\dots,\delta p_o,\delta p_1,\dots) &= V(q_o) + \frac{\delta p_o^2}{2M} \\ &+ \frac{1}{2}\sum_{\alpha\neq 0} \delta p_o\ \delta p_\alpha \frac{\partial^2 H}{\partial p_o \partial p_\alpha} + \frac{1}{3!}\sum_{\alpha\neq 0} \delta p_o^2\ \delta p_\alpha \frac{\partial^2 H}{\partial p_o^2 \partial p_\alpha} \end{aligned} \tag{3.19}$$

the coupling terms are

$$\frac{\partial^2 H}{\partial p_o \partial p_\alpha} = i\ \frac{\partial V}{\partial q_o} \langle\phi|[Q_\alpha,\frac{\partial Q_o}{\partial p_o}]|\phi\rangle \tag{3.20}$$

and

$$\frac{\partial^2 H}{\partial p_o^2 \partial q_\alpha} = \langle\phi|[P_\alpha,i[Q_o,[H,Q_o]]-[H,\frac{\partial Q_o}{\partial p_o}] + \frac{2i}{M}\frac{\partial Q_o}{\partial q_o}]|\phi\rangle \tag{3.21}$$

Thus if one demands

$$\frac{\partial^3 H}{\partial p_o^2 \partial q_\alpha} = 0 \qquad (3.22)$$

then one can evaluate $\partial Q_o/\partial p_o$ from that. If the other coupling term $\partial^2 H = \partial p_o \partial p_\alpha$ should be small then one has to demand that the $\partial Q_o/\partial p_o$ should be small, hence the validity condition is

$$p_o^2(q) \| \frac{\partial Q_o}{\partial p_o}(q) \| \ll 1 \qquad (3.23)$$

The solution of ATDHF which fulfills (3.23) best is the desired maximally decoupled collective path. This is, as shown in a simple model calculation[28], the one extending from the saddle point towards Hartree-Fock.

4. Path Integral Approaches

The basic feature of the path integral theory is that it expresses the time evolution of a many body system, subjected to two body interactions, by a coherent sum over the time evolution of the system in mean fields. Actually, one considers the multitude of (classical) trajectories of the system in all possible external one-body potentials (mean fields) and performs a functional integral over them with the weight given in terms of the two-body interaction. Of particular interest are those trajectories which satisfy the stationary phase condition since in many circumstances they dominate the time evolutions of the mean field. The actual conditions depend on the matrix element of the time evolution operator associated to the motion under consideration.

The proper object to study the stationary properties of a self bound system is the partition function, i.e. the trace of the time evolution operator over the complete space of the system

$$Z^{(T)} = \mathrm{Tr}\{\exp(-iHT)\} \qquad (4.1)$$

where H is the microscopic Hamiltonian of the many body system consisting out of a single particle term and the effective two body interaction V. If $Z^{(T)}$ is known the basic properties of the system can be extracted. For instance, the exact eigenenergies for stationary motion are given by the poles of the Fourier transform $\tilde{Z}$ of the partition function:

$$\tilde{Z}(E) = -i \int_0^\infty dT\, Z^{(T)}\, e^{iET} = \sum_n \frac{1}{E-E_n} \qquad (4.2)$$

In the approach of Reinhardt[15] one proceeds by expressing the $Z^{(T)}$ as a functional integral over all possible density matrices being periodic with period T, i.e. $\rho(t+T) = \rho(t)$. This reads

$$Z^{(T)} = (\mathrm{Det}V)^{1/2} \int D\rho \exp\{iS[\rho]\} \tag{4.3}$$

The DetV is the determinant of the two body interaction in the configuration time space. The $S[\rho]$ is the action corresponding to $\rho(t)$ and is given by

$$S[\rho] = \frac{1}{2}\int_0^T dt\ \mathrm{Tr}\{\rho(t)V\rho(t)\} - \sum_\nu n_\nu \varepsilon_\nu[\rho]T \tag{4.4}$$

Here the n_ν and $\varepsilon_\nu(T)$ are occupation numbers and Floquet indices of certain single particle states $|f_\nu(t)\rangle$. These are related to $\rho(t)$ by means of the eigenvalue equation

$$(i\frac{\partial}{\partial t} - h[\rho])|f_\nu(t)\rangle = -\varepsilon_\nu[\rho]|f_\nu(t)\rangle \tag{4.5}$$

where $h[\rho]$ is the single particle Hamiltonian associated to $\rho(t)$,

$$h[\rho] = T + \mathrm{Tr}\{V\rho(t)\} \tag{4.6}$$

and $|f_\nu(t)\rangle$ has the same periodicity as $\rho(t)$:

$$|f_\nu(t)\rangle = |f_\nu(t+T)\rangle \tag{4.7}$$

The eqs. (4.3)-(4.7) form a complete system and an exact description of the full partition function.

The virtue of the present PIA consists now in providing in a natural way a semi-classical approximation to the exact functional integral. Such an approximation is expected to be useful for systems performing a large amplitude collective motion because one expects the action $S[\rho]$ to change rapidly (in comparison to the magnitude of h) for small charges in the classical trajectories $\rho(t)$. Thus a stationary phase approximation to the integral (2.4) seems to be appropriate, which selects out of the multitude of trajectories $\rho(t)$ those which minimize the action. These particular trajectories can be shown to fulfil the selfconsistency condition in addition to eqs. (4.5)-(4.7):

$$\rho(t) = \sum_\nu |f_\nu(t)\rangle n_\nu \langle f_\nu(t)| \tag{4.8}$$

The total set of eqs. (4.5)-(4.8) is equivalent to a time dependent Hartree-Fock approximation (TDHF) with periodic trajectories:

$$\begin{aligned} i\frac{\partial}{\partial t}\rho(t) &= [h[\rho],\rho] \\ \rho(t+T) &= \rho(t) \end{aligned} \tag{4.9}$$

The $\rho(t)$ is composed as $\rho(t) = \sum g_\nu(t) g_\nu^*(t)$ with g_ν obeying the TDHF eqs. They are quasi-periodic and are related to the periodic $f_\nu(t)$ by $g_\nu(t) = \exp\{-i\varepsilon_\nu t\} f_\nu(t)$. For a given period T the total energy of the system is thus constant and can be evaluated as

$$E(T) = \mathrm{Tr}\{\rho T\} + \frac{1}{2} \mathrm{Tr}\ \mathrm{Tr}\{\rho V \rho\} \tag{4.10}$$

In order to obtain quantized stationary states one has to select from the multitude of periodic TDHF orbits with varying E those which correspond to the pole of the Fourier-transform Z(E) of Z^T. They are distinguished by the fact that the classical action satisfies the condition

$$S(T) - T \frac{\partial}{\partial T} S(T) = 2\pi n \qquad n = 0,1,2,\ldots \tag{4.11}$$

or explicitly

$$\int_0^T dt \sum_\nu n_\nu \langle f_\nu(t) | i \frac{\partial}{\partial t} | f_\nu(t) \rangle = 2\pi n \tag{4.12}$$

This prescription selects some energies E_n corresponding to the stationary states of the system. Actually Griffin et al.[31)] have come to the same quantization condition in a different approach.

5. The Generator Coordinate Method and Quantized Time-Dependent Mean Field Motion

The present section investigates the TDHF description of nonadiabatic collective vibrations by means of generator coordinate techniques and compares the results with those of the path integral approach of sect. 4. The derivation starts with the dynamical GCM ansatz (DGCM) as it is explained in Ref. 37: One assumes the motion to evolve along a collective path $\{|\phi_{qp}\rangle\}$, which is a manifold of Slater determinants $|\phi_{qp}\rangle$ representing a collective degree of freedom with coordinate q and conjugate momentum p. The corresponding stationary states are then given by

$$|\psi\rangle = \int dq\, dp\, f(q,p) |\phi_{qp}\rangle \,. \tag{5.1}$$

Both the superposition function f(q,p) and the manifold $\{|\phi_{qp}\rangle\}$ are to be determined by means of a variation of the total energy $\langle\psi|H|\psi\rangle$ with respect to f(q,p) and $|\phi_{qp}\rangle$ with the subsidiary condition of properly normalized $|\psi\rangle$. Since q and p are conjugate parameters describing basically the same collective motion $\{|\phi_{qp}\rangle\}$ is highly overcomplete. For slow motions the problems caused by this overcompleteness can be overcome by an adiabatic expansion in powers of p around the static path $\{|\phi_{q,p=0}\rangle\}$. It is plausible that for nonadiabatic motions an expansion around a finite momentum is the appropriate way.

Therefore we assume that p is not an independent integration variable but depends on q like the momentum along a classical trajectory, i.e. $p = p(E,q)$ is given by $\mathcal{H}(q,p) = E$ where $\mathcal{H}$ is the classical Hamiltonian $\mathcal{H}(q,p) = \langle\phi_{qp}|H|\phi_{qp}\rangle$ and H is the microscopic many-body Hamiltonian. Thus we reduce the DGCM to a simpler one-parameter GCM with the collective path $\{|\phi_q^{(E)}\rangle\} = \{|\phi_{q,p(E,q)}\rangle\}$:

$$|\psi\rangle = \int dq\; f^{(E)}(q)|\phi_q^{(E)}\rangle \; . \tag{5.2}$$

By construction we have energy conservation along the path: $\langle\phi_q^{(E)}|H|\phi_q^{(E)}\rangle = E = \text{const}$. The approximate DGCM ansatz (5.2) still is a fully quantal state but the dp-integration has been replaced by fixing p to a classical trajectory. Therefore we call eq. (5.2) a semiclassical approximation to DGCM. This approximation is best understood by the analogy to a well established procedure in the restauration of broken rotational symmetry: If one uses the Euler angles Ω instead of q then the p corresponds to the cranking frequency ω and eq. (5.1) is a Peierls-Thouless double projection. Eq. (5.2) is then the analogue of a Peierls-Yoccoz angular momentum projection out of a cranked HF state obtained by $\langle\delta\phi(\omega)|H-\omega J_x|\phi(\omega)\rangle = 0$ with $\langle\phi(\omega)|J_x|\phi(\omega)\rangle = [J(J+1)]^{1/2}$. The cranking enhances the angular momentum components near $\langle\phi|J^2|\phi\rangle = J(J+1)$ and thus improves the validity of the Peierls-Yoccoz projection.

Still, a full variation of $\langle\psi|H-E|\psi\rangle$ with regard to both, f and ϕ, is not feasible nor sensible. A further approximation is needed. Since we deal with collective motion we perform a Gaussian overlap expansion (GOA) up to second order in $(q-q')$

$$I^{(E)}(q,q') = \langle\phi_q^{(E)}|\phi_{q'}^{(E)}\rangle = \exp[-\tfrac{1}{2}\lambda^{(E)}(\bar{q})(q-q')^2] \; , \tag{5.3}$$

$$\begin{aligned} H^{(E)}(q,q')\; I^{(E)}(q,q') &= \langle\phi_{\bar{q}}^{(E)}|H|\phi_{\bar{q}}^{(E)}\rangle \\ &- \tfrac{1}{2}\, i(q'-q)\langle\phi_{\bar{q}}^{(E)}|\{P,\bar{H}\}|\phi_q^{(E)}\rangle - \tfrac{1}{8}(q-q')^2\langle\phi_{\bar{q}}^{(E)}|\{P,\{\bar{H},P\}\}|\phi_{\bar{q}}^{(E)}\rangle \; , \end{aligned} \tag{5.4}$$

with $P|\phi_q^{(E)}\rangle = (i\partial/\partial q)_{ph}|\phi_q^{(E)}\rangle$ and $\lambda^{(E)}(q) = 2\langle\phi_q^{(E)}|P^2|\phi_q^{(E)}\rangle$ and $\bar{H} = H - \langle\phi_{\bar{q}}^{(E)}|H|\phi_{\bar{q}}^{(E)}\rangle$ and $\bar{q} = (q+q')/2$. The GOA is actually a well established and frequently used approximation for collective motions in many-body systems. If one assumes consistently the various $\bar{q}$-dependent terms to be slowly varying compared to $I^{(E)}(q-q')$ one can rewrite $\langle\psi|H-E|\psi\rangle$ as a one-dimensional integral involving the collective wave functions $g^{(E)}(q) = \int dq'\; I^{1/2}(q,q') \times f^{(E)}(q')$ rather than the unphysical $f^{(E)}(q)$, see ref. 37.

$$\langle\psi|H-E|\psi\rangle = \int dq\ g^*(q)\{\langle\phi_q^{(E)}|H-E|\phi_q^{(E)}\rangle$$

$$- [1/4\,\lambda(q)](\partial^2/\partial q^2)\langle\phi_q^{(E)}|H|\phi_q^{(E)}\rangle - [1/4\,\lambda^{(E)}(q)]\langle\phi_q^{(E)}|\{P,\{\bar{H},P\}\}|\phi_q^{(E)}\rangle$$

$$+ :p-p(q,E),[1/\lambda^{(E)}(q)]\langle\phi_q^{(E)}|\{\bar{H},P\}|\phi_q^{(E)}\rangle:$$

$$+ \frac{1}{2} :(p-p(q,E))^2,[1/\lambda^{(E)}(q)^2] \times \langle\phi_q^{(E)}|\{P,\{H,P\}\}|\phi_q^{(E)}\rangle:\}g(q)\ , \tag{5.5}$$

and $\langle\psi|\psi\rangle = \int dq\ g^*(q)g(q)$. Here $p-p(q,E) = -i\partial/\partial q$ and it is acting on each function of q on its rhs and the normal ordering is defined as

$$:(p-p)^n,A(q): = 2^{-n} \sum_{k=0}^{n} \binom{n}{k}(\partial^k/\partial q^k)A(q)\partial^{n-k}/\partial q^{n-k}\ .$$

The g(q), representing a bound state, is to be normalizable

$$\int dq\ g^*(q)g(q) < \infty\ . \tag{5.6}$$

Actually eq. (5.5) is the result of a GOA up to second order in q-q' or equivalently p-p(q,E). This is an expansion of quantum fluctuations about the classical trajectory p(q,E).

Now we ask for the least order in p-p(q,E) which is necessary to obtain a definite result for $|\phi_q\rangle$ and g(q) if one performes the variations of $\langle\psi|H-E|\psi\rangle$ with regard to $|\phi_q\rangle$ and g(q). For this one should note that the second and third term on the rhs of eq. (5.5) originate from terms $\alpha(q-q')^2$ and are therefore of second order. From variation of the order $(p-p(q,E))^0$, i.e. $\langle H-E\rangle = \int dq\ g^*(q)g(q)\langle\phi_q^{(E)}|H-E|\phi_q^{(E)}\rangle$ we obtain an equation for the path $|\phi_q^{(E)}\rangle$. But the wave function g(q) remains undetermined and with it the collective dynamics. We therefore start with order $(p-p(q,E))^1$, i.e.

$$\langle\psi|H-E|\psi\rangle \simeq \int dq\ g^*(q)[\langle\phi_q^{(E)}|H-E|\phi_q^{(E)}\rangle + :p-p(q,E),\langle\phi_q^{(E)}|\{\bar{H},P\}|\phi_q^{(E)}\rangle]g(q)\ .$$

This yields from variation $\delta g^*(q)$

$$\begin{aligned}&\langle\phi_q^{(E)}|H-E|\phi_q^{(E)}\rangle g(q) - \frac{1}{2} i(\partial_q\ \phi_q^{(E)}\langle\{\bar{H},P\}|\phi_q^{(E)}|(g(q))\rangle \\ &-i\langle\phi_q^{(E)}|\{\bar{H},P\}|\phi_q^{(E)}\rangle\partial_q g(q) = 0\ ,\end{aligned} \tag{5.7}$$

and from variation $\delta|\phi_q^{(E)}\rangle$

$$g(q)\ ^2\langle\phi_q^{(E)}|(a_p a_h^+)^\perp H|\phi_q^{(E)}\rangle + i(\partial_q g^* g - g^*\partial_q g\ \langle\phi_q^{(E)}|(a_p a_h^+)^\perp\{\bar{H},P\}|\phi_q^{(E)}\rangle = 0\ , \tag{5.8}$$

where $(a_q^+ q_h)^{\perp}$ is any 1p-1h operator which is orthogonal to the path generator P. Actually eq. (5.7) can be employed to simplify eq. (5.8): If one multiplies eq. (5.7) from left with g^* and adds its complex conjugate one obtains, using energy conservation and hermiticity of H and P, $\partial_q g^* g - g^* \partial_q g = 0$, and thus from eq. (5.8)

$$g(q)\ ^2\langle\phi_q^{(E)}|(a_p a_h^+)^{\perp} H|\phi_q^{(E)}\rangle = 0\ . \tag{5.9}$$

Thus we have a coupled set of equations (5.9) and (5.7) which determine the path $\{|\phi_q\rangle\}$ and the collective wave function q(g), respectively.

Let us first investigate the consequences of eq. (5.9): It allows for a nontrivial $g(q) \neq 0$ only if the 1p-1h elements W_{ph} of H are orthogonal to $(a_p^+ a_h)^{\perp}$; that means W_{ph} has to be paralell to P, i.e. $W_{ph}|\phi_q^{(E)}\rangle = \alpha P|\phi_q^{(E)}\rangle$. Here $W = T + tr\{\rho_q^{(E)} V\}$ is the usual Hartree-Fock Hamiltonian for the Slater determinant $|\phi_q^{(E)}\rangle$ and the index ph indicates the 1p-1h part of W. $\alpha=\alpha(q)$ is some proportionality factor which has to be a real number. Indeed we have with $\partial/\partial q \propto W_{ph}$ the equation

$$\langle\phi_{q+\delta q}^{(E)}|H|\phi_{q+\delta q}^{(E)}\rangle - \langle\phi_q^{(E)}|H|\phi_q^{(E)}\rangle = i\partial q\langle\phi_q^{(E)}|\alpha^* W_{ph} H - \alpha H W_{ph}|\phi_q^{(E)}\rangle\ .$$

The lhs vanishes due to energy conservation of $|\phi_q^{(E)}\rangle$. Thus we need $\alpha^*=\alpha$ in order that the rhs vanishes. It is then convenient to choose a parametrization in which $\alpha=1$. This is no restriction because we can perform in any case a (real) scale transformation $t(q) = \int_0^q dq'\ \alpha(q')^{-1}$ such that $\tilde{\alpha}=\tilde{\alpha}(t)=1$. In the t-parametrization we obtain as equation for the path

$$i\partial_t|\tilde{\phi}_t^{(E)}\rangle = W_{ph}|\tilde{\phi}_t^{(E)}\rangle\ , \tag{5.10}$$

which determines the $|\tilde{\phi}_t^{(E)}\rangle$ at any t if one $|\tilde{\phi}_{t_0}^{(E)}\rangle$ is given. In fact, that is the well-known TDHF equation and the parameter t is the time along the trajectory $|\tilde{\phi}_t^{(E)}\rangle$.

We now can proceed to investigate the consequences of eq. (5.7): For a given path $\{|\tilde{\phi}_t^{(E)}\rangle\}$ one can solve eq. (5.7), using energy conservation, and one obtains $0 < \langle\tilde{\phi}_t^{(E)}|\{\tilde{H},\tilde{P}\}|\tilde{\phi}_t^{(E)}\rangle g(t) = \text{const.} < \infty$ where we consider from now on all quantities in the transformed coordinate t. The $\langle\tilde{\phi}_t^{(E)}|\{\tilde{H},\tilde{P}\}|\tilde{\phi}_t^{(E)}\rangle$ certainly will have some upper bound, thus one gets for all t

$$0 < \tilde{g}_{min} < \tilde{g}(t) < \tilde{g}_{max} < \infty\ . \tag{5.11}$$

At first glance the two conditions eq. (5.11) and eq. (5.6) seem to be contradictory, namely on the one hand that g(t) should be a smooth and nonvanishing function of t according to eq. (5.11), and on the other hand that $\tilde{g}(t)$ be normalizable, see eq. (5.6). However, they can be made compatible if we require that $|\phi_t^{(E)}\rangle$ is periodic with a period T, i.e. $|\tilde{\phi}_{t+T}^{(E)}\rangle = |\tilde{\phi}_t^{(E)}\rangle$. Indeed, this guarantees that $|\tilde{\phi}_t^{(E)}\rangle$ contains all the information within a finite t-interval and that it nevertheless is smoothly connected and finite everywhere.

Thus we have obtained from the present semiclassical approximation to the DGCM and the GOA to first order a precise prescription for $\{|\phi_q^{(E)}\rangle\}$ or equivalently $\{|\tilde{\phi}_t^{(E)}\rangle\}$ given in eq. (5.10): They are defined to be the strictly periodic TDHF orbits.

Actually the physical implications of this simple prescription are remarkable and can be discussed as follows: Let us consider for a moment not strictly but only quasi-periodic TDHF orbits $|\phi_t^{(E)}\rangle$. This means that the density matrix (and with it the physical information) is periodic, i.e. $\rho(t+T) = \rho(t)$, but $|\phi_{t+T}^{(E)}\rangle$ is only up to a phase identical to $|\phi_t^{(E)}\rangle$. Such a quasi-periodic initial value problem can be transformed to an equivalent eigenvalue problem with strictly periodic orbits $|\bar{\phi}_t^{(E)}\rangle$ at the price of introducing the Floquet-index $\varepsilon^{(E)}$:

$$(i\partial_t - W_{ph})|\bar{\phi}_t^{(E)}\rangle = -\varepsilon^{(E)}|\bar{\phi}_t^{(E)}\rangle \ . \qquad (5.12)$$

This equation has a continuous manifold of periodic solutions over a certain range of energies E. Accordingly $\varepsilon^{(E)}$ is a continuous function of the period T and hence, equivalently, of the energy E. The quasi-periodic $|\phi_t^{(E)}\rangle$ and the strictly periodic $|\bar{\phi}_t^{(E)}\rangle$ are related by a phase:

$$|\phi_t^{(E)}\rangle = \exp(-i\varepsilon^{(E)}t)|\bar{\phi}_t^{(E)}\rangle \ ,$$

$$\varepsilon^{(E)}T = \int_0^T \langle\bar{\phi}^{(E)}(t)|\,W_{ph} - i\partial_t|\bar{\phi}^{(E)}(t)\rangle dt \ .$$

Thus from the continuum of quasi-periodic wave functions only some are strictly periodic, i.e. those, for which $\varepsilon^{(E)}T = 2\pi n$. This means that demanding strictly periodic TDHF solutions rather than quasi-periodic ones incorporates already a quantization prescription. Only those quasi-periodic TDHF solutions are allowed which fulfil the quantization rule

$$\int_0^T \langle\bar{\phi}_t^{(E)}|i\partial_t|\bar{\phi}_t^{(E)}\rangle dt = 2\pi n \ , \qquad n=0,1,2,\ldots \qquad (5.13)$$

Actually these are exactly the equations which come out of the PIA, namely the quasi-periodic TDHF eigenvalue eq. (5.12) for the classical mean field motion and in addition the semiclassical quantization rule (5.13), by considering the classical action along the orbits. The minor modification that the PIA is formulated in terms of W whereas in the present considerations always the W_{ph} is used can be proven to be immaterial.

Thus the present section shows that the path integral approach together with its semiclassical quantization condition on the one hand and the DGCM with a semiclassical approximation and the GOA on the other hand yield identical results. This is also to be seen in the context of the next section where the PIA with adiabatic approximations is shown to lead to the quantized ATDHF theories as it comes out from the DGCM. Thus we have the remarkable result that both theories although being of different conceptual origin, although employing different techniques with different degrees of complexity, allow for equivalent approximations and lead then to identical results. Thereby it seems that the GOA is comparable to the saddle-point approximation in the PIA. It is an interesting task for future investigations to work out a one-to-one correspondence between these both approximations.

By now we have used the GOA expansion in the GCM up to order $(p-p(q,E))^1$. One should be able to go to order $(p-p(q,E))^2$ and study the quantum fluctuations. This will probably yield correction terms similar to those obtained in a second-order loop expansion[15)] in the PIA.

6. Periodic Quantization and Quantized ATDHF

Concerning the quantization of TDHF the results of PIA and of the previous section can be summarized in a two step procedure:

i) search periodic TDHF solutions for a range of periods T and corresponding energies E.

ii) select particular energies E_n by the quantization condition.

Actual numerical calculations in realistic cases have not yet been established. It is even not known whether periodic TDHF solutions indeed exist at all. To judge the method it might therefore be useful to establish a connection to quantized ATDHF, which is known to be numerically applicable and indeed has very similar objectives as PIA. In fact this connection has been discussed recently in Ref. 32. Without going into details the results can be reviewed as follows:

The path-integral-approach (PIA) considers a multitude of TDHF solutions $\phi_T(t)$, the density matrix of which is periodic with the period T, i.e. $\rho(t+T) = \rho(t)$. Actually there may be various distinct solutions for a given period T being associated e.g. to different multipole vibrations. However we assume in the following that these different solutions have clearly different characteristics such that one can always identify those associated to the mode one is interested in. We furthermore assume that within certain limits there exists for any T a $\phi_T(t)$ associated to the considered mode and that they are continuously connected. This means that the trajectory $\phi_{T+\delta T}(t)$ is infinitesimally close to $\phi_T(t)$ if δT is small enough. The multitude $\{|\phi_T(t)\rangle\}$ of those solutions (associated to a certain collective model) we shall call a branch. We will assume that a branch contains an HF point and that all quantities calculated from $\phi_T(t)$ will be distinguishable in T and also distinguishable in the energy $E = \langle\phi_T(t)|H|\phi_T(t)\rangle$ and that there is a unique relation between E and T. Hence any Slater determinant, which is member of a branch, is uniquely characterized by giving E and t, i.e. $|\phi_E(t)\rangle$. This labeling in E and t, rather than in T and t, has the advantage that the energy provides a much more sensitive discrimination, ranging from E_{HF} to the upper end of the branch, whereas T is compressed at $T_{min} = \omega_{RPA}^{-1}$ at the lower end of the spectrum (harmonic limit).

Furthermore, this labelling has the convenience that E and t are something like canonically conjugate quantities. This establishes immediately a bridge between a branch $\{|\phi_E(t)\rangle\}$ and a path $\{|\phi_{qp}\rangle\}$ since both multitude are characterized by two real parameters which can be understood as classical canonical conjugate variables. Within the branch $\{|\phi_E(t)\rangle\}$ a periodic trajectory $|\phi_E(t)\rangle$ represents an actual classical motion. This is characterized within the path $\{|\phi_{qp}\rangle\}$ by $q = q(t,E)$ and $p = p(t,E)$ which obey the classical equations of motion. Hence the members of a branch and of a path can be related by

$$|\phi_E(t)\rangle \propto |\phi_{q(E,t),p(E,t)}\rangle$$

with a reverse relation $E = E(q,p)$ and $t = t(q,p)$ such that $\{|\phi_{qp}\rangle\}$ has the characteristics of a collective path, with canonical labelling, i.e.

$$\langle\phi_{qp}|\overleftarrow{\partial}_q\overrightarrow{\partial}_p-\overleftarrow{\partial}_p\overrightarrow{\partial}_q|\phi_{qp}\rangle = i$$

In practice such a transformation may be constructed explicitly such that q(t) and p(t) obey classical canonical equations.

In order to establish a connection of a branch to an adiabatic path we assume the labelling q,p to be such that it has something like turning points, i.e. that it has two times t_a and t_b at which $p(E,t_a) \sim p(E,t_b) \sim 0$. We furthermore assume that the states $|\phi^{PIA}_{q',p'=0}\rangle = |\phi^{PIA}_{q'}\rangle$ are time even, which simplifies the calculational techniques. Actually the path $\{|\phi_{q'p'}\rangle\}$ is fully identical to $\{|\phi_E(t)\rangle\}$ and it may have any complicated q'- and p'-dependence. An adiabatic collective path, however, as it is used in ATDHF, is assumed to have the special structure of a separation ansatz

$$|\phi^{ATDHF}_{qp}\rangle = \exp[ipQ(q)^{ATDHF}]|\phi^{ATDHF}_q\rangle$$

where $Q^{ATDHF}(q)$ is a 1p-1h operator with respect to $|\phi_q^{ATDHF}\rangle$ corresponding of course to the derivative with respect to p at p=0

$$Q^{ATDHF}(q)|\phi_q^{ATDHF}\rangle = (i\frac{\partial}{\partial p})_{1p-1h}|\phi_{qp}^{ATDHF}\rangle_{p=0}$$

Apparently an adiabatic $\{|\phi_{qp}^{ATDHF}\rangle\}$ is a useful object if it can be constructed in such a way that it is as close as possible to the $\{|\phi_{qp}^{PIA}\rangle\}$.

If one requires that $\phi_{qp}^{PIA} = (1+ipQ^{PIA}+\ldots)|\phi_q^{PIA}\rangle$ and $\phi_{qp}^{ATDHF} = (1+pQ^{ATDHF}+\ldots)|\phi_q^{ATDHF}\rangle$ are identical to order p^1 one obtains exactly the ATDHF eqs. (3.4a), (3.4b) or equivalently (3.8). It is remarkable that there is such a strong connection between PIA and ATDHF. One also can derive the ATDHF-validity condition from PIA by requiring the deviation between ϕ_{qp}^{PIA} and ϕ_{qp}^{ATDHF} to be as small as possible in order p^2. For details see Ref. 32.

Concerning the quantization one can proceed as follows. The classical action along the periodic TDHF trajectories is then given by eq. (4.11) with

$$S(T) = \int_0^T dt\,\langle\phi_{E(T)}(t)|i\partial_t - H|\phi_{E(T)}(t)\rangle$$

If we transform this to q(t) and p(t) we obtain after some algebra

$$S(T) = \int_0^T dt(\dot{q}(t)p(t)-\mathcal{H}(q(t),p(t))$$

with q and p obeying classical canonical equations with $\mathcal{H}(q,p)$ as classical Hamiltonian.

In the adiabatic limit we expand

$$\mathcal{H}(q,p) = \frac{p^2}{2M(q)} + V(q)$$

If one now evaluates eq. (4.11) and transforms from an integral over t into one over q with $dt = \dot{q}dq$ one gets

$$2\int_a^b \sqrt{\frac{2M(q)}{\hbar^2}(E-V(q))}dq = 2\pi n$$

Apart from an additive term $\pi/2$ on the rhs we would get the same expression if we evaluated the eigenenergies of the collective quantized ATDHF-Hamiltonian by WKB methods, neglecting however the zero point energies Z(q). If we performed a Bohr-Sommerfeld quantization involving the (q,p) we would obtain an identity.

If one considers the fluctuations in PIA a lengthy calculation shows that the action S is changed to

$$S+\Delta S = \int_0^T dt(\dot{q}(t)p(t)-\tilde{\mathcal{H}}(q(t),p(t))$$

where

$$\tilde{\mathcal{H}}(q,p) = \mathcal{H}(q,p) - Z^{PIA}(q)$$

with

$$Z^{PIA}(q) = \frac{\lambda_0(q)}{4M_0(q)} + \frac{\partial^2 V_0(q)}{\partial q^2} \cdot (4\lambda_0(q))^{-1}$$

The PIA expression differs from the corresponding one of quantized ATDHF in that it is lacking the residual two body interaction in the zero point energies which, however, is not expeced to be a severe limitation. Again the PIA corresponds to the Bohr-Sommerfeld quantization of the quantized ATDHF-Hamiltonian including this time the zero point energies.

In summary one can conclude: There is a remarkably close connection between the PIA and quantized ATDHF: If one extracts an adiabatic collective path from the periodic PIA orbits with turning points one obtains exactly the adiabatic TDHF equations derived by Villars[9)] and Goeke and Reinhard[10,13)] described in sect. 3 of this lecture. If one solves the ATDHF collective Hamiltonian by semiclassical WKB methods one obtains basically the PIA quantization. These relationships shed also an interesting light on ATDHF: The ATDHF path appears to be an approximation to a very special <u>multitude</u> of TDHF trajectories with <u>varying energies</u>, namely the periodic TDHF orbits.

7. Time Dependent Generator Coordinate Method

If one considers deep inelastic collisions and the associated quantities evaluated by means of TDHF, one realizes e.g. that TDHF describes well the mean value of the particle number distribution of e.g. the projectile-like fragment. The width of the distribution, however, is by an order of magnitude too small in TDHF.

Two different effects are believed to contribute to the large spreading widths, the first one is of a statistical nature and involves random coupling of the collective degrees of freedom of the system to the intrinsic ones. Many different approaches have been developed[21-27)]. It has not yet explicitly been demonstrated if these theories indeed give an increase of the width compared to TDHF or not. In fact there are arguments[33,34)] that one single particle potential alone hinders the system to develop a large spreading width. Hence a superposition of several time dependent mean fields would be preferable since this is expected to have a large effect on the 2-body density matrix.

It is the aim of this section to pursue the approach of calculating RPA-like correlations as a function of time into a heavy ion reaction[19,20]. To this end we shall proceed from a time dependent generalization of the generator-coordinate variational principle (TDGCM). In the representation of TDGCM we have for the wave function of the total system

$$\psi(t) = \int dq\ f(q,t)|\phi_q(t)\rangle \qquad (7.1)$$

where $|\phi_q(t)\rangle$ represents the collectively distorted time dependent Slater determinants and $f(q,t)$ is the time dependent superposition function. The q is a collective variable associated to the mode whose zero point correlations are under consideration, see Fig. 16. The assumptions required in order that the deformed sin-

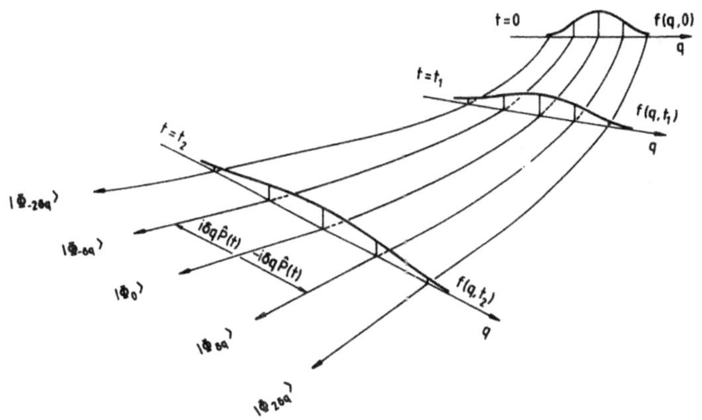

Figure 16

gle Slater determinants $|\phi_q(t)\rangle$ obey the TDHF equations are made very transparent in the TDGCM. They are basically the Gaussian overlap assumption characteristic of most GCM models, and the neglect of the feedback of the correlations on the Slater determinant equations of motion. We note that the TDHF mean path is itself a large amplitude (semi-classical) time dependent collective path and that we are looking for correlations in collective motions which are orthogonal to the path. We will see that the TDGCM takes into account the effect of the full history of the mean path onto the correlations. We want to point out that this is something very different from a local RPA around the TDHF state at each time separately. In TDGCM a local RPA is used only once, namely as to define an initial condition for the dynamical evolution of the correlations during the reaction. In this paper we are going to consider the particular case of the TDGCM for harmonic (small amplitude) motion in the correlation channels. Of course, the TDGCM model can also be presented without making the harmonic assumption which is usually justified if the amplitude of the correlations is small. In the general case we have a large amplitude theory of the dynamical evolution of correlations perpendicular to the path. This is not an eigenvalue problem as in stationary RPA or GCM but an initial value problem. The TDGCM aims to determine the time evolution of an initially given oscillation (orthogonal to a TDHF trajectory) and the corresponding ground state correlations. The objective is to find an equation of motion for $|\phi_q(t)\rangle$ and $f(q,t)$ in eq. (7.1) by varying $\langle\psi(t)|H-i\frac{\partial}{\partial t}|\psi(t)\rangle$ with respect to f and $|\phi\rangle$. This variation requires some simplifications which are fulfilled in the Gaussian overlap approximation (GOA)

$$\langle\phi_q(t)|\phi_{q'}(t)\rangle = \exp\{-\frac{\lambda(\bar{q},t)}{2}(q-q')^2\} \tag{7.2}$$

with $\bar{q} = \frac{1}{2}(q+q')$ and

$$\lambda(q,t) = \langle\phi_q(t)|\frac{\overleftarrow{\partial}}{\partial q}\frac{\overrightarrow{\partial}}{\partial q}|\phi_q(t)\rangle \tag{7.3}$$

The GOA allows to define a collective wave function

$$g(q,g) = \int dq'\,(\frac{\lambda(q,t)}{\pi})^{1/4}\exp\{-\frac{\lambda(q,t)}{2}(q-q')^2\}f(q',t) \tag{7.4}$$

such that for any operator A one obtains:

$$\begin{aligned}\langle\psi(t)|A|\psi(t)\rangle = \int dq\, g^*(q,t)\{&\langle\phi_q(t)|A|\phi_q(t)\rangle - \frac{1}{4\lambda}\frac{\partial^2}{\partial q^2}\langle\phi_q(t)|A|\phi_q(t)\rangle \\ &+ :(-i\frac{\partial}{\partial q})\langle\phi_q(t)|\frac{\overleftarrow{\partial}}{\partial q}A - A\frac{\overrightarrow{\partial}}{\partial q}|\phi_q(t)\rangle/\lambda: \\ &+ :(-\frac{\partial^2}{\partial q^2}-\frac{\lambda}{2})\langle\phi_q(t)|\frac{\overleftarrow{\partial}^2}{\partial q^2}A - 2\frac{\overleftarrow{\partial}}{\partial q}A\frac{\overrightarrow{\partial}}{\partial q} + A\frac{\overrightarrow{\partial}^2}{\partial q^2}|\phi_q(t)\rangle:\}g(q,t)\end{aligned} \tag{7.5}$$

If one inserts for A the $H-i\frac{\partial}{\partial t}$ and varies with respect to $|\phi_q(t)\rangle$ one obtains in lowest order

$$(H_{ph} - i\frac{\partial}{\partial t})|\phi_q(t)\rangle = 0 \tag{7.6}$$

where H_{ph} is the 1p-1h part of the total Hamiltonian with respect to $|\phi_q(t)\rangle$. This is exactly the time dependent Hartree-Fock equation for each $|\phi_q\rangle$. The variation with regard to g(q) yields an equation for g(q,t) which describes the time evolution of the superposition function g(q,t) or f(q,t). In case of harmonic motions in q orthogonal to the TDHF trajectories the g(q,t) can be written as

$$g(q,t) = N(t)\exp\{-\frac{q^2}{2\alpha(t)}\} \tag{7.7}$$

with

$$i \frac{\partial}{\partial t} \alpha(t) = - \frac{1}{2\lambda(o,t)} \langle \phi_o(t) | \overleftarrow{\partial}_q^2 H - H \overrightarrow{\partial}_q^2 | \phi_o(t) \rangle / \lambda(o,t) \qquad (7.8)$$

where all values are to be taken at q=0.

The spreading width $\Delta^2 A$ of a 1-body operator can also be given very simply as

$$\begin{aligned} \langle \psi | \Delta^2 A | \psi \rangle = \langle \phi_o | \Delta^2 A | \phi_o \rangle + \frac{1}{2} \left(\frac{\alpha\alpha^*}{\alpha_r} - \mu\right) \langle \phi_o | \overleftarrow{\partial}_q A + A \overrightarrow{\partial}_q | \phi_o \rangle^2 \\ + \frac{i\alpha_i}{\alpha_r} \mu \langle \phi_o | \overleftarrow{\partial}_q A + A \overrightarrow{\partial}_q | \phi_o \rangle \langle \phi_o | \overleftarrow{\partial}_q A - A \overrightarrow{\partial}_q | \phi_o \rangle + \frac{1}{2} \left(\frac{1}{\alpha_r} - \frac{1}{\mu}\right) \mu^2 \langle \phi_o | \overleftarrow{\partial}_q A - A \overrightarrow{\partial}_q | \phi_o \rangle^2 \end{aligned} \qquad (7.9)$$

This can be evaluated straightforwardly after performing two successive TDHF calculations with varying δq-initial conditions, and after solving the differential equations for $\alpha(t)$.

Altogether we see that in the harmonic approximation all information on $|\psi(t)\rangle$ is carried by the three quantities $|\phi_o(t)\rangle$, $\frac{\partial}{\partial q}|\phi_o(t)\rangle$ and $\alpha(t)$. See the complete set of eqs. (7.7)-(7.9). In practice we handle the $\frac{\partial}{\partial q}|\phi_o(t)\rangle$ as finite difference $\frac{\partial}{\partial q}|\phi_o(t)\rangle = (|\phi_{\delta q}(t)\rangle - |\phi_o(t)\rangle)/\delta_q$. Thus it remains to propagate two neighboured TDHF trajectories $|\phi_o(t)\rangle$ and $|\phi_{\delta q}(t)\rangle$ and to work out additionally the time evolution of the width $\alpha(t)$ by solving eq. (7.8). Obviously this procedure can easily be implemented into any existing TDHF code. The extra expense of TDGCM consists mainly in carrying TDHF twice. The numerical applicability and the relevance of the TDGCM theory has been checked by means of a one dimensional TDHF code with quartet symmetry and using the α+α-system. The force used corresponds to a single particle Hamiltonian of the form

$$h[\rho] = - \frac{\hbar^2}{2m} \frac{d^2}{dx^2} + \frac{V_o}{\sqrt{\pi\sigma'}} \int dx' \exp\{- \frac{(x-x')}{\sigma}\} \rho(x') + t_3 \rho^2(x)$$

with V_o = -50 MeV, σ = 2 fm, t_3 = 88 MeV fm^2. The TDHF solutions show fusion up to an incoming momentum of p = 0.28 fm^{-1}.

Since we have only one spatial coordinate there are only few collective modes, associated to q, to be excited. Besides the unperturbed motion there are basically three modes possible whose characteristics are given in Fig. 17. Corresponding to the three modes, we have six observables. We consider the following three of them

$$P_{rel} = \int_0^\infty dx\ j(x) - \int_{-\infty}^0 x\ j(x)$$

$$R_+^2 = \int_0^\infty dx\ x^2 \rho(x) + \int_{-\infty}^0 dx\ x^2 \rho(x)$$

$$R_-^2 = \int_0^\infty dx\ x^2 \rho(x) - \int_{-\infty}^0 dx\ x^2 \rho(x)$$

and in addition the difference in particle number

$$N_{rel} = \int_0^\infty dx\ \rho(x) - \int_{-\infty}^0 dx\ \rho(x)$$

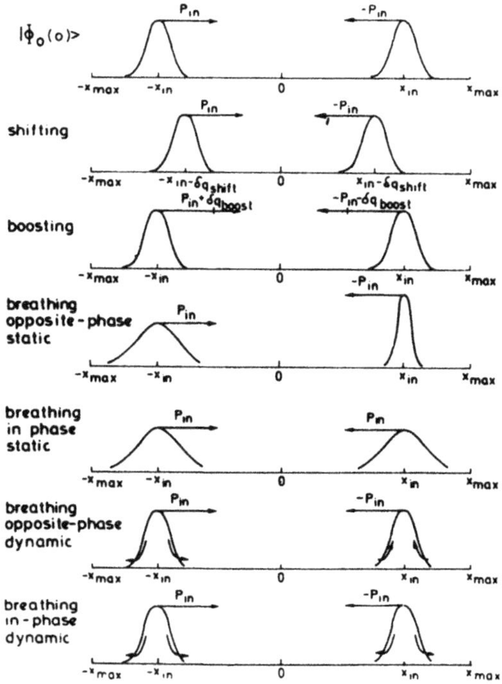

The various initialization modes exemplified by drawing the initial density profile as a function of coordinate x. The arrows with p_{in} indicate the initial momentum of the clusters. The little arrows in part 3f indicate the breathing flow of matter inside the clusters.

Figure 17

These are the operators to be inserted in place of A in eq. (7.5). See for illustration Figs. 18 and 19. There the spreading $(\Delta^2 R_-^2)$ is evaluated by the suitable collective mode of breathing-out-of-phase. The trajectories are shifted in order to have the same exit time out of the interaction region. One sees that the coupling of the channels yields approximately the same curve for $(\Delta^2 R_-^2)$.

Now, having shown the numerical reliability of TDGCM and knowing how to assure phase stability, we can evaluate every combination of initialization and measurement. The results are given, for various impact momenta p_{in}, in Table 1 and the values are extracted from $\psi(t)$ immediately after the separation of the ions. First, we observe strong contributions from the collective correlations to the spreading widths. Moreover we see that the dominant contribution always comes from that initial mode which is related to the measurement; this feature could be called "channel-memory". The cross channel contributions are much smaller. One cannot exclude that many of them may add up to another substantial enhancement, but a larger manifold of channels can only be treated statistically.

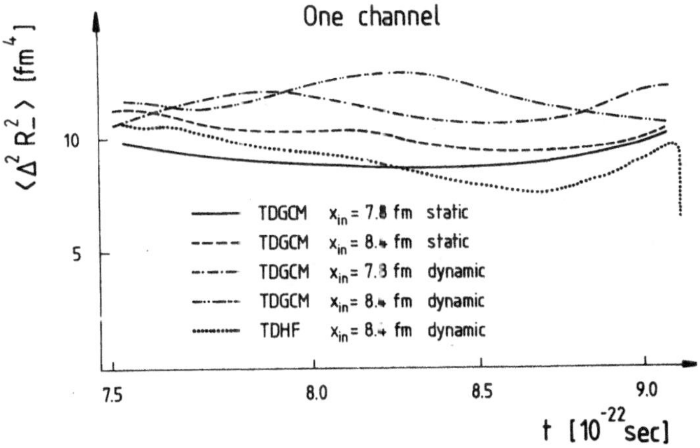

Figure 18

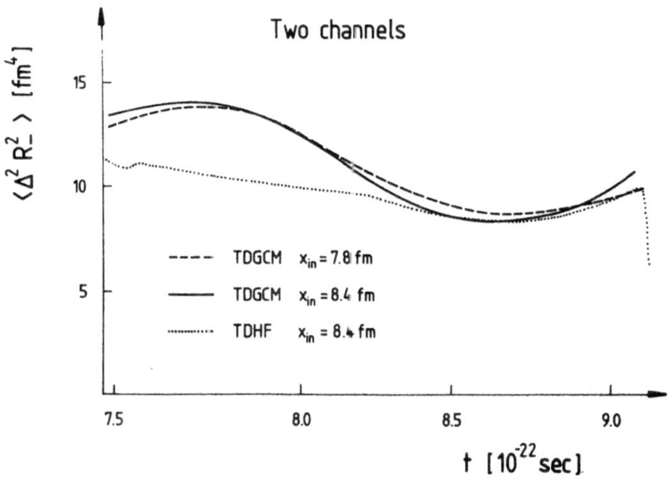

Figure 19

In Table 1 we see also that the correlation effects are strongest for low momenta, near the fusion window at $p_{in} = .28\ fm^{-1}$, and tend to decrease for higher impact momenta. In Fig. 20 we have plotted the TDGCM result for the spreading with of the internal excitation energy E^+ (after reaction) as a function of p_{in}. We see that the correlations indeed produce a significant enhancement of Δ^2E^+, particularly for low p_{in} near the fusion window. For larger p_{in} the contribution decreases but seems to level off at a ratio of about 2, leaving perhaps a strong effect also for fast processes.

The increase of the energy width by an order of magnitude in TDGCM compared to TDHF is a remarkable result. Although the present calculations are only one dimen sional and we have considered only a few modes, it shows that the sort of correlations treated in TDGCM is simply necessary for a correct description of the width

Table I

The spreading width of various measuring operators are given for TDHF and for a couple of explicitly considered correlated time-dependent states treated by means of TDGCM

Initialization	Measurement			
	$\Delta^2 p_{rel}$(fm^{-2})	$\Delta^2 R_+^2$ ((fm^4)	$\Delta^2 R_-^2$ (fm^4)	$\Delta^2 N_{rel}$
p_{in} = 0.30 fm^{-1} E_{lab}/A = 7.5 MeV				
TDGCM:				
shift + boost	*2.40* (760%)	18.03 (9%)	10.01 (< 1%)	0.50 (< 1%)
breathing (+)	0.42 (50%)	*35.60* (116%)	10.01 (< 1%)	0.50 (< 1%)
breathing (−)	0.29 (< 1%)	16.51 (< 1%)	*13.14* (31%)	0.52 (4%)
TDGCM (sum)	2.54 (807%)	37.1 (124%)	13.14 (31%)	0.52 (4%)
TDHF	0.28	16.50	10.00	0.50
p_{in} = 0.35 fm^{-1} E_{lab}/A = 10.2 MeV				
TDGCM:				
shift + boost	*0.88* (203%)	7.81 (36%)	6.42 (< 1%)	0.36 (< 1%)
breathing (+)	0.31 (6%)	*9.60* (68%)	6.42 (< 1%)	0.36 (< 1%)
breathing (−)	0.20 (< 1%)	5.73 (< 1%)	*7.86* (22%)	0.33 (−9%)
TDGCM (sum)	0.90 (209%)	11.6 (103%)	7.86 (22%)	0.33 (−9%)
TDHF	0.29	5.73	6.42	0.36
p_{in} = 0.40 fm^{-1} E_{lab}/A = 13.3 MeV				
TDGCM:				
shift + boost	*0.63* (110%)	5.79 (19%)	4.84 (< 1%)	0.24 (< 1%)
breathing (+)	0.31 (3%)	*6.04* (25%)	4.84 (< 1%)	0.24 (< 1%)
breathing (−)	0.30 (< 1%)	4.85 (< 1%)	*5.02* (4%)	0.24 (< 1%)
TDGCM (sum)	0.64 (113%)	6.13 (26%)	5.02 (4%)	0.24 (< 1%)
TDHF	0.30	4.85	4.84	0.24

The TDGCM is initialized using the TDHF width. One notices that for a given measuring operator only the correlations of the corresponding mode (given in italics) contribute essentially. The important point is that this correspondence is not destroyed in the interaction area. Apparently the treated RPA correlations have in some cases a substantial effect on the spreading width which, however, decreases with increasing impact momentum, or with decreasing interaction time. The numbers in brackets indicate the increase of the TDGCM value relative to the TDHF one. The values of TDGCM (sum) sum up all diagonal terms of the three collective modes considered.

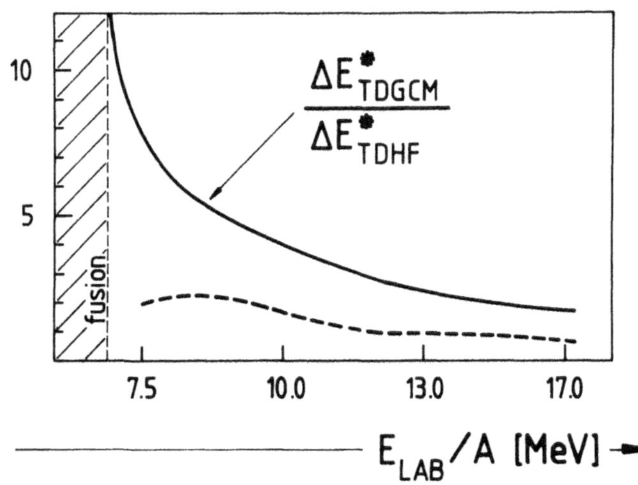

Figure 20

of one body operators. These correlations are actually created by a coherent superposition of different mean fields associated to the collective coordinate q. By a simple argument of Feldmeier[5)] it can be shown that particularly in the case of the particle number width in an HI reaction such a superposition is the preferable mechanism to enlarge the TDHF width, compared to the inclusion of collision terms. If by the use of a collision term a particle is transferred from one ion to the other it is bound to occupy a high-lying state since the potential well is not changed. If we, on the other hand, allow a change of the potential well a it is done in TDGCM by a superposition of many of them, then the additional particle can occupy a lower-lying state in a broader potential. Hence the superposition is energetically favoured. This corresponds also to the findings of Wolschin[5)]. Thus the present mechanism described in TDGCM is probably indispensible for a correct description of widths and preferable to inclusion of collision terms.

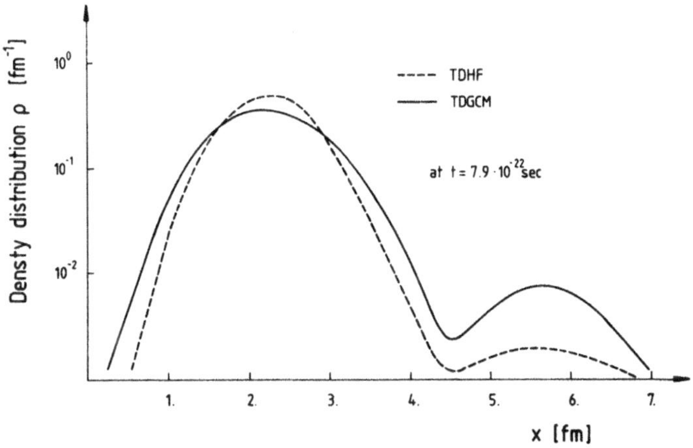

Figure 21

There are some other aspects to be mentioned: As we see at Fig. 20 there are dramatic things happening at the fusion window. Hence it is very probable that findings of TDHF of an angular momentum window are changed if one considers correlations in this section. This may also be true of the results of TDHF concerning precompound emission of particles. This happens in a low density region which is expected to be most affected by correlations. If one evaluates for example for the breathing mode considered the density distribution, see Fig. 21, one realizes big changes in the tail. It has not been checked if the addition of various modes reduces the effect. However, one has to be careful in interpreting TDHF results in those limiting cases.

8. Summary

The present lecture series can be summarized by a few qualitative statements:

The basic success of TDHF for heavy ion collisions consists in providing a good insight into the geometry of shapes involved in an HI reaction, in describing well the energy flow from the entrance channel to internal excitations (heat) for deep inelastic reactions and also for fusion events. Furthermore, the time evolution of one body operators turns out to be well described. All this is achieved without free adjustable parameters solely on the basis of an effective interaction and a variational solution of the time dependent Schrödinger equation.

The disadvantage of TDHF consists in the difficulty to extract quantum mechanical information from it. There are no ways to describe barrier penetration phenomena in low energy fusion or fission events. TDHF does not give phase shifts or even clearly defined reaction channels. TDHF gives a continuum of vibrational excitations rather than a discrete spectrum.

The quantized ATDHF theory is an approach, which is fully quantum mechanical from the beginning and which is able to find the maximally decoupled collective path of a system. By means of a validity condition the limits and the range of applicability can be formulated precisely. It has now frequently been applied to the collision of heavy ions, fusion below and above the barrier, ground state rotational bands, phase shifts of elastic HI-scattering, etc. A three dimensional code has been developed which, with a suitable force, was able to reproduce well the astrophysical S-factors of the $^{16}O+^{16}O \rightarrow$system.

The functional integral approaches are conceptually a progress since they provide means to extract quantum mechanical information from TDHF without explicitly making the adiabatic approximation. They also can be used to formulate a complete reaction theory. For bound states they result in a multitude of periodic TDHF orbits and a quantization condition. Unfortunately, numerical calculations in realistic systems are missing and to not seem to be available in the near future.

Actually the basic result of PIA, namely the periodic quantization, can also with noticeably less technical requirements be derived by means of a semiclassical approximation to the GCM with a pair of conjugate parameters. This yields also periodic TDHF trajectories where the strictly periodic ones determine the stationary energies.

There is a strong relationship between PIA and quantized ATDHF. The adiabatic approximation incorporated in PIA yields the ATDHF equation and the corresponding collective path, which means that the ATDHF path is a well defined approximation to the multitude of periodic TDHF trajectories. The semiclassical approximate solution of the quantized ATDHF collective Hamiltonian reproduces the quantization condition of PIA also when fluctuations are taken into account.

If one sticks to the semiclassical nature of TDHF and is interested in inclusive cross sections then correlations due to collision terms and due to RPA vibrations on top of TDHF go beyond the TDHF approach. In the first sort of approaches excited single particle levels of the TDHF-like potential well are fractionally occupied. In the second case, i.e. time dependent GCM (TDGCM), several Slater determinants associated to different mean fields are superimposed. Actual one dimensional calculations show that the latter theory is able to explain the large widths of one body operators not reproduced by TDHF. This is particularly relevant for the description of particle number fluctuation in HI reactions, which is probably due to RPA correlations and not due to statistical two body collisions. However, calculations for this particular example are missing yet since they are bound to involve pairing correlations. Calculations in more than one dimension, allowing also for finite impact parameters, are also missing.

One of the most promising aspects of future work is the development of the quantized ATDHF theory to the description of coupled channel phenomena by means of angular momentum projected collective paths. This is presently under investigation.

References

1. P. Bonche, S. Koonin, J. Negele, Phys. Rev. C13 (1976) 1226.
2. R.Y. Cusson, J.A. Maruhn, W.A. Meldner, Phys. Rev. C18 (1978) 2589.
3. K.T.R. Davies, K.R. Sandhya-Devi, M.R. Strayer, Phys. Rev. C20 (1978) 1372.
4. J.W. Negele, Rev. Mod. Phys. 54 (1982) 913.
5. "Time Dependent Hartree-Fock and Beyond", Proc. Int. Meeting at Bad Honnef, June 7-11, 1982, ed. by K. Goeke and P.-G. Reinhard, Lecture Notes in Physics, Springer-Verlag, 1982.
6. G. Holzwarth, T. Yukawa, Nucl. Phys. A219 (1974) 125.
7. D.J. Rowe, R. Bassermann, Can. J. Phys. 54 (1976) 1941.
8. T. Marumori, Prog. Theor. Phys. 57 (1977) 112.
9. F. Villars, Nucl. Phys. A285 (1977) 269.
10. K. Goeke, P.-G. Reinhard, Ann. Phys. 112 (1978) 328.
11. M. Baranger, M. Vénéroni, Ann. Phys. 114 (1978) 123.
12. P.-G. Reinhard, K. Goeke, Nucl. Phys. A312 (1978) 121.
13. P.-G. Reinhard, K. Goeke, Phys. Rev. C20 (1979) 1546.
14. P.-G. Reinhard, J.A. Maruhn, K. Goeke, Phys. Rev. Lett. 44 (1980) 1740.
15. H. Reinhardt, Nucl. Phys. A346 (1980) 1.
16. S. Levit, J. Negele, Z. Paltiel, Phys. Rev. C21 (1980) 1603.
17. M. Kuratsuji, T. Suzuki, Phys. Lett. 92B (1980) 19.
18. J.P. Blaizot, M. Orland, J. Phys. 41 (1980) L53.
19. P.-G. Reinhard, K. Goeke, R.Y. Cusson, Z. Phys. A295 (1980) 45.
20. P.-G. Reinhard, R.Y. Cusson, K. Goeke, Nucl. Phys. A398 (1983) 141.
21. C.Y. Wong, H.H. Tang, Phys. Rev. Lett. 40 (1978) 1070.
22. H.S. Köhler, Nucl. Phys. A343 (1980) 315.
23. P. Grangé, H.A. Weidenmüller, G. Wolschin, Ann. Phys. 136 (1981) 190.
24. R. Balian, M. Vénéroni, Ann. Phys.
25. S. Ayik, Z. Phys. A298 (1980) 83.
26. A.F.R. de Toledo Piza, see Proc. of Ref. 5.
27. H.C. Pauli, see Proc. of Ref. 5.

28. K. Goeke, P.-G. Reinhard, D.J. Rowe, Nucl. Phys. A359 (1981) 408.
29. D.J. Rowe, A. Ryman, J. Math. Phys. 23 (1982) 732.
30. H. Flocard, P.H. Heenen, D. Vautherin, Nucl. Phys. A339 (1980) 336.
31. K.K. Kan, J.J. Griffin, P.C. Lichtner, M. Dworzecka, Nucl. Phys. A332 (1979) 109.
32. K. Goeke, P.-G. Reinhard, H. Reinhardt, Nucl. Phys. A378 (1982) 474.
33. H. Feldmeier, see Proc. of Ref. 5.
34. G. Wolschin, see Proc. of Ref. 5.
35. K. Goeke, P.-G. Reinhard, Ann. Phys. 124 (1980) 249.
36. K. Goeke, F. Grümmer, P.-G. Reinhard, Phys. Lett. B124 (1983) 21.
37. K. Goeke, F. Grümmer, P.-G. Reinhard, Ann. Phys. 150 (1983) 504.
38. P.-G. Reinhard, J. Friedrich, K. Goeke, F. Grümmer, D.H.E. Gross, to be published.
39. K. Goeke, P.-G. Reinhard, J. Phys. G6 (1978) L245.
40. K. Goeke, P.-G. Reinhard, J.N. Urbano, in preparation.

SHELLS : MODEL AND THEORY

A.P. ZUKER
Centre de Recherches Nucléaires et Université
Louis Pasteur de Strasbourg
Groupe de Physique Nucléaire Théorique
BP 20 67037 Strasbourg Cedex, France

INTRODUCTION AND SUMMARY

These lectures are devoted to examine the theoretical bases of the Shell Model. We shall be led to reformulate the theory of effective interactions so as to incorporate naturally whatever regularities and simplicity emerge from the data. The importance of linked cluster mechanisms will be stressed, and it will appear that much we learn from the behaviour of the Schrödinger equation in the full Hilbert space can be put to good use in diagonalizing large but finite matrices. This is due to the factorization properties of the amplitudes in the wave-functions which are at the origin of linked cluster theory and make it possible to take advantage of the low rank (1+2 body) of the forces to devise accurate truncations.

After some preliminary remarks to define the context of these talks, we shall review in Lecture 1 some basic ideas about the shell model, show some of its successes and then examine the difficulties to be overcome in a rigorous microscopic theory.

Lecture 2 is devoted to introduce the notion of dressed states or quasiconfigurations as a physical picture on which to build effective interaction theory. A very elementary introduction to (quasi-)degenerate perturbation theory will follow, and then we shall show that in second order we can already obtain accurate diagonalizations for some systems.

Lecture 3 deals with Linked Cluster Theory (LCT), the term we have chosen to describe the general formulation in terms of coupled non linear integral equations that is related to (quasi-)degenerate perturbation theory in the same way the coupled cluster (exp(s)) method is related to non degenerate Perturbation Theory (PT).

The first sections are devoted to a diagrammatic derivation of the coupled cluster equations. Then it is shown how the classical approximations, RPA and LOBT (Low Order Brueckner Theory) emerge naturally and how a reading of the history of nuclear matter follows from the Bochum truncation scheme. LCT is then derived in a way that stresses that degenerate theory is the continuation of non degenerate theory by other means. After that the RPA is reexamined to establish the connection between the equations for the Ground State (GS) and the excitation spectra. The last section is devoted to an elementary derivation of the form of the G-matrix that shows how the repulsive core of the interaction is changed into a density dependent saturating mechanism.

Lecture 4 deals with the monopole field and shows the empirical and semiphenomenological evidence that suggested the formulation of LCT. The emphasis is put on the need to enforce good saturation properties as a necessary and (almost) sufficient condition to obtain good spectroscopy with realistic forces. The results suggest that non nucleonic effects must appear as a density dependent modulation of the two body force rather than as genuine higher rank forces.

Throughout, emphasis has been put on conceptual rather than formal development. Still, the diagrammatic derivation of the linked cluster equations is very compact and it is recomended that Lecture 3 be read pencil in hand keeping an eye on [PZ81a]. Although much of the material is new, all of it is the outcome of much rethinking on the fundamental ideas and contributions in the field. In this sense it is an introduction and a review but also a proposal and an invitation to new research.

PRELIMINARY REMARKS TO DEFINE THE HISTORICAL AND PHYSICAL CONTEXT OF THESE LECTURES.

The success of the Shell Model has always been a puzzle : how could a system interacting strongly through a potential containing a repulsive core, behave experimentally as if made of free or weakly interacting particle ?

This problem could be faced with an attitude of mathematical rigour by trying to understand how to deal directly with the many body Schrödinger equation. It was Brueckner who pioneered this approach in the 50's and following his lead it was assumed or rather decided, that the foundations of the Shell Model would be firmly established by the time a calculation could yield the binding energy of nuclear matter. The difficulties encountered by the theory were such, that it has taken some 30 years before reliable estimates of this quantity could be obtained.

In the meantime Kuo and Brown were the first to implement an approach to nuclear spectroscopy which assumes that the main ingredient in the (weak) effective forces needed in shell model calculations, is a Brueckner G-matrix. Although the initial results were encouraging it became apparent that it was necessary to go beyond this lowest approximation to achieve a smooth connection between the N-N potentials and the level schemes. An intense activity followed and a new branch of nuclear physics developped : the theory of effective interactions. In 1975 a topical conference was held at Tucson [BA75] that probably left in practically all the participants the dismal feeling that it was impossible to get good spectra in any rigorous way and the project was abandoned by most nuclear physicists[†]. As these

[†]To become an important branch in Quantum Chemistry !

lectures are largely devoted to show that the project is perfectly viable and interesting, we would like to start by quoting one of the few hopeful remarks of Kirson in the summary talk to the Tucson conference (the whole talk is highly recommend reading) :

In the light of the irreductible imprecision inherent in the perturbation theory approach, it would perhaps be wise to adopt a more qualitative attitude and to search for understanding in terms of simple physical concepts which can be associated (pictorially) with certain terms in the perturbation series. As emphasized by Harvey, a major part of the renormalization of effective operators appears to be associated with self-consistent-field effects. Köhler and Zuker have independently suggested that our favourite two-body renormalization, core polarization, may in fact simply reflect a change in the self-consistent field due to the addition of two valence particles, and in that sense is hardly a legitimate two-body effect. Similarly, we are all familiar with the picture of single-particle energies changing smoothly with nucleon number as particles are added to a shell. However, all our calculations are done with fixed single-particle energies. Perhaps we should reinterpret some apparent two-body terms as arising from this smooth change in the single-particle energies. Bertsch has shown how Pauli-blocking effects can change the effective two-body force towards the end of a shell, something which shows up in our theory as an effective three-body force. Given that exact numerical precision is out of our reach, it is possible that new, qualitative, points of view such as these could be useful in refining our understanding both of effective interaction theory and of nature.

And this not so hopeful comment during one of the discussions : (Kallio)

I want to bring in some criticism of this whole approach. To my mind this theory of folded diagrams is reaching such a degree of complication that of the two possibilities

a) you can perform practical calculations
b) you cannot,

the case (b) would be preferable because just explaining what you have done in (a) is next to impossible. I would feel it preferable to break your initial symmetry and hence decrease its degeneracy so as to be able to stick to non-degenerate Bethe-Goldstone theory with deformation.

With the benefit of hindsight, these remarks can be interpreted as a diagnostic of the three diseases that almost killed the theoretical efforts at the time :

i) Disregard of the influence of the selfconsistency properties on the spectra.

ii) Disregard of the need of going beyond two body forces.

iii) Use of poorly defined perturbative prescriptions.

The three diseases are intimately connected and can be cured by a single powerful prescription provided it is used carefully :

A shell model calculation must be thought as the last step in a process of exact solution of the full many body Schrödinger equation.

Let us make a preliminary exploration of the use of this guideline by considering in turn the 3 problems.

I) Selfconsistency is a word borrowed from Hartree-Fock theory, but it can be used more generally to indicate the mechanism that ensures saturation, i.e. the equilibrium of the system at a given energy and radius. Traditionally it was assumed that these bulk properties had little influence on the spectra and that a force producing poor saturation could produce good spectra. There is some truth in these statements but we must be careful not to transform this sound phenomenological idea into a theoretical assumption. Let us take the sd shell as an example. The spectrum for the two particle states (A=18), is usually calculated in a single particle oscillator basis chosen so as to produce a closed shell (A=16) which has the radius experimentally observed in ^{16}O ($\hbar\omega \approx 15$). If the G-matrix is calculated in such a basis for an interaction that does not saturate at that radius, trouble will follow in estimating a better approximation to the force through a theory that is in principle exact. The equations will drive the solutions towards the correct saturation point, which is not the observed one and bad spectra or even fictitions divergence in some perturbative expansions will result. The Sauer-Vary-Wong [VSW73] effect which precipitated the crisis has probably this origin.

II) If we consider again the sd shell we may ask why we should have a fixed two body interaction that reproduces well ^{18}O ($\hbar\omega \approx 15$) rather than ^{38}Ca ($h\omega \approx 11$), at the other end of the shell. The natural answer is that the interaction must change smoothly from one end to the next. One could expect this variation to be described by a 3-body force, but this is not the case. One of the most important new results we are going to present at these lectures is that the " two body " force must have a smooth non analytic variation on the number of particles, which should be interpreted as a density dependence. It is quite obvious to think of this effect as due to a rearrangement of the selfconsistent field. But then it is important to realize that it is not only the radius but the energy of the systems that must propagate smoothly from ^{16}O to ^{40}Ca. Except for the non analytic behaviour of the variation these statements seem selfevident. There is however one thing we have not mentioned : once this correct saturating behaviour is forced on the naïve G-matrices strikingly good spectroscopic results are obtained (see lecture 4).

In retrospect we may say that the " naïve " Kuo Brown matrix elements for ^{18}O are much better than they look and that the real trouble in the original calculations came from the lack of recognition of the density dependence of the G matrix. Too much time and effort was spend in studying ^{18}O, a rather complicated system, under the delusion that all there was to be known about the " two body " force was contained in the two particle system.

III) Although from the point of view of a shell model calculation the closed shells ^{16}O and ^{40}Ca are trivial, from the point of view of the theory of effective interactions they are not. Both nuclei are particular solutions of the general equations that define the effective Hamiltonian and there is no point in assuming that perturbation theory may be of relevance in going from one end of the shell to the next. The idea one solves a complicated set of coupled integral equations for ^{16}O and then applies perturbation theory to define the effective interaction is wishful thinking.

Linked cluster perturbation theory is of enormous historical interest and some variants of it may also be of great practical use as we shall show in lecture 2. However, in solving the Schrödinger equation we are confronted with integral equations, that can be interpreted as sums of an infinite class of diagrams. In the nuclear case the first equation of this kind defines the G-matrix. Now : if we are interested in integral equations rather than individual diagrams it is better to have a formulation in which the latter emerge naturally from expansion of the former, rather than proceeding artificially the other way round. It is very hard to guess what has to be summed beyond the G-matrix.

The idea that the closed shells (and single particle or single hole states) at the ends of the model space should be calculated on same the footing as all other states between them has an important consequence : a satisfactory theory exists and has been implemented ; the exp(s) or coupled cluster method of Coester and Kümmel [KLZ78, PZ81a]. It defines a set of coupled non linear integral equations from which Goldstone's linked cluster perturbation theory can be obtained by Neumann expansion. Unfortunately, in their original algebraic form these equations are rather forbidding and they were not understood by enough people to be accepted as the natural cornerstone of effective interaction theory when the crisis set in the field. In particular the significance of Zabolitzky's calculations in the closed shell nuclei (^{4}He, ^{16}O and ^{40}Ca) was not appreciated for a long time.

Lectures 2 and 3 will be devoted to rederive in a simple diagramatic notation the coupled cluster equations and to extend them to the general degenerate case. Schematically the different forms of the theory can be represented as follows :

NLCPT=	⟶	LCPT =
Goldstone		= Brandow
↓		↓
NLCT =	⟶	LCT =
Coester Kümmel		= these lectures

The arrows indicate increasing generality. LCT stands for Linked Cluster Theory, N for non degenerate and P for perturbation. Except for LCT that is new, the content of lectures 2 and 3 are meant as a guided introduction to Quasiconfiguration Theory (QT) [PZ81a,b]. Perhaps the reader has the feeling that there are too many names for the same theory and the point deserves an explanation.

In the second paragraph of this introduction we mentioned Brueckner attitude as one of mathematical rigour by which we meant a direct attempt to get exact solutions, without free parameters. There is another attitude, of physical rigour : if a strongly interacting system looks so much like a weakly interacting one, let us not try to give a proof, but let us rather try to take advantage of the fact, in extracting some new infomation. The shell model in its simplest and most elegant form is an example of this attitude and in lecture 1 we shall show some results. Landau's theory of ^{3}He is the other paradigm in the field.

The point we want to stress, is that in formulating a mathematically rigorous theory, physical rigour can be put to good use : the theory must be designed so as to take maximum advantage of the empirical knowledge we have of the final solutions. This is the idea behind quasiconfiguration theory : it is a physically appealling way a writing down the LCT equations. Conversely, if in doing phenomenological work we keep in mind the need of relating it to theory we may obtain hints about the form it must take : this was the guiding attitude in the work about the monopole effect that unearthed the explicit need of density dependence (lecture 4) and suggested the way to write LCT (lecture 3).

Two final comments :

1) QT deals with the Schrödinger equation as if it were a matrix secular problem. As a consequence the equations may be of relevance in diagonalizing finite matrices. This is an interesting possibility, since so far little attention was paid to the linked cluster properties of shell model matrices. It is explored in lecture 2, where we show how second order PT can approximate with acceptable precision very large diagonalizations in the fp shell.
2) It is now known that an exact solution of the Schrödinger equation with two body forces will not reproduce nuclear data. This is not a good reason not to attempt such calculations. On the contrary ; the discrepancy with experiment will give a measure of the non nucleonic effects necessary and the more discrepancies we have, the more potential information on new phenomena.

LECTURE 1. THE SHELL MODEL

Fig. 1

At the origin the shell model was the assumption that the ground states of nuclei could be described as systems of non interacting particles filling the lowest orbits of an oscillator well whose degeneracies were split by a spin orbit force. Exactly filling a major shell (s,p,sd,fp etc.) or some subshells of given j would give rise to particularly stable structures : the closed shell nuclei. The spins of the odd-even ground states were those of the last orbit being occupied. Excited states could be obtained by diagonalizing " some " interaction in the states of that orbit.

Definitions : a configuration is the set of states $m=[n_1,n_2,\dots n_m]$ having n_1 particles in orbit 1, n_2 particles in orbit 2 etc.

N_i is the occupancy, $2(2j_i+1)$, of orbit i. If only neutrons are occupied $N_i=2j_i+1$. In fig. 1 N(Nν) refer to total (neutron) occupancies at closed shells. Example : the closed shell for A=40 is the configuration

$$\left[s^4_{1/2}p^4_{1/2}p^8_{3/2}s^4_{1/2}d^8_{3/2}d^{12}_{5/2}\right]=\left[4,4,8,4,8,12\right]=\left[N_1,N_2,\dots N_6\right]$$

The word configuration mixing is used for diagonalizations involving several configurations, not only the lowest one in the oscillator ordering.

To decide which configurations mix is an art rather than a science. As a rule, shell model calculations are performed for all the configurations that can be constructed for a given set of subshells. Refering to Fig. 1, an A=45 calculation would assume an inert A=40 core and include not only $f^4_{7/2}p_{3/2}$ as shown, but $f^{n_1}_{7/2}p^{n_2}_{3/2}$, $n_1+n_2=5$ (we hasten to say this is a very bad calculation on physical grounds and that we do not recommend it by any means).

Although the oscillator field plus spin orbit accounts for some global trends and gives the order in which levels are filled, it is misleading to think of it as some approximation to the Hamiltonian. If we look at the spectrum of ^{41}Ca we would expect to see a $7/2^-$ ground state and some single particle excited states ($3/2^-$, $1/2^-$

$5/2^-$) in the first 5 or 6 MeV. Then at " $1\hbar\omega$ " (10 MeV) some states would appear corresponding to the promotion of a (ds) particle to an (fp) orbit. Actually we see some 100 levels. The reason is that the centroïd of a configuration is not well approximated by a single particle field and we must use the proper definition :

$$\frac{1}{D}\sum_{\alpha}\langle[n_1\ldots n_m]\alpha|H|[n_1\ldots n_m]\alpha\rangle = \sum K_i n_i + \sum n_i(n_j+\delta_{ij})(1+\delta_{ij})^{-1}V_{ij} \qquad (I.1)$$

where $D=\binom{N_1}{n_1}\binom{N_2}{n_2}\cdots\binom{N_m}{n_m}$ is the dimensionality of the configuration, K_i is the expectation value of the kinetic energy and V_{ij} is a centroïd defined in eq. (4.2)) and α the extra set of quantum numbers that the states define. If now we consider only configurations of the type $m=[N_1\ldots N_n\ n_i\ldots n_m]$, i.e. a closed shell plus some particles moving in valence orbits, we can specialize eq. (I.1) to

$$\langle m|\bar{H}|m\rangle = \underbrace{\sum K_i n_i + \sum n_i\cdot n_j V_{ij}}_{U_0} + \underbrace{\sum n_r(K_r+n_i V_{ir})}_{U_1} + \underbrace{\sum n_r\cdot n_s V_{rs}}_{U_2} \qquad (I.2)$$

$n_\alpha\cdot n_\beta = n_\alpha(n_\beta-\delta_{\alpha\beta})(1+\delta_{\alpha\beta})^{-1}$, ij= occupied ; rs= half occupied orbits.

We see there is a constant term U_o giving the energy of the closed shell, a term U_1 that acts as a single particle field for states r and a two body term U_2. The quantity $K_1 + \sum_i n_i V_{ir} = \varepsilon_r$ corresponds to the energy of the single particle state r, above the closed shell. Eq. (I.2) defines the 1+2 particle monopole field in which the nucleons move, it has a richer structure than a simple one particle (central field) approximation. It can go a long way in explaining why there are so many levels in the neighbourhood of the ^{41}Ca ground state. We shall return to the monopole field in lecture 4. As of now it is introduced only to stress that the naïve shell model picture may define a perfectly good coupling scheme ($\bar{H}$ in eq. (I.2) commutes with the central field) even if the harmonic oscillator well + spin orbit can't be used to estimate unperturbed energies reliably.

Now we would like to examine some of the predictions of the naïve shell model, and see how well they agree with experiment.

In figs. 2 we show some experimental spectra for two particle systems. The heavy lines correspond to the T=1 states expected from a $f_{7/2}^2$ configuration, the thin ones to the T=0 states and the dashed ones to some other states. Some striking features are immediately apparent. The spectra of ^{42}Sc and ^{54}Co (corresponding to having two holes on a ^{56}Ni ($f_{7/2}^{16}$) core) are very similar. If the forces were of strict rank two they would be identical. The spectrum of ^{50}Ti is practically identical to that of ^{42}Ca (T=1 states of ^{42}Sc). Again this is to be expected on simple grounds : in ^{50}Ti we are depositing two protons on a core (closed neutron shell in $f_{7/2}$) of ^{48}Ca and the dynamics is that of a two particle state. There is a severe experimental catch on this argument though : the two lowest dashed lines in the ^{42}Sc spectrum represent 0^+ and 2^+ states that share much of $f_{7/2}^2$ strength. In particular, a

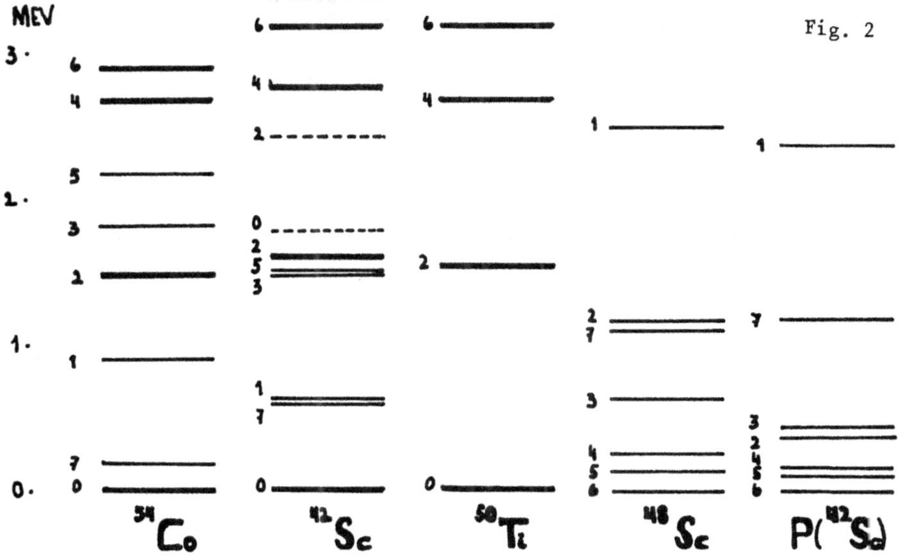

Fig. 2

(dp) reaction on ^{41}Ca populates with equal strength the two 2^+ states indicating that the " model " $f^2_{7/2}$ state must be at the centroïd position of the two and can't be identified with any of them. But then : why is the spectrum of ^{50}Ti identical to that of ^{42}Ca when no " intruders " such as the 0^+ and 2^+ dashes are present in the former ? More on this in the second lecture.

Finally we examine ^{48}Sc, which can be obtained by changing a neutron in the closed shell of ^{48}Ca into a proton : a particle-hole excitation in a sense. The particle-hole energies ω^λ are related to the 2-particle energies E^J by a Pandya transformation

$$\omega^\lambda(\mathrm{ph}) = - \sum \begin{Bmatrix} j & j & \lambda \\ j & j & J \end{Bmatrix} E^J(2p) \qquad \text{(I-3)}$$

and the last column shows the nice agreement obtained when the ^{42}Sc energies are used.

From this quick examination of some typical shell model states we can conclude that the naïve picture is very good, has predictive power even quantitatively and that the riddles it proposes should not invalidate the idea of pure $f^n_{7/2}$ configurations interacting via two body forces but rather establish it as a very sound first approximation in a more refined treatment.

As our ultimate aim is to go from some fundamental theory to the spectra it is appropriate to foresee the kind of difficulties we may encounter :

i) the hard core in the N-N potential

ii) the non nucleonic effects

iii) " phase transitions "

iv) coupling schemes and configuration mixing.

Let's examine them separately.

i) The existence of a strong repulsion at short distances in the N-N potential was thought for years to be the main obstacle in our understanding of the success of the shell model. The problem is now under fairly good control and it will be studied in some detail in lectures 3 and 4.

ii) The quality of numerical calculations of the Schrödinger equation with two body forces is now such that there is no doubt that even the best potentials fail to reproduce the observed properties of nuclear systems. For A=3 and 4 the calculations are exact and for A=16 and 40 they are good enough to establish that something is needed beyond two body forces

Fig. 3

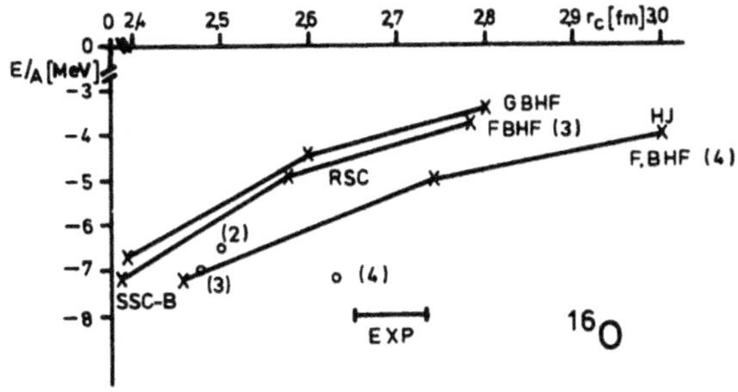

Figure 3, extracted from Zabolitzky's calculations [KLZ78] is a good example of a " Coester line ", i.e. a line over which the results for different potentials move in the energy vs density (or radius in this case) plot, without ever intersecting the experimental point. Loosely speaking we call " non-nucleonic " anything that has to be changed in or added to the non relativistic Schrödinger equation with twobody forces. This may be an abuse of language. For one thing the potentials themselves are non-nucleonic and for another, the possibility that relativistic kinematics could play a role cannot be excluded.
This situation is typical of nuclear physics, it makes life harder and at the same time more interesting. In calculating the properties of liquid Helium for example, one uses some potential (Lennart-Jones or Assize). If exact calculations (see Kalos' lectures) show discrepancies with the data we can speak of a nuisance but we do not expect to learn from it anything fundamental about the forces : in the end everything must come from the Coulomb potential as it is quite unlikely that the structure of the ^{3}He or ^{4}He nuclei could play a role in the problem.
In calculating the properties of nuclei, things are quite different, and whatever we learn about the discrepancies with the data may be useful in formulating the more fundamental theory we do not have yet. One comment is worth making in this respect. Two nucleon forces have been of good service and they certainly provide a very good and solid ground for proceeding. After all, the corrections necessary to the conventional picture are smallish. Therefore, we

think it is very dangerous to take the view that nucleons are dead and that nuclear physics should be done with quarks. For one thing we would not how to do it, except in crude and phenomenological forms that quantitatively have no hope of competing with the conventional view. On the other hand, this way of pretending we start from first principles we do not really have, would hinder the discovery of something interesting about these very principles. The approach we would advocate is intermediate. What is needed, we think, is some phenomenological form of Hamiltonian, including non-nucleonic species of course, that provides the N-N force as effective interaction in the two particle case, and a little something else in the multiparticle case, so that nuclei could move away from the Coester lines to their saturation points.

For example, we could risk an analogy between the nucleon-nucleon and atomic He potentials. In both cases we are reducing complex systems to single entities. The He potentials have a strong repulsion starting at about ~2Å (10^{-8} cm), a distance enormously larger than the size of the He nuclei and the force is entirely generated by the Coulomb field of the nuclei and the orbiting electrons. For nucleons, the 3 quarks in a little bag would play the role analogous to that of the He nuclei in the atomic case, and the real orbiting electrons would be replaced by virtual mesons. It should be noticed that the sizes of the little bags advocated by Brown and Rho [BR83] are very much the same as the sizes of the N-N hard core (~.5 fm), and one could think of the hard core as a Pauli effect, the quarks refusing to be squeezed beyond a certain point. So far the analogy is not working wonders, but there one aspect where it becomes useful. QCD may well be responsible for binding the deuteron, but what we expect of it is to bind the nucleon. The energies in play are larger by a factor 10^3. In this case, there is an even greater difference between the ionization energy of the atoms and the binding of the He molecules (eV vs K$\equiv 10^{-4}$ eV). So what we are advocating, or rather asking the experts to invent, is a phenomenological model to produce the N-N force and a little something else ; in which the nucleons will already come in with masses close to their observed ones, form factors, hard cores and what not. At this stage in the development of nuclear physics we can't afford to be too fundamental, nor can we throw away all we have learnt with the " conventional " view.

iii) Phase transitions are not supposed to take place in small systems. Still, there is a point in asking what is it that may happen in a nucleus, that would be called phase transition in nuclear matter. The normal phase is no doubt a liquid. Something similar to superconductivity occurs in nuclei but we won't discuss it here and we shall ignore the possibility of ordered or crystalline phases. This leaves us with gases of α particles or eventually some mixture of different nuclear species.

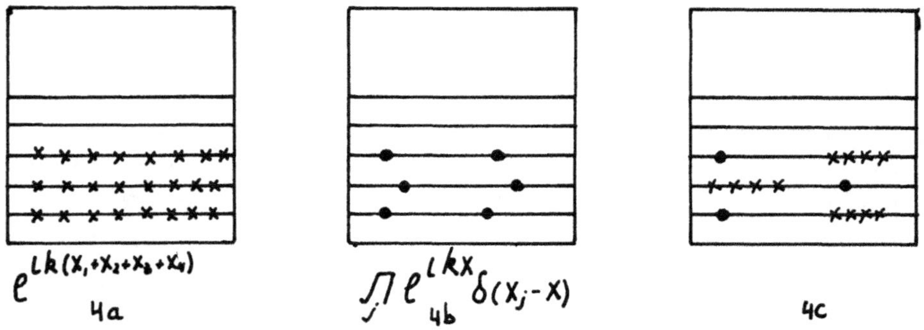

Fig. 4

The figure 4a shows a sample of liquid nuclear matter : each line represents several possible momentum states of equal modulus $|k|$, over which we spread particles. In fig. 4b we have the same particles coalescing into α clusters, which for simplicity are taken to have a δ internal wave-function, while 4c shows an intermediate situation in which some of the particles cluster while others do not. If the number of particles is made arbitrarily large we have a well defined order parameter : the number of clusters/the number of particles. If it becomes of order 1/4 we have a phase transition. What is important to notice from our point of view is the very small overlap between a wave-function of type 4a and a wave-function of type 4b or 4c. If we think in terms of shell structure, to construct the cluster correlations we need very high momentum excitations. Correspondingly we would expect the α gas to be at a very high energy with respect to the normal phase. But we have very simple physical arguments to expect the opposite : while nuclei are bound by an average 8 MeV/particle, the α particle is bound by 7 MeV/particle. Therefore somewhere around A MeV in a nucleus of mass A, we may find something similar to an α-gas. In light nuclei states of this type are observed. In ^{12}C for example the 0^+ level at 7 MeV is made of 3 loosely bound α particles. In ^{16}O the 0^+ at 6 MeV can be interpreted as an α revolving around ^{12}C, while at some 18 MeV there is a rotational band made of 4 α particles. From a shell model point of view these states have a 4 particle-4 hole (4p-4h) and 8p-8h character respectively and naïve shell model energetics would estimate their positions at some 4 or 8 " $\hbar\omega$ ". Even if we take for " $\hbar\omega$ " a <u>very</u> small value, 5 MeV say, we have major trouble in understanding how they come so low if we insist on thinking about them in terms of shell model configurations.

What we would like to suggest here is the possibility that beyond their spectroscopic interest, these particular levels may suggest the existence of " phase transitions " at moderate excitation energies, i.e. regions in the compound nucleus where states are predominantly of " gas cluster " rather than liquid (shell model) character. If this is the case we shall have to change the idea that microscopic descriptions of nuclei are synonymous with shell model calculations. In particular we may expect important consequences on the density of levels (or density matrix, or partition function) which is the quantity of interest at those energies. In professor's French talks we have learnt that under rather general circumstances Bethe's expressions for the densities can be expected to follow rigorously from spectral distribution theory. Therefore we have to bear in mind that possible deviations from Bethe's densities may indicate the necessity of some other form of counting levels due to something we may call a phase transition.

iv) While phase transitions of the type we have described above, represent an external challenge to the shell model, configuration mixing is the internal challenge. In the former case we may have to change our view on the dominance of shell structure while in the latter we have to learn how to live with shell structure. As D in eq. (I.1) shows, even if we have a single configuration the number of states may become extremely large and numerical diagonalizations difficult or even impossible. A fortiori, the problem becomes worse if mixing between different configurations occurs. Even if diagonalizations were everywhere feasible we would always prefer to know on which physical (rather than numerical) grounds the lowest and most collective states happen to be... the lowest and most collective. This raises the problem of coupling schemes, which can be stated through two questions :

1°) Why is it we choose a closed shell as natural candidate for the ^{16}O ground states ?

2°) What do we choose for some other nuclei, say A=20 ?

The answer to the first question is simple but rather propound : if we could not choose a closed shell we would have no shell model. In other words we would have another phase. For example, it could have happened that 2p-2h excitations could come so low and be so strongly coupled to the closed shell that a superconducting (BCS) wave-function would have been necessary. But this is simply not the case : ^{16}O is a normal Fermi liquid.

Now we come to the second question. From experience we can give the answer. ^{20}O (a T=2 state) is as well represented by a $d^4_{5/2}$ configuration as ^{16}O by a closed shell, while ^{20}Ne (T=0) is close to an α particle revolving around ^{16}O. As it happens $(sd)^4$ (or $[d^{n_1}_{5/2} d^{n_2}_{3/2} s^{n_3}_{1/2}]$, $n_1+n_2+n_3= 4$) is a sufficiently large space to accomodate such a cluster state, at the price of very strong (and very opaque) configuration mixing. Another completely equivalent way of describing the system is to say that the ground state rotational band belongs to the (8,0) representation of SU3. The advantage of a shell model calculation in the sd shell is that it will produce naturally the correct answer for both ^{20}O and ^{20}Ne. The advantage of having a coupling scheme such as the SU3 model of Elliot is that the physics is more transparent and the mathematics simpler and more elegant for ^{20}Ne, but of little relevance for ^{20}O. Actually, for this nucleus we have Racah's seniority (states constructed with pairs of particles coupled to J=0 come lowest) as a good coupling scheme. The problem is not so much how to know which coupling scheme to choose but to have mathematically convenient and physically relevant coupling schemes. Unfortunately they are not available and we would not hesitate to say that their existence poses the major question of nuclear spectroscopy nowadays. Even in the examples of A=20 above, the interest of the perfect operation of seniority and SU3 is hindered by the lack of " communication " between the two schemes : either one or the other works, but we have nothing in between. We can diagonalize some linear combination of the Casimir operators for the two algebras to define intermediate situations, but these diagonalizations would be as difficult to perform as in the case of an arbitrary force. It should be noticed that except for SU3, seniority (SU2 for one kind of particles or R5 for neutrons and protons) and SU4 (Wigner) we have no other algebras that define possible coupling schemes. The IBM model of Iachello and Arima [IA79] is very interesting in this context because it is defined by a general algebra, SU6, which contains three possible subalgebras SU3, SU5 and O5 , that correspond respectively to rotational, vibrational and transitional nuclei. As these are the three broad types of behaviour that have emerged from the traditional classifications, the IBM certainly deserves all the attention it has got in recent years. Unfortunately its microscopic foundations remain uncertain. If it becomes possible to unearth them, the nature of shell model calculations will change radically, and we would like to close this lecture by explaining how.

When a large scale diagonalization is performed in a space of dimensionality 10^3 or 10^4 (for given J and T) we are not interested in 10^3 or 10^4 eigenstates but at most in a half dozen of them. The rest could and should better be treated by the spectral distribution methods developed by French and his school. Then, we could ask : why go through a diagonalization ? Simply because we have very little idea of what the lowest states look like. Therefore we select a space large enough go as to be sure the states we want will emerge as linear combinations. If we had some good starting guess, we could do the diagonalization by other methods which will be described at length on the following lectures. Now, the short range attractive nature of the effective force, plus the fact it is basically of rank 1+2 and acts on a space of antisymmetric functions, limits the possible solutions. The matrices could well be enormous but their eigenvectors are far from arbitrary (as would be the case with forces of arbitrary rank). Therefore the solutions must fall on some broad classes, we call coupling schemes. Different coupling schemes represent in a sense different phases but the smallness of the systems will preclude sharp transitions : we shall show at the end of lecture 2 how the ground state of ^{12}C can be obtained perturbatively from a $p_{3/2}^{8}$ closed shell, while it is also well described as an SU3 deformed state.

Let us now put the idea of coupling scheme in proper context by summarizing some important points.

The wave-function obtained from a shell model diagonalization is to the exact wave-function for that nucleus, the same thing a single Slater determinant is to the exact wave-function in a closed shell nucleus.

The exact wave-function of a closed shell nucleus is very different from a closed shell and has only a small overlap with it, as it will be explained later. Similarly, the result of a shell model diagonalization is only a small part of the exact wave-function. In both cases, what we expect of the unperturbed solution (closed shell or shell model state) is to provide a starting point from which we could evolve to the exact eigenstate in " some perturbative " sense. In the following lectures we shall clarify this notion, which implies that the exact results could be well approximated by some algorithm that is much simpler than numerical diagonalization.

If now we restrict attention to the shell model space itself we see what is requested of a coupling scheme :

A coupling scheme should provide a (set of) state(s) that is (are) to the exact solution(s) in the model space, the same thing a closed shell is to the exact solution of a closed shell nucleus in the full space.

LECTURE 2 : EFFECTIVE INTERACTIONS, QUASICONFIGURATIONS AND PERTURBATION THEORY (PT).

(2.I) General formulation of effective interaction theory.

We want to formalize the idea that the basic states entering a shell model calculation are " dressed " by excitations coming from the whole Hilbert space. Let us assume we have a Hamiltonian

$$H = H_o + V \qquad H_o|i\rangle = \varepsilon_i, \qquad H_o|j\rangle = \varepsilon_j \tag{2.1}$$

where H_o is not necessarily a one body operator, but some conveniently defined operator (the monopole field for example) diagonal in the basis. The states i belong to the " model " or i-space and the states j are outside of it. The secular problem can be thought of in term of a matrix, which we write in block form

$$\left[\begin{array}{c|c} \begin{array}{l}\varepsilon_i + \langle i|V|i\rangle \\ \quad \langle i|V|i'\rangle\end{array} & \langle i|V|j\rangle \\ \hline \langle j|V|i\rangle & \begin{array}{l}\varepsilon_j + \langle j|V|j\rangle \\ \quad \langle j|V|j'\rangle\end{array} \end{array}\right] \longrightarrow \left[\begin{array}{c|c} X & 0 \\ \hline 0 & X \end{array}\right] \qquad (2.1')$$

the arrow indicates the block diagonal form we want to achieve. We have written only one typical diagonal of off-diagonal matrix element in each block.

The decoupling between the blocks can be achieved by defining the following transformations(*)

$$\langle\bar{i}| = \langle i|+\langle i|\hat{A}|J\rangle\langle J| \qquad (2.2)$$

$$|\bar{j}\rangle = |j\rangle-|I\rangle\langle I|\hat{A}|j\rangle \qquad (2.3)$$

$$\langle\bar{i}|\bar{j}\rangle = 0 \quad \langle\bar{i}|i'\rangle = \delta_{ii'} \quad \langle j|\bar{j}'\rangle = \delta_{jj'} \qquad (2.4)$$

Einstein's summation convention applies to repeated capital indeces. Strict orthogonality between the dressed states is manifestly respected by the transformation, whatever the nature of the general (many body) operator $\hat{A}$. Inside the blocks, however, orthogonality does not obtain automatically, but for our purposes the biorthogonal basis defined in eq. (2.4) will be sufficient.

The operator $\hat{A}$ will be defined by requesting perfect decoupling between blocks :

$$\langle\bar{i}|H|\bar{j}\rangle = 0, \qquad \text{which leads directly to}$$

$$\varepsilon_{ij}\langle i|\hat{A}|j\rangle = \langle i|V|j\rangle+\langle i|\hat{A}|J\rangle\langle J|V|j\rangle-\langle i|V|I\rangle\langle I|\hat{A}|j\rangle$$
$$-\langle i|\hat{A}|J\rangle\langle J|V|I\rangle\langle I|\hat{A}|j\rangle \quad , \quad \text{with} \quad \varepsilon_{ij}= \varepsilon_i- \varepsilon_j \qquad (2.5)$$

This non linear integral equation for the operator $\hat{A}$ can be Neumann expanded by initializing

$$\langle i|\hat{A}^{(1)}|j\rangle= \frac{\langle i|V|j\rangle}{\varepsilon_{ij}} \quad , \qquad (2.6)$$

and looking for solutions of the form

$$\langle i|\hat{A}|j\rangle = \hat{A}^{(1)}+ \sum_{n=2}^{\infty} \langle i|\hat{A}^{(n)}|j\rangle \qquad (2.7)$$

Inserting in eq. (2 5) and regrouping terms, order by order, we obtain the recursion formula

(*) Notice that in [PZ81a,81b] $\hat{A} = \underline{\hat{V}}$. We prefer the new notation.

$$\varepsilon_{ij}\langle i|\hat{A}^{(n)}|j\rangle=\langle i|\hat{A}^{(n-1)}|J\rangle\langle J|V|j\rangle-\langle i|V|I\rangle\langle I|\hat{A}^{(n-1)}|j\rangle-$$

$$-\langle i|\hat{A}^{(\lambda)}|J\rangle\langle J|V|I\rangle\langle I|\hat{A}^{(\mu)}|j\rangle \quad , \qquad \lambda+\mu= n-1, \quad n>2 \tag{2.8}$$

which defines a linked cluster form of quasidegenerate perturbation theory. (By quasi-degenerate we mean that the energies ε_i may be different, as opposed to Brandow's theory [BR 67] which demands strict degeneracy). Linked cluster properties are demonstrated and extensively discussed later. A thorough study of the perturbation expansion can be found in [PZ81a], for our purpose however, we shall only need the expansion of the dressed wave-function to second order :

$$\langle\bar{i}| = \langle i|+ \frac{\langle i|V|J\rangle\langle J|}{\varepsilon_{iJ}} + \frac{\langle i|V|J\rangle\langle J|V|J'\rangle\langle J'|}{\varepsilon_{iJ}\varepsilon_{iJ'}} - \frac{\langle i|V|I\rangle\langle I|V|J\rangle\langle J|}{\varepsilon_{iJ}\varepsilon_{IJ}} \ . \tag{2.9}$$

If we assume eq. (2 5) has been solved, we can be certain that exact eigenstates in the (dressed !) model space will have the form

$$\langle\phi_\alpha|= \sum\bar{C}_i^\alpha\langle\bar{i}|= \sum\bar{C}_i^\alpha\left[\langle i|+\langle i|\hat{A}|J\rangle\langle J|\right] \tag{2.10}$$

and we can solve the Schrödinger equation

$$\langle\phi_\alpha|(H-E_\alpha)= 0 \implies \sum\bar{C}_i^\alpha\left[\langle\bar{i}|H|i'\rangle-\delta_{ii'}E_\alpha\right] =$$

$$= \sum_{ii'} \bar{C}_i^\alpha\left[\langle i|V|i'\rangle+\langle i|\hat{A}|J\rangle\langle J|V|i'\rangle+\delta_{ii'}(\varepsilon_i-E_\alpha)\right] = 0. \tag{2.11}$$

From which we conclude that everything happens as if we had truncated the space and replaced the original interaction V by an effective interaction

$$V \longrightarrow V+ \hat{A}|J\rangle\langle J|V = V_{eff} \tag{2.12}$$

It should be noticed that V_{eff} is manifestly non-Hermitian, and consequently the ket $|\phi_\alpha\rangle$ has no meaning. To define some orthogonality conditions we can solve the " fictitions " eigenvalue problem

$$(H-E\ )|\Phi_\alpha\rangle = 0, \qquad \text{with } |\Phi_\alpha\rangle= \sum C_i^\alpha|i\rangle \tag{2.13}$$

We are certain the spectra in (2.13) and (2.11) are identical, as the secular determinant is invariant under the exchange of rows and columns. Therefore, assuming non degeneracy we obtain the useful biorthogonality property

$$\langle\phi_\alpha|\Phi_{\alpha'}\rangle= \delta_{\alpha\alpha'} = \sum_i C_i^\alpha\bar{C}_i^{\alpha'} \tag{2.14}$$

Formally we have achieved a solution to the problem of difining properly a model space and its associated effective interaction. The lack of Hermiticity should not be a source of trouble ; in [PZ81a] it is shown explicitly how to restore it. In our formulation we have emphasized the possibility of interpreting a shell model calculation as the outcome of an exact diagonalization in the full Hilbert space. Nothing is " truncated ". It is in the nature of things that the definition of dressed states (2 2) be ambiguous, since the block diagonalization (2.1') can be achieved up to an arbitrary (not necessarily unitary !) transformation in the i or j-spaces. What is essential is to have a unique solution to the eigenvalue problem and this we have. The fact that shell model calculations can be successful with Hermitian two-body interaction is an important empirical clue but it does not imply that the effective forces are either strictly Hermitian or strictly two-body. Still the clue is so strong that it is our main task to show that V_{eff} in eq. (2.12) is actually a rather simple object. It should be understood that the block decoupling in (2 1') and the resulting equations are well defined mathematically but in principle V_{eff} is a many body operator of arbitrary complexity. Whatever simplicity obtains must be a consequence of a good choice of model space. Here again we have much empirical evidence but the situation is not really satisfactory : the choices are either based on some tacit agreement between experts and/or on the dangerous idea that the bigger the model space the better the calculation. More on this in lecture 3.

(2.II) Quasiconfigurations in Perturbation Theory.

Before we move on to examine the general form of the coupled integral equations we shall work out a few examples in low order perturbation theory. So far we have made no specification concerning the states $|i\rangle$ and their dressed counterparts $|\bar{i}\rangle$.

If states $|i\rangle$ belong to some shell model configuration, we call the set of states $|\bar{i}\rangle$ a QUASICONFIGURATION. The word is meant as a reference to Landau's idea of quasiparticle. Equations (2.11) show transparently in what sense we are working with an interacting system which formally behaves as another (presumably more weakly) interacting system which has the same labels i, i.e. the same sates as the original one. If the original interaction V was acting on fermions, V_{eff} is now acting on quasiparticles that are again fermions. There is no point however in trying to identify new fermion operators : to all intents and purposes they are the same as the original ones. It is V_{eff} that takes care of whatever changes may be necessary. In going from the original form of the secular problem to the new one we are essentially going from a Schrödinger to a Heisenberg (or intermediate) representation. Perturbation theory is very useful in clarifying the mechanisms involved.

To do so we introduce the notion of process in which some configuration in i-space jumps to j-space and comes back again to i-space, eventually after a number of jumps in both spaces. Each jump corresponds obviously to going to a higher order in PT. To simplify matters we consider only three generic orbits p(particles) f(valence) and h(holes). A model space configuration has the form [N,n,0], where N is the total number of holes, n the number of valence particles while the particle states are empty.

Possible processes in second order are

$$[N,n,0] \xrightarrow[W_{fpfh}]{} [N-1,n,1] \xrightarrow[W_{fhfp}]{} [N,n,0] \tag{2.15}$$

$$[N,n,0] \xrightarrow[W_{fpff}]{} [N,n-1,1] \xrightarrow[W_{fffp}]{} [N,n,0] \tag{2.16}$$

$$[N,n,0] \xrightarrow[W_{ppff}]{} [N,n-2,2] \xrightarrow[W_{ffpp}]{} [N,n,0]\ ,\ \text{etc} \tag{2.17}$$

We have written under each arrow the amplitude (two body matrix element) with which the jump takes place. We assume the Hamiltonian to be written in the usual second quantization form

$$H = \sum K_{ij} a_i^+ a_j - \sum_{\substack{i\leq j\\ k\leq \ell}} W_{ijk\ell} a_i^+ a_j^+ a_k a_\ell \qquad i,j,k,\ell \equiv pp_z, ff_z, hh_z, \tag{2.18}$$

where p_z, f_z, h_z are the spin, isospin projections, which are tacitly assumed when we write a matrix element such as W_{fpfh}.

The form of V_{eff} in second and third order is obtained by taking the overlaps of $<i|$ in eq. (2-9) with $H|i'>$, from which we get

$$\begin{aligned} <i|V_{eff}|i'> &= <i|V|i'> + (\varepsilon_{iJ})^{-1}<i|V|J><J|V|i'> + \\ &+ (\varepsilon_{iJ}\varepsilon_{iJ'})^{-1}<i|V|J><J|V|J'><J'|V|i'> + \\ &(\varepsilon_{iJ\ IJ})^{-1}<i|V|I><I|V|J><J|V|i'> \end{aligned} \tag{2.19}$$

Notice the manifest lack of symmetry in the i, i' indeces. For second order it is correct to restore Hermiticity by taking the half sum of $<i|V_{eff}|i'>$ and $<i'|V_{eff}|i>$, for higher orders things are somewhat more complicated : see [PZ81a] for details.

The particular processes (2.15) to (2.17) give the following contributions to V_{eff}

$$\begin{aligned} &\left(\varepsilon_{f^n} - \varepsilon_{f^{n-1}p}\right)^{-1} <f^n|W_{fhfp} a_f^+ a_h^+ a_f a_p|f^{n-1}p><f^{n-1}p|W_{fpfh} a_f^+ a_p^+ a_f a_h|f^n> \\ &= \varepsilon^{-1}<f^n|a_f^+ a_f a_f^+ a_f|f^n>W_{fhfp}^2\ , \qquad \text{and similarly} \\ &\varepsilon^{-1}<f^n|a_f^+ a_f^+ a_f a_f^+ a_f a_f|f^n>W_{fffp}^2,\quad \varepsilon^{-1}<f^n|a_f^+ a_f^+ a_f a_f|f^n>W_{ffpp}^2 \end{aligned} \tag{2.20}$$

To obtain these expressions we assume the denominators to be constant for a given set of intermediate states, which can always be done by proper choice of H_o. Then we invoke closure to eliminate the sum over intermediate states and we contract out the operators for the p and h orbits, a straightforward operation. Notice the appearence of 2+3 body effective forces for the second process.

Effective transition operators can also be calculated, but then we have to be careful with the norms, and the non-Hermiticity problem has to be faced. The important point however is that one body bare operators may become, already in second order, two body effective operators. This can be understood most simply by considering the second process again and jump out of the i-space with a transition operator and come back with H (or vice versa).

Diagrammatically we can represent the three processes in the standard Goldstone form, by assuming (arbitrarily) the valence lines are particle (or hole) lines, which leads for process (2 16) to

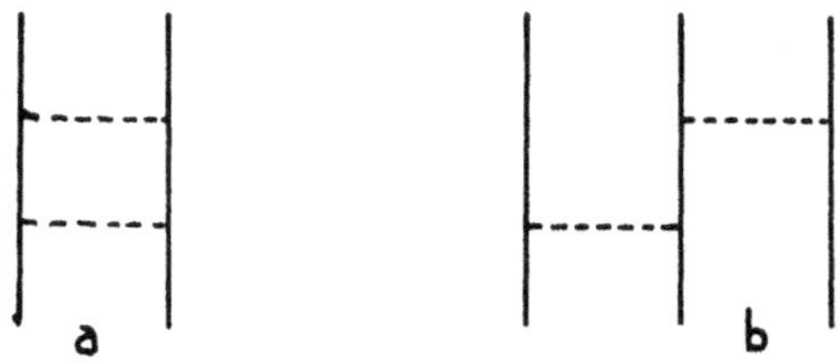

Fig. 5

In the figure the unlabeled lines correspond to f-particles. The fact that valence orbits are not really particle nor hole orbits in the diagrammatic sense, since they are partially empty (or occupied) becomes a problem in third order. It we choose a process of the type

$$f^n \xrightarrow[W_{ffff}]{} f^n \xrightarrow[W_{fhfp}]{} f^n ph^{-1} \xrightarrow[W_{fpfh}]{} f^n \quad , \tag{2.21}$$

the conventional diagrammatic counterpart of which would be fig. 6(a) (for the two body part)

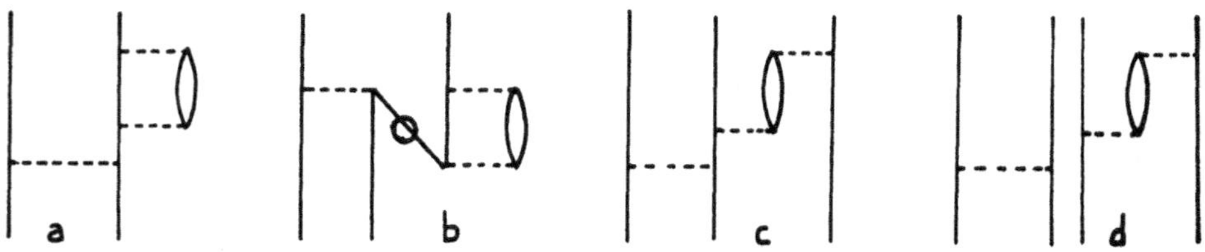

Fig. 6

As in fig. 5, arrows are drawn only for p or h lines. Diagram a poses a problem if interpreted according to Goldstone's rules, as the first energy denominator would be of the form ε_{iI}, which does not make sense : the only denominators coming out of eq. (2.5) are of the form ε_{ij} (or ε_{Ij}, ε_{IJ}, etc). When Brandow formulated the first linked cluster version of degenerate PT, the problem appeared in a different way but the essence is the same : denominators of the form $\varepsilon_{ii'}$ vanish and therefore they can't appear. In our case the solution is simple since eq. (2.19) gives us explicitly the right answer. Brandow's solution was to invent " folded diagrams " of which fig. 6(b) is an example. Assuming H_o is a strict rank-1 operator, by transforming the internal f-particle line into a hole line, i.e. by folding it, (as shown by a circle) one recovers the right energy denominator of eq. (2.19) and the correct minus sign. This happens because a general diagrammatic rule is that internal hole-lines carry a minus sign, a result that in this case eq. (2.8) gives automatically since each contribution with an intermediate i-state is negative. Fig. 6(c) gives the 3-body contribution (folding not shown), while fig. 6(d) is the rank-4 operator. As it is unlinked it shall be cancelled by a similar diagram coming from a process in which the order of the jumps will be reversed (see [PZ81a] for an example).

Folding and non hermiticity are the characteristic features of quasidegenerate PT. We have already hinted there is no reason to worry about non-hermiticity. Folding is a more complicated matter. If we add the fact that energy denominators should not be strictly one body (the monopole field is an absolute need if we want reasonable results, as we shall see soon) we are left with the feeling that diagrammatic methods may be helpful in guiding the eye but very difficult to formulate with sufficient precision and generality to make them useful. To this we may add the comment already made, that diagrams seldom have individual meaning and they have to be resumed into an integral equation. In that case it is better to start directly with the integral equations, and understand in a general context the nature of folding and other problems. But before doing that we discuss two situations in which second order PT gives remarkable results.

(2.III) The $f^n_{7/2}$ quasiconfigurations and f-shell spectroscopy.

The general formulation we have presented is meant to apply to the full Hilbert space but we can just as well think of it as a way to diagonalize a large matrix. Block decoupling in (2.1') is a preliminary operation done with the hope (or the conviction) that simpler methods are available than brute force numerical diagonalization. The latter becomes inevitable in the model space, mainly because the states are degenerate and no a priori choice about the structure of those that come lowest can be made. Strict degeneracy however happens only for states in a given configuration. As many configurations are present in most model spaces, it is worth asking whether it is really necessary to treat them all on the same footing and whether it is possible to define some smaller model space.

The first example we consider is that of the fp shell (see fig. 1). From the comments made after fig. 2 we expect phenomenologically to be able to define successfully an $f_{7/2}$ model space. If the other configurations $(f_{7/2}^{n_1} f_{5/2}^{n_2} p_{3/2}^{n_3} p_{1/2}^{n_4})$ could be used to dress perturbatively the $f_{7/2}^n$ states, impossibly large calculations (dimensions 10^6 and more) could be replaced by very small ones (dimensions 20, for typical values of nJT). This programme has been fully carried out in [PZ81b] and we are only going to make here some comments and show some significant examples.

Technically the problem involves only processes of the type (2.16) and (2.17), which we rewrite schematically

$$f^n \longleftrightarrow f^{n-1} r, \quad f^n \longleftrightarrow f^{n-2} r^2 \qquad \begin{array}{l} r = \text{generically } f_{5/2} p_{3/2} p_{1/2} \\ f = f_{7/2} \end{array} \tag{2.21}$$

One significant feature on which hinges the success of the calculations is the use of the monopole field in the energy denominators i.e.

$$\varepsilon\left(f^{n_1} r^{n_2}\right) = n_1 \varepsilon_f + n_2 \varepsilon_r + \frac{n_1 (n_{1-1})}{2} V_{ff} + \frac{n_2 (n_{2-1})}{2} V_{rr} + n_1 n_2 V_{fr} \tag{2.22}$$

To give an idea of the importance of the two body terms we mention that in ^{41}Ca, $\varepsilon_f - \varepsilon_{p_{3/2}} = 2$ MeV while in ^{56}Ni $\varepsilon(f^{16}) - \varepsilon(f^{15} p_{3/2})$ reaches some 6 MeV : including only the single particle terms would lead to disastrous results. The interaction used [KB 68] had to adjusted so as to give a good monopole field (see lecture 4) and no other significant changes are necessary to give excellent spectroscopy.

The conventional $f_{7/2}^n$ calculations [MC64,GI63] use a pure two body interaction extracted from the experimental spectrum of ^{42}Sc (other choices are possible but the results do not depend strongly on them). They give a fairly good description of the level schemes, but a number of systematic discrepancies indicate there is room for improvement :

1) Cross conjugate symmetry. With a strict two body force the spectra obtained by adding n particles to a ^{40}Ca core must be identical to those obtained by removing n particles from ^{56}Ni. In general for even-even nuclei the similarity is striking, but discrepancies appear for odd-odd and odd-even nuclei. Figs. 7, 8, 9 illustrate the situation in the pairs ^{44}S - ^{52}V, ^{46}Ti - ^{50}Cr, ^{47}Ti - ^{49}V.

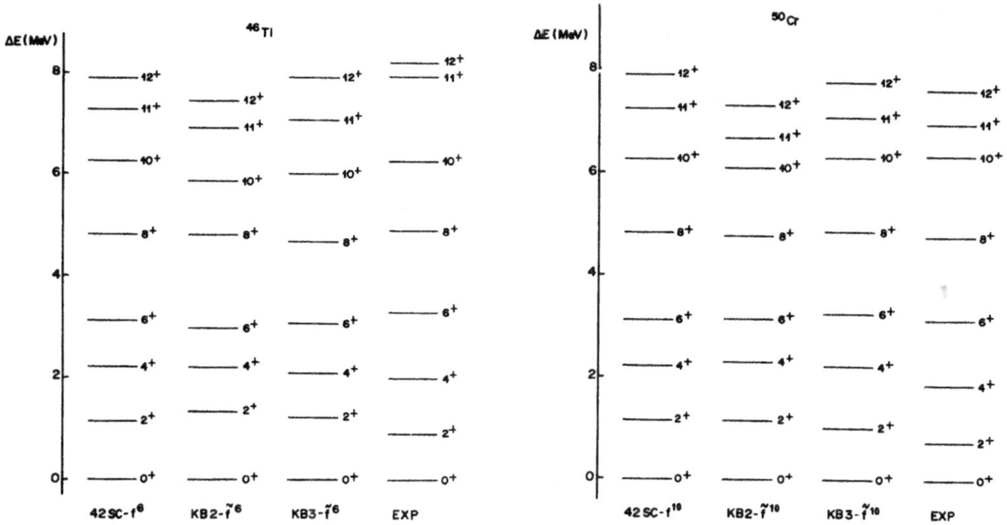

ΔE(MeV)
^{46}Ti
42SC-f^6
KB2-$\tilde{f}^6$
KB3-$\tilde{f}^6$
EXP
ΔE(MeV)
^{50}Cr
42SC-f^{10}
KB2-$\tilde{f}^{10}$
KB3-$\tilde{f}^{10}$
EXP

Fig. 7

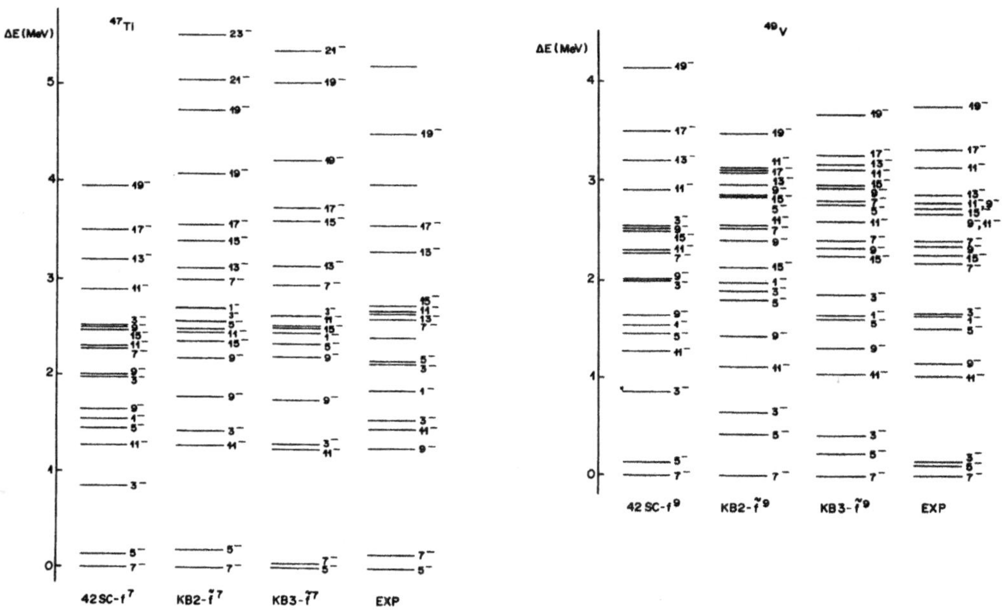

ΔE(MeV)
^{47}Ti
42SC-f^7
KB2-$\tilde{f}^7$
KB3-$\tilde{f}^7$
EXP
ΔE(MeV)
^{49}V
42SC-f^9
KB2-$\tilde{f}^9$
KB3-$\tilde{f}^9$
EXP

Fig. 8

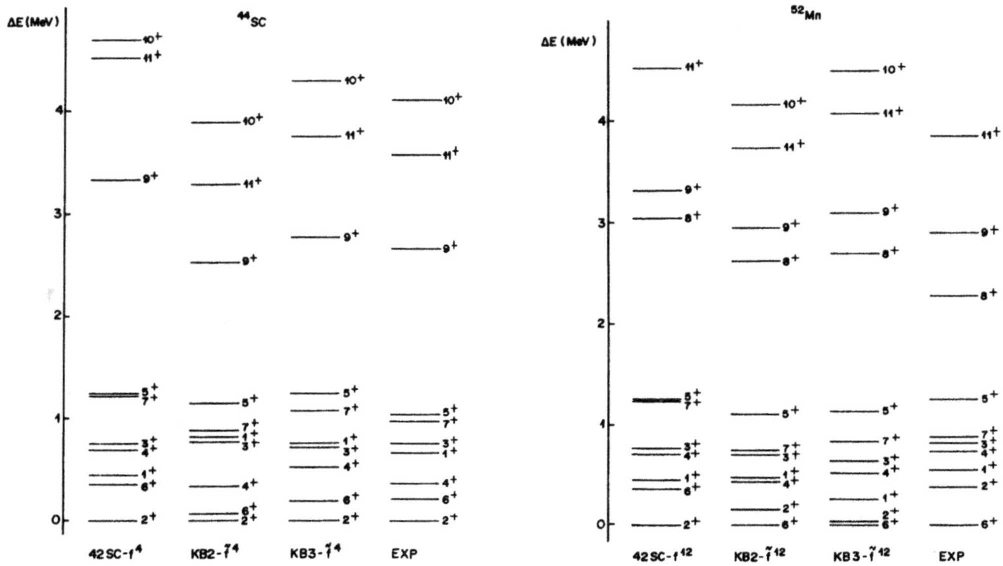
ΔE (MeV)
44SC
0
1
2
3
4
42SC-f4
KB2-f̃4
KB3-f̃4
EXP
52Mn
ΔE (MeV)
0
1
2
3
4
42SC-f12
KB2-f̃12
KB3-f̃12
EXP

Fig. 9

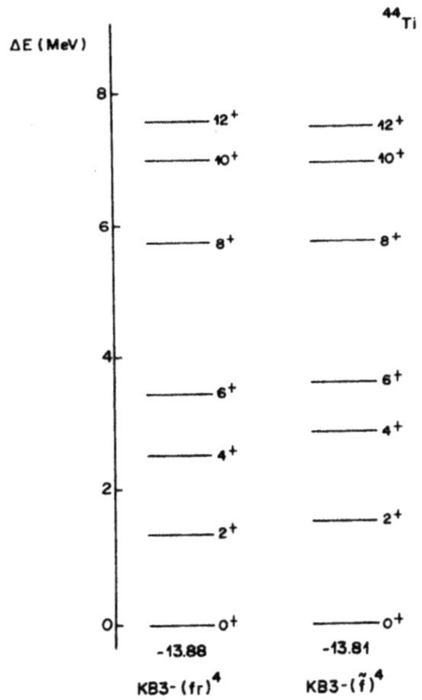
44Ti
ΔE (MeV)
0
2
4
6
8
-13.88
-13.81
KB3-(fr)4
KB3-(f̃)4

Fig. 10

2) The spectrum of ^{42}Ca is not well given by the interaction we used because we have insisted on <u>not</u> including renormalization effects due to jumps from the sd shell. As these jumps are associated to intruder states (dashed lines in fig. 2 for ^{42}Ca) that couple strongly to the model space, a perturbative treatment is dangerous to use and we prefered to ignore those processes, which are only important at the beginning of the shell. By the time we reach ^{50}Ti intruder states have disappeared completely, and it is extremely interesting to find that our calculations give a perfect spectrum for ^{50}Ti, i.e. one that looks very much like the <u>experimental</u> spectrum of ^{42}Ca. Furthermore the BE2 transition rates are well given by the calculation, while the naïve hypothesis of two particles on a closed shell would have produced a severe underestimate.

3) The quadrupole moments in the simple $f^n_{7/2}$ model are systematically rather large and positive, while experimentally they are large and negative, e.g. 7.32 vs -30.4 ± 9 efm^2 for ^{50}Cr. The quasiconfiguration result is −19.7. It is good to remember that the quadrupole operator is now of rank 2. It is also of interest to notice we can go <u>perturbatively</u> from a positive value to a negative one, which shows there is not much point in thinking of an oblate to prolate phase transition.

4) Direct comparison between perturbative and exact diagonalizations are only possible in very few cases due to the very rapid increase of dimensionalities which go roughly as $(10)^{n-2}$ at the beginning of the fp shell, to reach 5.10^7 at mid-shell. Fig. 10 shows the case of ^{44}Ti. Clearly the quasiconfiguration results are more than adequate from a numerical point of view. A very important point is that the amount of bare model state ($f^n_{7/2}$) in the exact result is of the order of 50 % only, a quantity that will strongly decrease for large n, while the quality of the perturbative calculation will remain high or even improve. More on this in lecture 3.

We remind that a full account of the calculations is given in [PZ81b].

(2.IV) <u>The ground state of ^{12}C and perturbation theory</u>.

Now we examine the 0^+ states in the $(p_{3/2}p_{1/2})^8$ configurations. The naïve shell model version would predict a $p^8_{3/2}$ closure. There are very good reasons however, to think that the state belongs to a rotational band (e.g. very large BE2 from the first excited 2^+ state) and that an SU3 representation is a better description. To settle the problem we shall study the matrix in a exact diagonalization

	$p_{3/2}^8$	$p_{3/2}^6p_{1/2}^2(01)^{JT}$	$p_{3/2}^2p_{1/2}^2(01)^{JT}$	3p3h	4p4h
	0	1	2	3	4
0	0	8.42	-2.70		
1		7.62	3.44	6.47	7.68
2			13.55	-3.86	-1.48
3				15.72	-5.79
4					11.95

(2.23)

Δ =-9.16 MeV

Exact $|\bar{0}\rangle = |0\rangle - .96|1\rangle + .24|2\rangle + .27|3\rangle + .31|4\rangle$

PT 1 - 1.19 .38 .42 .49

$$\Delta^{(2)} = -\frac{(8.42)^2}{7.62} - \frac{(2.7)^2}{13.35} = -9.85 \text{ MeV}$$

The matrix elements are calculated with the Cohen and Kurath interaction [CK65] Δ is the correlation energy, i.e. the gain with respect to the unperturbed value. On the face of it, the very highly correlated wave-function and the value $V_{01}/\varepsilon_{01} > 0$, would seem to preclude any meaningful perturbative treatment. However the energy to second order $\Delta^{(2)}$ is not that bad and the wave-function to 3^d order has perfect phase agreement with the exact one. Even the amplitudes are reasonable : as a variational guess this wave-function would be excellent.

Tu understand what is happening we shall solve the secular equation.:

$$\begin{aligned}
\langle 0|H|\bar{0}\rangle &= \Delta = \quad A_1V_{10} + A_2V_{20} \\
\langle 1|H|\bar{0}\rangle &= A_1\Delta = V_{10} - A_1\varepsilon_1 + A_2V_{12} + A_3V_{13} + A_4V_{14} \\
\langle 2|H|\bar{0}\rangle &= A_2\Delta = V_{20} + A_1V_{12} - A_2\varepsilon_2 + A_3V_{23} + A_4V_{24} \\
\langle 3|H|\bar{0}\rangle &= A_3\Delta \quad + A_1V_{13} + A_2V_{23} - A_3\varepsilon_3 + A_3V_{34} \\
\langle 4|H|\bar{0}\rangle &= A_4\Delta \quad + A_1V_{14} + A_2V_{24} + A_4V_{34} - A_4\varepsilon_4
\end{aligned} \qquad (2.24)$$

First we extract A_3 and A_4 as a function of A_1, A_2 and Δ

$$A_3(A_1,A_2,\Delta) = A_3\left(\frac{P(\Delta^3)}{Q(\Delta^3)}\right) =$$

$$= \left[\frac{V_{34}(A_1V_{14}+A_2V_{24})}{(\varepsilon_3+\Delta)(\varepsilon_4+\Delta)} + \frac{A_1V_{13}+A_2V_{23}}{(\varepsilon_3+\Delta)}\right]\left[1 - \frac{V_{34}^2}{(\varepsilon_3+\Delta)(\varepsilon_4+\Delta)}\right]^{-1} \qquad (2.25)$$

and the equivalent expression for A_4. P and Q are 3^d order polynomials in Δ. We are making use of the fact that A_1 and A_2 are linear functions of Δ. The equation for the A_1 amplitude now becomes

$$A_1\varepsilon_1 - V_{10} = A_2V_{12} - \Delta A_1 + R(\Delta^3)S^{-1}(\Delta^3) \qquad (2.26)$$
$$0 \neq .69 \; -10.84 \; 6.69 \qquad , \text{ if } A_1\varepsilon_1 = V_{10} \text{ (PT)}$$

where R and S are again polynomials, and the numbers below indicate the values obtained by initializing perturbatively. The equations are not satisfied but things are not that bad : the 3.5 MeV deficit is the difference of two largish numbers. So the goodness of the perturbative initialization does not depend so much on V/ε being a small number but on a cancellation which can be achieved with extraordinary numerical efficiency if we notice that eq. (2.26) is of the form

$$T(\Delta^2) = R(\Delta^3)S^{-1}(\Delta^3) \qquad (2.26')$$

i.e. that we are equating a rapidly varying function of Δ with a very flat one, as the figure shows

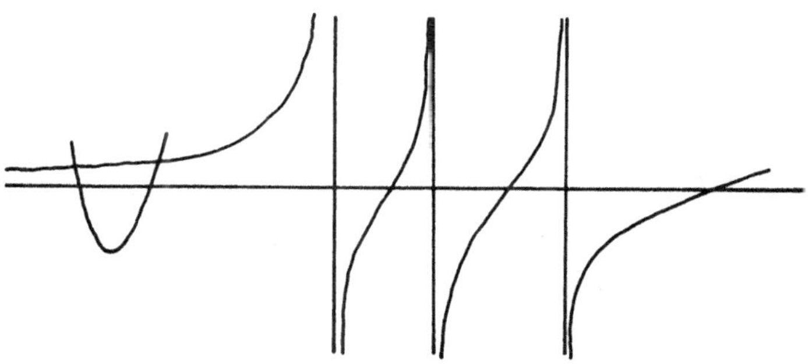

Fig. 11

A change of only 15 % (trivial to guess if we assume R/S is flat) in Δ drives it to the correct result. This very encouraging situation may be quite general as the figure hints : when the dimensionality of the matrix grows we simply have more spikes to the right.

There is something magic about <u>second order</u> PT. In this particular case the series converges poorly and may even diverge (3^d and 4th orders give no improvement), but second order is a very good initialization to an algorithm that suggests quite strongly, by inspecting eq. (2.25), that some PT-like expansion is plausible if we use very carefully adapted, order dependent, energy denominators.

We can obtain a striking confirmation of the good behaviour of second order PT by working out the expression for the process

$$p_3^{n_3} p_1^{n_1} \xrightarrow[W_{3311}]{} p_3^{n_3-2} p_1^{n_1+2} \qquad 3,1 \equiv 3/2,\ 1/2 \qquad (2.27)$$

Instead of resorting to the exact matrices we'll use simpler average values

$$\bar{\Delta}(n_1n_3)= \frac{1}{D}\sum_{\alpha\alpha'} \left[<n_1n_3\right]\alpha|V\left[n_1+2\ n_3-2\right]\alpha'>|^2\ \frac{1}{\varepsilon(n_1n_3)} =$$

$$= \frac{1}{\varepsilon(n_1n_3)}\ \frac{n_3^{(2)}(N_1-n_1)^{(2)}}{N_1^{(2)}\ N_3^{(2)}} \sum(W_{3311})^2, \quad \text{with } D= \sum_{\alpha}, \ n^{(2)}= \frac{n(n-1)}{2} \tag{2.28}$$

When the configuration reduces to a single state the average involves no approximation. Of course for ε we can't take the detailed expectation values we have used above, but the monopole field gives results that are only slightly different (better !). The interesting point is that

$$\Delta(8,1) = \frac{1}{2}\ \Delta(8,0), \tag{2.29}$$

indicating a loss of half the correlation energy in going from ^{12}C to ^{13}C, a result that checks beautifully with the exact calculations.

LECTURE 3. LINKED CLUSTER THEORY (LCT).

(3.I) Linked cluster and factorization properties of the many body states.

When we consider second order processes (2.21), no explicit mention is made of configurations $f^{n-\mu}r^{\mu}(\mu>2)$, although they are not " neglected ". They enter in a subtle way that is at the heart of many body theory. To explain the mechanism we consider the non degenerate case, which in our language amounts to having a single model state, which we assume to be a Fermi sea (Slater determinant) we call $|0>$.

The exact wave-function in its most general form is then

$$|\bar{0}> = |0>+<0|\hat{A}|J><J|=|0>+ \sum_{\alpha i} A_{\alpha i}|\alpha i>+ \sum_{\substack{\alpha\beta \\ ij}} A_{\alpha\beta ij}|\ \alpha\beta ij>+ \ldots=$$

$$= |0>+ \sum_{n,x} A(n,x)|n,x> , \qquad |\alpha i>= a^{+}_{\alpha}a_{i}|0> \quad \text{etc.} \tag{3.1}$$

Here $\alpha\beta\ldots(ij\ldots)$ orbits are above (below) the Fermi sea. The energy is

$$<0|H|\bar{0}>=<0|K+V|0>+ \Sigma<0|K+V|\alpha i>A_{\alpha i}+ \Sigma<0|V|\alpha\beta ij>A_{\alpha\beta ij}$$

$$= \underbrace{\underset{(1)}{\sum_i K_{ii}} + \underset{(2)}{\sum_{i<j} W_{ijij}}}_{E_o} + \underbrace{\underset{(3)}{\sum_{\alpha ij}} (K_{\alpha i}+ \underset{(4)}{W_{\alpha jij}})\ A_{\alpha i}}_{\Delta_1} - \underbrace{\underset{(5)}{\sum_{\substack{\alpha<\beta \\ i<j}} W_{\alpha\beta ij}A_{\alpha\beta ij}}}_{\Delta_1} = <0|\bar{0}>E=E= \tag{3.2}$$

= 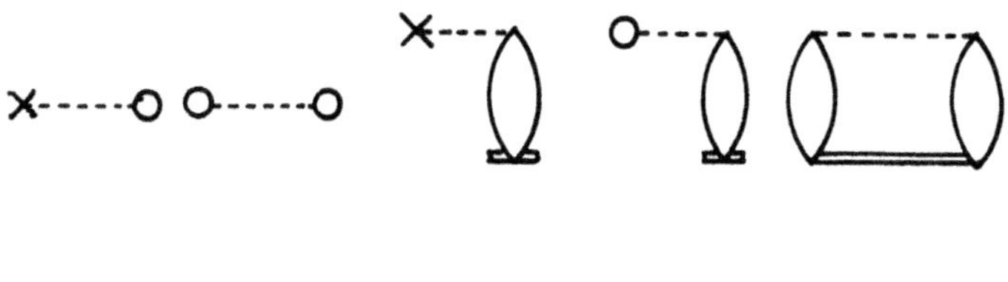 $=E_o+\Delta_1+\Delta_2=E_o+\Delta$

Before explaining the diagrammatic notation we notice that only the $A_{(1,x)}$ and $A_{(2,x)}$ amplitudes are present. Therefore, if we have a way of estimating these amplitudes, we have an estimate for E. Second order PT is one possible way, and from eq. (2.9) we get

$$\langle i|H|\bar{i}\rangle \tilde{=} \langle i|H|i\rangle + \frac{\langle i|V|J\rangle\langle J|V|i\rangle}{\varepsilon_{iJ}} \tilde{=} E_o + \Delta \tilde{=} E \qquad (3.3)$$

Although under favorable circumstances this may be a good approximation to E, it does not mean eq. (2-9) is a good approximation to the wave-function. This can be understood quite generally by considering again the general problem

$$|\bar{i}\rangle = |i\rangle + |J\rangle\langle J|\hat{A}|i\rangle \implies E = \langle i|H|\bar{i}\rangle = \langle i|H|\bar{i}\rangle + \langle i|V|J\rangle\langle J|\hat{A}|i\rangle \qquad (3.4)$$

But then we should be able to obtain the energy by overlapping with any state i.e.

$$\langle j|H|\bar{i}\rangle = \langle j|V|i\rangle + \langle j|H|J\rangle\langle J|\hat{A}|i\rangle = \langle j|\bar{i}\rangle E = \langle j|\hat{A}|i\rangle\left[\langle i|H|i\rangle + \langle i|V|J\rangle\langle J|\hat{A}|i\rangle\right], \qquad (3.5)$$

obtained by simply replacing eqs. (3-4) in the middle equality. By trivial reordering, using the separation $H=H_o+V$, we get

$$\varepsilon_{ij}\langle j|\hat{A}|i\rangle = \langle j|V|i\rangle + \langle j|V|J\rangle\langle J|\hat{A}|i\rangle - \langle j|\hat{A}|i\rangle\langle i|V|i\rangle - \langle j|\hat{A}|i\rangle\langle i|V|J\rangle\langle J|\hat{A}|i\rangle \qquad (3.6)$$

which is eq. (2.5) specialized to the non degenerate case. Therefore, to be consistent, an approximation to the wave-function must obey eq. (3.6) for all states $\langle j|$, and to understand what happens we start probing the exact wave-function with different basic states. The first equation we have is (3.2) a detailed form of (3.4). Next we probe with some p-h state

$$\langle m\varepsilon|H|\bar{0}\rangle = A_{m\varepsilon}E = \langle m\varepsilon|H|0\rangle + \sum\langle m\varepsilon|H|\alpha_i\rangle A_{\alpha i} + \sum\langle m\varepsilon|H|\alpha\beta ij\rangle A_{\alpha\beta ij}$$
$$+\langle m\varepsilon|H|\alpha\beta\varepsilon ijm\rangle A_{\alpha\beta\varepsilon ijm} \qquad (3.7)$$

which can be extended schematically to any state of np-nh

$$\langle n|H|\bar{0}\rangle = A(n)E = \sum_x \langle n|H|n+\mu,x\rangle, \qquad \mu=-2,-1,0,1,2 \qquad (3.8)$$

The equations in full algebraic detail soon become very clumsy. To give an example : H can take some 2p-2h state $\langle\alpha\beta ij|$ to states $|\alpha'\beta'ij\rangle$ (particle particle rescattering, p-p), $|\alpha\beta i'j'\rangle$ (h-h), $|\alpha'\beta i'j\rangle$ (p-h), $|\alpha'\beta ij\rangle$ (particle rescattering, p, can be mediated by $K_{\alpha\alpha'}$ or by $W_{\alpha k\alpha' k}$ terms in H), $|\alpha\beta i'j\rangle$ (h). To keep track of all the possible terms in a compact form a diagrammatic method is desirable. In eq. (3.2) we have already made use of it and it is obviously more economical than the algebraic expression.

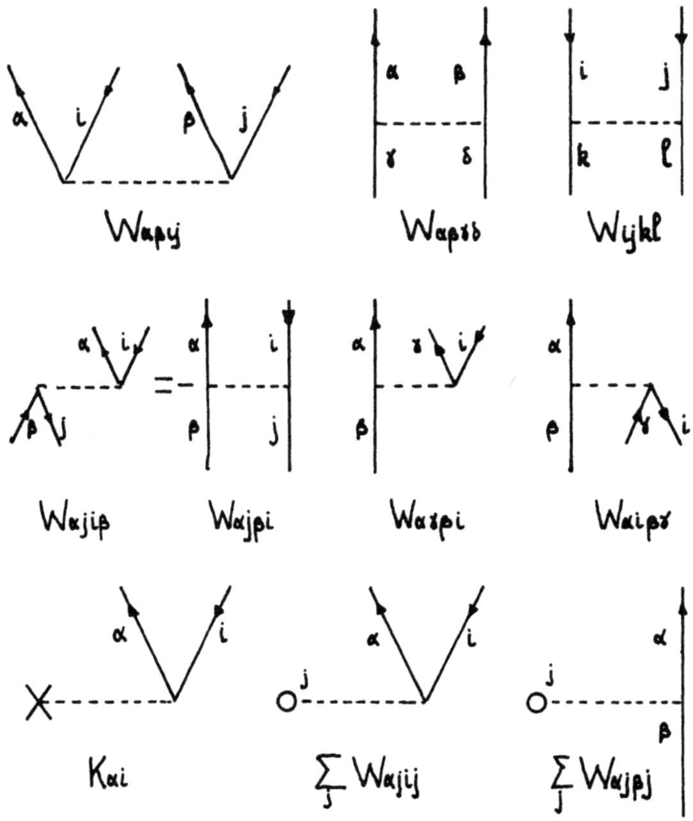

Fig. 12

In fig. 12 we show the usual (Goldstone) form of the matrix elements which are taken to be antisymmetric and normalized. In fig. 13 we introduce a notation for the amplitudes $A_{\alpha\beta\ldots ij\ldots}$. In fig. 14 we compress in a single diagram all the processes mentioned above, whereby a 2p-2h state is scattered by the potential into all possible 2p-2h states. The first of the diagrams stands for $\sum_{\alpha'\beta'} W_{\alpha\beta\alpha'\beta'} A_{\alpha'\beta' ij}$ the others are left as an exercise. The parenthesis in the last diagram indicate it will disappear from the family : either it will cancel for off diagonal rescatterings ($\alpha \to \alpha'$, $i \to i'$ etc.), as part of a selfconsistency condition or, for diagonal rescatterings it will be included in a single particle term as shown in fig. 15, where ε stands for

$$\varepsilon \equiv \varepsilon_{\alpha\beta\ldots ij\ldots} = \varepsilon_\alpha + \varepsilon_\beta + \ldots - \varepsilon_i - \varepsilon_j \ , \qquad \varepsilon_r = \sum_i W_{riri} + K_{rr} \tag{3.9}$$

Notice in fig. 14 that not all possible processes have been drawn. In particular the line r in parenthesis can be any of the lines of the diagram.

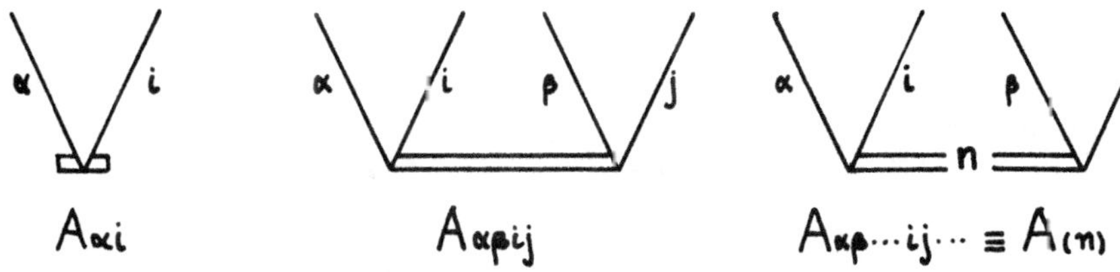

Figure 13

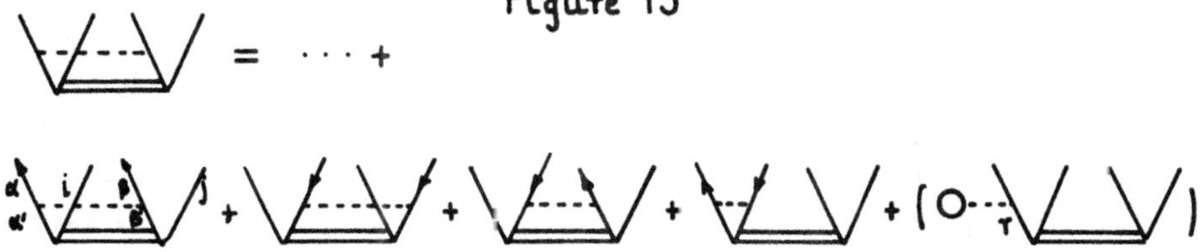

Figure 14

Figure 15

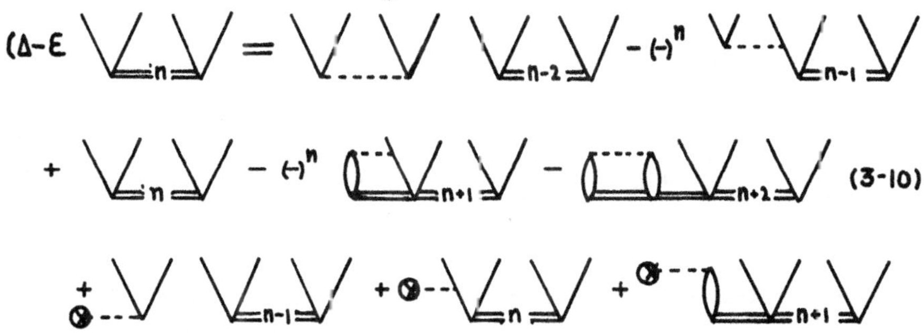

Figure 16

Now we are in a position to write eq. (3.8) quite generally in diagrammatic form. The resulting equation is shown in fig. 16 (eq. (3.10)).

In the last line each insertion may either correspond to a potential (circles) or a kinetic energy (crosses). A complete set of rules for dealing with this expression is given in [PZ81a] but for our purposes it is sufficient to say that each internal hole line carries a minus sign and that antisymmetry against interchange of particle or hole lines must be enforced. The equation for the $A_{(1)}$ amplitude is

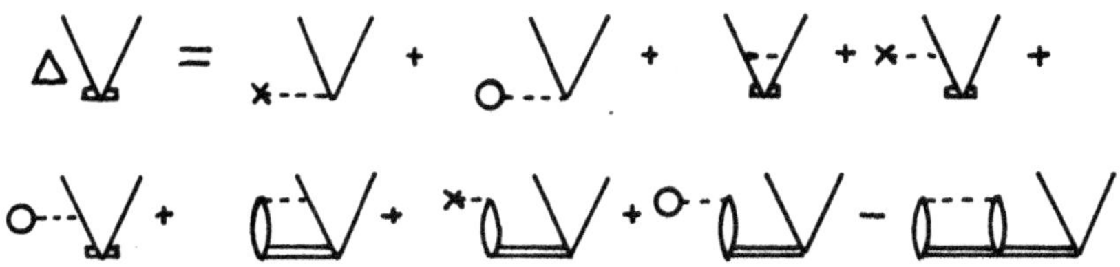

(3.11)

First let us notice that if we neglect the $A_{(2)}$ and $A_{(3)}$ amplitudes and request vanishing of the $A_{(1)}$ amplitude we are left with the Hartree-Fock condition that p-h jumps from the Slater determinant $|0>$ are forbidden by cancellation of the off diagonal kinetic and potential terms (first two terms in rhs, the only ones left). Then we notice that as Δ is given in terms of $A_{(1)}$ and $A_{(2)}$ in eq. (3.2), the ℓhs of eq. (3.11) factorizes. For large number of particles N, this is a very dangerous situation as Δ should be linear in N, while most quantities in the rhs are constants. So we have to find something to cancel the ℓhs. This can be done very simply by requesting that the $A_{(2)}$, $A_{(3)}$ and higher amplitudes have the following form (which with little effort is seen to solve the riddle after equation (3.3))

$$A_{(2)} = A_{(1)}A_{(1)} + A^{L}_{(2)}$$

$$A_{(3)} = A_{(1)}A_{(1)}A_{(1)} + A^{L}_{(2)}A_{(1)} + A^{L}_{(3)}$$

$$A_{(4)} = A_{(1)}A_{(1)}A_{(1)}A_{(1)} + A^{L}_{(2)}A_{(1)}A_{(1)} + A^{L}_{(2)}A^{L}_{(2)} + A^{L}_{(3)}A_{(1)} + A^{L}_{(4)} \tag{3.12}$$

etc.

The superscript L stands (or will stand) for LINKED, i.e. non factorable, but since $A^{L}_{(2)}$, $A^{L}_{(3)}$ and $A^{L}_{(4)}$, in principle, can be anything they like, there is no approximation involved. If we examine what happens with $A_{(3)}$ in eq. (3.11) we have

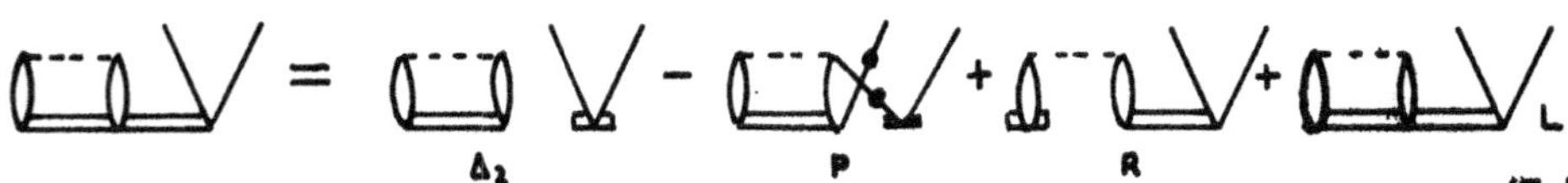

(3-13)

The first term cancels clearly the Δ_2 contribution in eq. (3.11). However we must remember that no two p or h lines may be identical in $A_{(3)}$ and this Pauli restriction is not obeyed by the Δ_2 term which is summed over all possible orbits. Therefore we cancel the Pauli violating terms in Δ_2 by the P term in which the dots indicate that the corresponding lines must be identical. The relinking term R shows that the interaction may keep a diagram linked even if the amplitude has factorized. The story is the same for the $A_{(1)}$ case. Once all the replacements are made in eq. (3.11) we are left with equations among fully linked terms, having all the same dependence on the number of particles.

The prescription (3.12) is equivalent to the exp(s) ersatz in which the wavefunction is written as

$$|\bar{0}\rangle = e^{S}|0\rangle \quad , \qquad S = \sum_n A^L_{(n)} \tag{3.14}$$

If we keep in S the $A_{(1)}$ terms only, we change a Slater determinant into another one (Thouless theorem) and as mentioned after (3.11), if we further request $A_{(1)}$ to vanish we obtain the Hartree-Fock equations. In the general case, in which we keep higher amplitudes we can still request vanishing of $A_{(1)}$. Then eq. (3-11) reads

0 = (3-15)

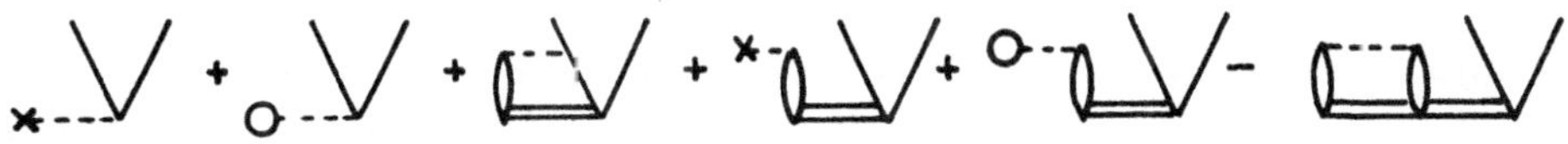

where we omit the L subscript since from now on all amplitudes are linked. To understand the meaning of this equation we have to know something about $A_{(2)}$ and $A_{(3)}$. The procedure to follow in obtaining the equations is exactly the same as for $A_{(1)}$: start from the general expression (3.10), replace each amplitude by its form (3.12), and classify terms as in (3.13). The general result if the following :

<u>Linked Cluster Theorem</u> : Equation (3.10) for $A_{(n)}$ reduces to a fully linked coupled, non linear integral equation for the amplitudes $A^L_{(m)}$, $m \leq n+2$.

A full, rather involved proof, is given in [LKZ78] and details of the diagrammatic derivation for $A_{(2)}$, $A_{(3)}$ and $A_{(4)}$ can be found in [PZ81a]†.

[The result is important for two reasons. One already stated after eq. (3.11), is that it ensures that all the terms in each equation have the same behaviour in the total number of particles. The other reason is of great physical significance. If we refer to eq. (3.13) we see that the unlinked piece is an $A_{(1)}$ amplitude times one of the terms of eq. (3.2), which we could call the $A_{(o)}$ equation. In fact all the terms of eq. (3.2) are hidden somewhere in the $A_{(1)}$ equation and this is how cancellation takes place. In general, for the $A_{(n)}$ equation it is all the preceeding

---

†The remarks in brackets are meant to give a qualitative feeling. It is advisable to read them in conjunction with [PZ81a] (neither before, nor after : in conjunction).

linked equations that are hidden in (3..10) and get " expelled " in the cancellations. In other words : the linked equations contain only new information about the higher amplitudes. The basis of all truncation schemes we can imagine is that, somehow, the amount of new information we get or we need in $A_{(n)}$ diminishes rapidly with n. Intuitively this can be understood in terms of the superposition principle. Starting from a " center " $|0\rangle$ the Hamiltonian acts as a field that reaches each np-nh state by choosing all possible paths through mp-mh states (m<n) and collecting each time the corresponding $A^L_{(m)}$ factor. This gives all the unlinked contributions to $A_{(n)}$ and $A^L_{(n)}$ is then interpreted as the residual contribution due to passage through ℓp-ℓh states with $\ell \geqslant n$.

Since we are primarily interested in the energy, for a rank 1+2 force we only need knowledge of $A_{(1)}$ and $A_{(2)}$. As we have coupled equations, these quantities depend on all the other $A_{(n)}$ and the whole question is : how strongly ? Figure 11 in (II.4) may provide some hint. The behaviour of the solutions for the ground state does not seem to depend very drastically on whether we have a 5 or 100 dimensional matrix. The power of the linked cluster mechanism is that it is adapted to driving very efficiently the system to the exact solution for the lowest state. Eq. (3.10) and its linked cluster version are equivalent to the full Schrödinger equation, and if we solve them in full we should obtain all the solutions. But we are not interested in them. As of now we want the lowest one(s) and we hope or expect to have a fair idea by just solving in some approximation for the lowest amplitudes.]

Refering again to [PZ81a] for details we write the $A_{(2)}$ and part of the $A_{(3)}$ equations

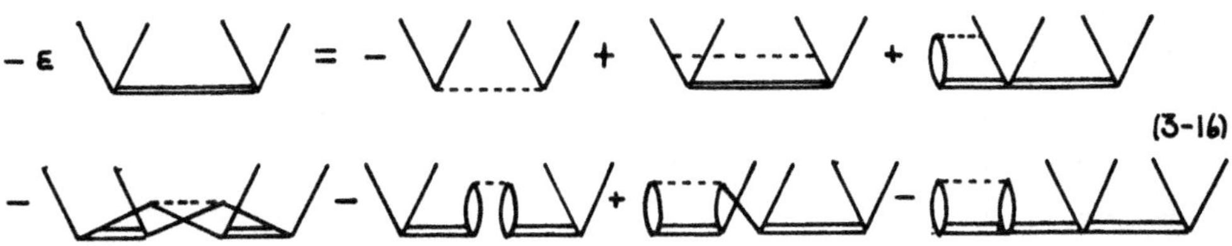

(3-16)

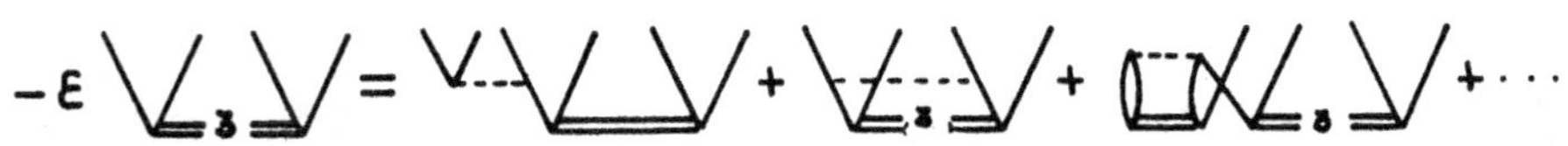

(3.17)

In writing eq. (3.16) we have assumed that the $A_{(1)}$ amplitudes vanish i.e. that eq. (3.15) is satisfied.

At first sight the equations are rather formidable (for $A_{(3)}$ we include only the very small fraction that will be discussed later). To give a preliminary reading we shall show how under some drastic assumptions the $A_{(2)}$ equation reduces to what we call " the classics " : the Random Phase Approximation (RPA) and Lowest Order Brueckner Theory (LOBT).

In principle theories start with the idea that the Hartree-Fock (HF) ground state should be a fairly good approximation which could be improved by perturbation theory (PT). For Fermi fluids this idea does not work in two cases. The HF approximation (equivalent to an optimized 1st order PT) makes no sense for potentials with repulsive cores and LOBT is the first step in solving this problem. For the Coulomb force (electron gas) HF makes very good sense but second order PT diverges (infrared catastrophe). The cure is here the RPA.

(3.II) RPA : preliminary discussion and example of diagrammatic expansions.

Although the RPA originated in the study of the electron gas, as a theory for the excitation spectrum of closed shell nuclei it is widely used in nuclear physics, where for reasons I do not understand,very well little attention is paid to the legitimacy of the approximation or to its direct influence on ground state properties.

The infrared divergence due to the behaviour of the Coulomb force at low momentum transfers (q^{-2}), can be eliminated once it is realized that the 2p-2h states that mix directly with the Fermi sea interact coherently among themselves. The 1p-1h spectrum exhibits strong collective states (the plasmons) whose frequency ω is pushed up with respect to the unperturbed ($\omega_o=k^2$) value for the p-h excitations of momentum $\vec{k}$ that build them. If the 2p-2h states are now constructed by recoupling diagonal p-h states, the divergence in 2^d order PT disappear since the force is now " screened " for low energies and the bulk of the interaction is exhausted by jumps from the Fermi sea to these high lying states. Diagrammatically, things are illustrated in eq. (3.18)

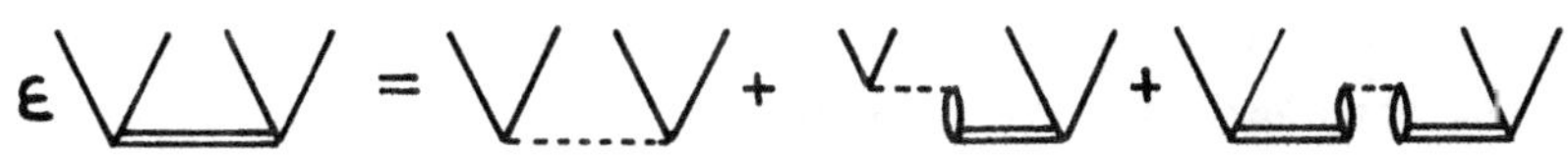
(3.18)

which is the RPA piece of eq. (3.16). Here we have used the exchange notation (see fig. 12) for the ph interaction vertex and out of the 4 possible ph rescatterings among the 2p-2h lines we have kept only two i.e.

$$\alpha\beta ij \rightarrow \alpha'\beta i'j \quad \text{or} \quad \alpha\beta' ij'$$

(3.19)

$$\text{but not} \quad \alpha\beta ij \rightarrow \alpha'\beta ij' \quad \text{or} \quad \alpha\beta' i'j$$

The approximation described above (Tamm-Dancoff) corresponds to the neglect of term B and to Neumann iteration of the term F (Forward). It is obvious that if instead of using a basis $\alpha\beta ij$ we had used a basis

$$\alpha\beta ij \equiv (\alpha i)(\beta j) \equiv \phi_{\omega k}\phi_{\omega' -k} \tag{3-20}$$

where $\phi_{\omega k}$ are p-h states diagonal in H, we would have no F term and Tamm-Daucoff in seen to be equivalent to 2^d order PT in the new basis. Historically and physically the reason to include the B (backward) term is that it generates diagrams of the same dominant behaviour on the density as term F and eq. (3-18) becomes exact in the high density limit.

To illustrate a very nice feature of the Goldstone diagrams obtained from the integral equation (3-18) we consider the result of the first iteration

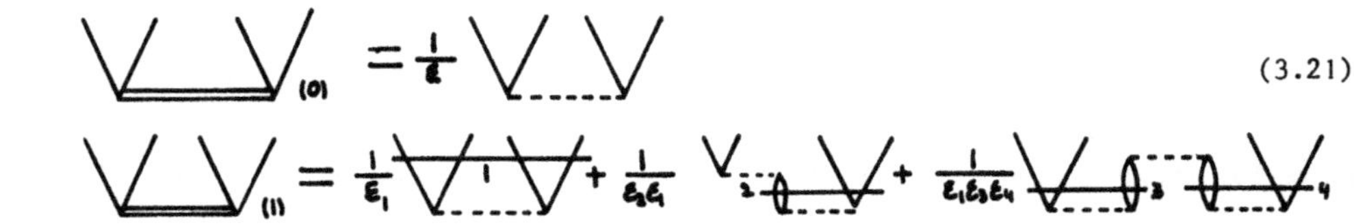

(3.21)

The superscrits (o) and (1) denote intialization and first iteration, not order in PT. The bars indicate the content of the energy denominators. In the Goldstone expansion we would have two diagrams with denominators $(\varepsilon_3+\varepsilon_4)\varepsilon_3$ and $(\varepsilon_3+\varepsilon_4)\varepsilon_4$ corresponding to different " time ordering " of the last term in (3-21). The integral equation yields only one with <u>local</u> denominators i.e. that do not go accross the whole diagram. By summing the two Goldstone diagrams we get

$$\frac{1}{\varepsilon_1}\left(\frac{1}{\varepsilon_3\varepsilon_{34}} + \frac{1}{\varepsilon_4\varepsilon_{34}}\right) = \frac{1}{\varepsilon_1\varepsilon_3\varepsilon_4}\left(\frac{\varepsilon_3+\varepsilon_4}{\varepsilon_{34}}\right) = \frac{1}{\varepsilon_1\varepsilon_3\varepsilon_4} \quad , \tag{3-22}$$

the value shown in (3-21). We shall encounter another example of the way the integral equations dispose automatically of the problem of resuming a class of diagrams that differ in their time ordering. (The P term in eq. (3.23)).

At the end of this lecture we shall return to the RPA to show the connection between ground state and excitation properties.

(3.III) Brueckner-Bethe-Goldstone theory, the Bochum truncation scheme and a short history of nuclear matter.

The screening of the repulsive core bears some formal analogy to the Tamm-Daucoff approximation discussed above in that each potential line (which is now infinite or very large) has to be associated to a ladder summation in which two particles (instead of a p-h pair) rescatter among themselves. The great difference now is that coupling between the $A_{(n)}$ equations becomes essential. In eqs. (3-2) or (3-15) the HF terms do not make sense and some knowledge of $A_{(2)}$ is essential before we could speak of an approximation to the GS energy. The original Bethe-Goldstone programme consisted in replacing each bare interaction line in a diagram according to the prescriptions

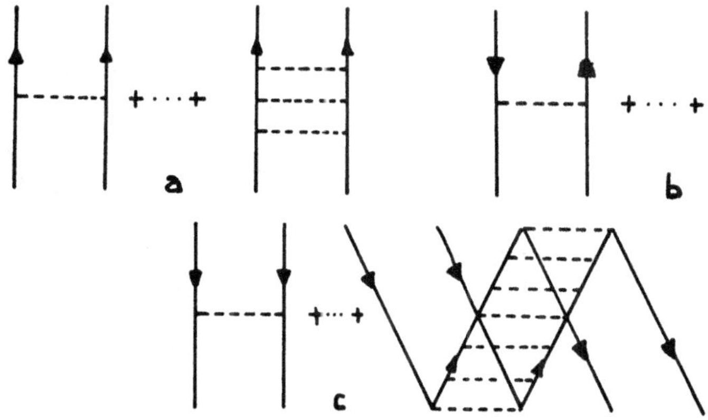

Fig. 17

The idea is attractive and we shall show later explicitly why this sort of operation yields finite results (3.VI) but to find out what has to be summed and how, turned out to be a very long story.

The coupled cluster equations of Coester and Kümmel contain a very natural and elegant rephrasing of the whole history of nuclear matter, that goes under the name of Bochum truncation scheme :

Start with the $A_{(2)}$ equation, neglect $A_{(3)}$ and $A_{(4)}$ and keep only those diagrams in the $A_{(2)}$ amplitudes that allow for the associations of fig. 17 (screening). Next introduce in the $A_{(2)}$ equation the $A_{(3)}$ amplitudes to screen some extra terms and solve the $A_{(3)}$ equations by neglecting $A_{(4)}$, $A_{(5)}$ and the $A_{(3)}$ processes that can't be screened. Etc.

The first step leads to the following equation

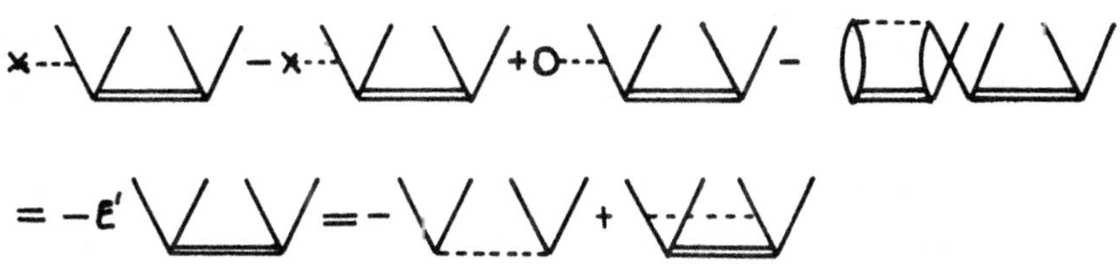

(3.23)

$$\varepsilon' A = G \; , \quad Q = \Sigma |\alpha\beta\rangle\langle\alpha\beta| \quad ; \quad G = V - V \frac{Q}{\varepsilon'} G$$

which is LOBT (G is the Brueckner G-matrix). The remarkable thing here is the very natural appearence of the Pauli term for the hole lines which can be used to screen the potential insertion as shown in fig. 18a

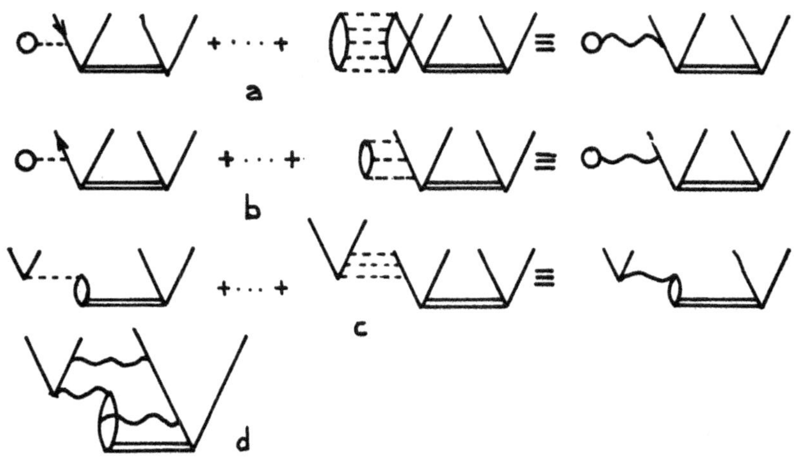

Fig. 18

The corresponding Pauli insertion on a particle line does not screen anything. In that case what is needed is the association shown if fig. 18b, but to generate this sort of ladders the $A_{(3)}$ equation is necessary, and for a long time, discarding altogether potential insertions on hole lines was the standard of nuclear matter calculations [BBP63]. It is obvious from eq. (3-17) that the Neumann iteration carries energy denominators that involve 3p-3h lines and therefore the G-matrix in fig. 18b is said to be " off shell " because it depends on the energies of all the lines, while G in fig. 18a does not and acts as a true on shell potential.

The definition of LOBT [BBP63] was the result of profound physical considerations and diagrammatic acrobacies (due to the multiplicity of time orderings, see eq. (3-22)), and it is a very nice feature of the coupled cluster equations to produce it automatically.

The next step involves the screening of the particle insertions, as shown in fig. 18b. But the $A_{(3)}$ equation also permits the screening of a p-h vertex as shown in fig. 18c and therefore, for consistency it should be introduced in the extended $A_{(2)}$ equation. However eq. (3-17), (called Bethe-Faddeev if we restrict rescatterings to particle lines) does not stop there, as processes of the type 18d appear that are of the same order of magnitude as 18a,b and have to be summed to all orders [BR67]. It has to be remembered that eq. (3-15) has to be satisfied simultaneously. If we do not introduce the $A_{(3)}$ amplitudes we can keep only the first three diagrams, the standard procedure in Brueckner-Hartree-Fock calculations. If $A_{(3)}$ comes in, the full equation is necessary. The coupled equations (3-15), (3-16) and (3-17) have been solved only once so far by Zabolitzky [ZA74,KLZ78] for ^{4}He, ^{16}O and ^{40}Ca. This calculation, which at the present time sets the standards in nuclear physics seems to have converged to something close to the exact solutions. (Selected pieces of the $A_{(4)}$ equation have been introduced also).

The Bochum truncation scheme retraces the main steps of the history of nuclear matter and clarifies the nature of the selfconsistency idea in the coupled equations, in a way that is far more economical (and complete) than the original derivations of Bethe and his school, starting from the fully expanded form of Goldstone's PT. It must be said however that the truncation scheme owes its use (and probably its existence) to the huge amount of physical insight that these derivations had demanded.

Much formal work remains to be done in giving the correct reading to the coupled equations, and other truncation schemes will have to be devised. In particular cancellations occur in the calculations that may a posteriori suggest a much simpler way of doing them.

From a Shell Model point of view it is interesting to speculate on the fact that formidable as they are in the full Hilbert space, coupled cluster calculations including up to $A_{(3)}$ (or $A_{(4)}$) amplitudes are a reasonably simple task when applied to a finite matrix, even if the dimensionalities are large and we can safely predict that this will be in a few years the standard method of dealing with enormous shell model problems. Assuming of course we really want to diagonalize big matrices. The point will be explored now.

(3.IV) Degenerate Linked Cluster Theory (LCT)

We have to remember now that our real task is to define a model space and solve eq. (2.5) rather than the simpler eq. (3.6) which corresponds to a model space of dimension 1. As stressed repeatedly, we want to obtain directly a set of integral equations from which degenerate PT may emerge. The full project has never been carried out and it is again to the Bochum school we owe the existing developments and calculations for excitations in closed shell nuclei [EM81,EZ81] for 1 and 2 particle systems [EZ79,OF76]. Many other attempts are presently being made in the field of quantum chemistry (see for example [LI78],[KU82]) but I feel there is room and need for a rather different approach that stresses the very close connection between degenerate and non degenerate theory. The problem of names is rather vexing : exp S was abandoned, probably because it made linked cluster properties appear as an ersatz rather than a fundamental fact. The current name, coupled cluster formalism, insists too much on the equations being coupled and too little on the clusters being linked. As it has gained acceptance it is better to keep it, at least for the non degenerate case but we propose to call the general formalism Linked Cluster Theory (LCT) to establish continuity with the standard historical name for Goldstone and Hugenholz work (LCPT). Probably LCCT would make everybody happy : Linked Coupled Cluster Theory.

The presentation that follows may be a bit baffling and the reader may wonder at first where is the theory to be found. For one thing there will be no equations (although in sections (3.V) and (3.VI) the RPA and the G-matrix will be studied in some detail). The main reason is that all the necessary equations are already written, and the only news will come from the way in which they should be solved. However, there is another surprising aspect in the results that follow : their simplicity. The only help I can offer in this respect is to say that it took quite a while to make things look so simple. In my view LCT is the answer to a number of riddles :

i) what to make of the non-hermiticity problem ?

ii) what is the status of folding in a general theory ?

iii) how do we choose the model space ?

iv) how strongly state dependent the effective interactions are ?

v) what is selfconsistency and what connection there is between global and spectral properties ?

iv) how does density dependence emerge from fundamental theory ?

It is well known that more often than not, riddles can appear as formidable before, and simple after, we know the answers. Let us look then at the answers bearing in mind the preceeding remarks.

Consider the situation depicted in fig. 19

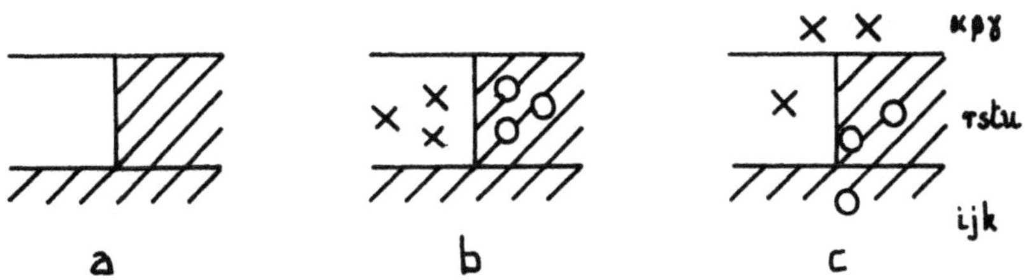

Fig. 19

The model space is made of all possible states for which the hole orbits are fully occuppied (core), n particles sit in valence orbits while the particle orbits are empty. In 19(a) we choose some arbitrary determinant in model space to serve as reference state $|ref\rangle$.

Fig. 19(b) shows another member of the model space, while 19(c) is a state outside (out state).

The most general solution to the Schrödinger equation has the following form

$$(1+\hat{A}_1+\hat{A}_2+\hat{A}_3+\ldots)(\hat{B}_o+\hat{B}_1+\hat{B}_2+\hat{B}_3+\ldots)|ref\rangle =$$

$$(\hat{C}_o+\hat{C}_1+\hat{C}_2+\ldots)|ref\rangle = |\overline{ref}\rangle \tag{3.24}$$

The operator $\hat{A}_k$ are rank k operators associated to an amplitude A_k. Acting on a model state they go to an out-state.

Example = $\hat{A}_3 = A_{\alpha\beta ritu}a^+_\alpha a^+_\beta a^+_r a_i a_t a_u$, goes from state (a) to state (c) in fig. 19.

The operators $\hat{B}_k$ are rank k operators associated to an amplitude B_k. Acting on a model state they go to another model state.
Example $\hat{B}_k = B_{rstuvw}\, a^+_r a^+_s a^+_t a_u a_v a_w$, goes from state (a) to state (b) in fig. 19. The operator $\hat{B}_o$ is a constant.

FUNDAMENTAL RULE.

The operators $\hat{C}_k$ acting on |ref> cover the whole Hilbert space. Everything we have said in the preceeding section about the $A_{(n)}$ amplitudes, their factorization properties and the equations they obey applies to the amplitude C_k. The whole theory rests on this fundamental remark.

FACTORIZATION RULES.

Amplitudes C_k obey decomposition rules of the type (3.24). The linked part C^L_k is identical to either a B^L_k or an A^L_k amplitude. Schematically we can write

$$C_k = B_o \prod_\ell A^L_\ell \circledast \prod_i B^L_i \circledast \prod A^L_m B^L_j + C^L_k$$

$$A_k = \prod_\ell A^L_\ell + A^L_k \;, \quad B_k = \prod_i B^L_i + B^L_k$$

$$\Sigma\ell = \Sigma i = \Sigma m + j = k \qquad (3.25)$$

The symbol $\circledast$ means that whenever $C^L_k = A^L_k (B^L_k)$ the second (the first and the third) term(s) vanishes (vanish) in the first equation.

DIAGRAMMATIC RULE.

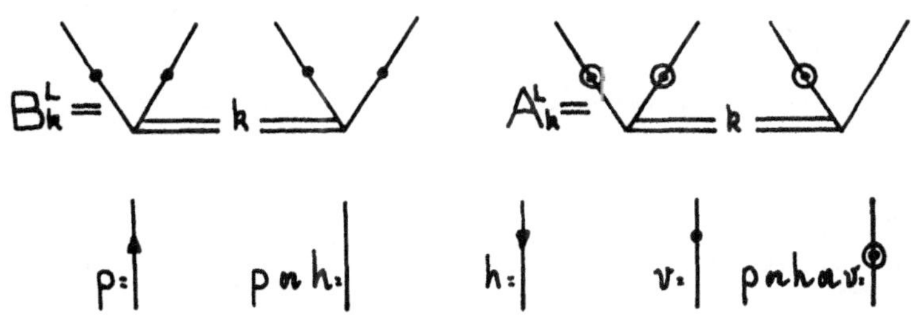

Fig. 20

The B_k amplitudes only contain valence (v) lines. The A_k amplitudes contain at least 1 non valence line. Valence lines are treated as particle (hole) lines if they create (anhilate) a particle when acting on |ref>.

REF-INDEPENDENCE RULE.

The operators $\hat{A}_k$ and the amplitudes B_i are <u>obviously</u> independent of the choice of $|ref>$, which is an arbitrary determinant in model space, provided B_o does not vanish. The operators associated with B_i <u>obviously</u> change if we change $|ref>$ but these changes cannot affect the calculations.

This rule is very profound and it is a consequence of eq. (3.24) which applies to <u>ONE</u> eigenstate, as the fundamental rule tells us that the A_k and B_k amplitudes are coupled. Therefore if we change eigenstate we change B_k and if we change B_k <u>in principle</u> we change A_k. The quality of the model space will depend on the degree of decoupling between A and B amplitudes. This point will be discussed below in CONNECTION WITH QT AND EQ (2.5). CHOICE OF MODEL SPACE.

THE SECULAR PROBLEM IN MODEL SPACE.

Amplitudes B_k correspond to states that are in principle degenerate or quasi-degenerate. The integral equations they obey are therefore <u>homogeneous</u> as the energy denominators ε affecting the inhomogeneous term vanish, or are very small. Using the definitions

$$|i> = B_i^{-1}\hat{B}_i|ref> \qquad |\bar{i}> = B_i^{-1}(1+\sum_\ell \hat{A}_\ell)\hat{B}_i|ref> \tag{3.26}$$

we can set up the secular problem

$$<i'|H\sum B_i|\bar{i}> = B_{i'}E = <i'|H(1+\sum_\ell A_\ell)\sum \hat{B}_i|ref> =$$

$$= \sum B_i <i'|H_{eff}|i> \quad ; \quad H_{eff} = H + H\sum_\ell A_\ell \tag{3.27}$$

FORM OF THE EFFECTIVE INTERACTION.

We illustrate the form of H_{eff} by means of an example. State $|i>$ is taken (with no loss of generality : see REF-INDEPENDENCE RULE) to be $|ref>$ and $B_i = B_o$. State $<i'|$ is a 3p-3h state with respect to $|ref>$. We choose a process involving the A_4 amplitude shown in eq. (3.28)

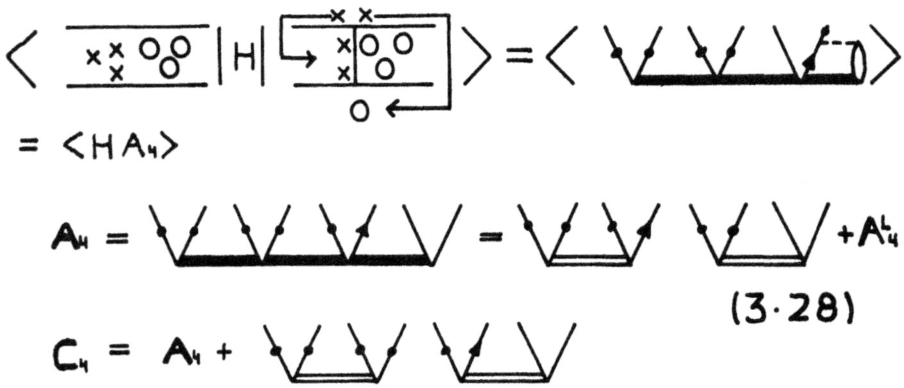

(3.28)

The arrows in the first equation indicate what H does. The contribution to H_{eff} is HA_4 and it contains the relinking of an unlinked term and the HA_4^L contribution. We have introduced heavy lines to distinguish A_4 from C_4 which contains a further factorization involving a B_2A_2 term. We have omitted factorizations in terms of A_1 or B_1 amplitudes (but they are there, see below : GENERALIZED SELFCONSISTENCY).

COUPLING BETWEEN A AND B AMPLITUDES.

Let us write eq. (3.23) for some $A_{\alpha\beta rs}$ amplitude

(3.29)

The coupling to B_{turs} amplitudes is manifest as the intermediate lines may well correspond to valence particles tu.

If we are in a truly degenerate regime all the B_2 amplitudes have roughly the same magnitude and it is not difficult to understand that $A_{\alpha\beta rs}$ does not depend on the choice of |ref>. If some B_2 amplitude happens to dominate in the wave-function, this argument does not hold anymore, but then we should do better to take as |ref> the dominant state and wonder whether a non degenerate regime is not adequate. At any rate the REF-INDEPENDENCE RULE holds for the exact A_2 and it may well be violated by a truncated equation such as (3.29). But then the very nature of this equation will guarantee, not only independence with respect to the choice of |ref> but decoupling of the A and B amplitudes. The reason is that $A_{(2)}$ or equivalently G in LOBT is insensitive to jumps in the vicinity of the Fermi surface and therefore the B amplitudes, whatever they are, will contribute little in eq. (3.29). The influence of |ref> comes through the Pauli principle : the particle states must be orthogonal to the hole states. Although it does not matter much which hole states appear in |ref> it matters a lot how many of them there are. Eq. (3.29) is formally the same for ^{16}O or ^{40}Ca, the only thing that changes is that 24 particle orbits become hole orbits. It is through this mechanism that density dependence will appear as we show in section (3.VI).

In fig. 20a we show how $A_{rs\alpha\beta}W_{\alpha\beta tu}$ screens the matrix element W_{rstu}

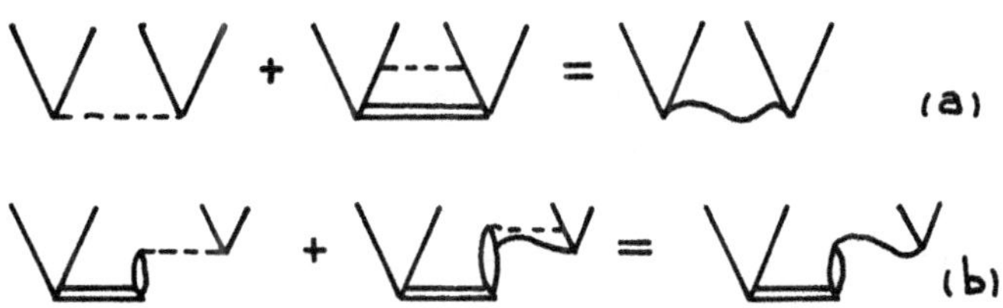

Fig. 20

In fig. 20b we have introduced the next ingredient in the effective interaction : the process called " core polarization ". The wiggles stand for G (i.e. εA). Notice that the remarks on fig. 18 apply here as we have to solve the $A_{(3)}$ equations to screen the isolated potential line in fig. 20b. (The only difference is that the G-matrix happens to be on shell because the energy denominators of the valence spectator lines vanish). Therefore we must bear in mind the possible necessity of having to solve the Bethe-Faddeev equation in full, and we are certainly not entitled to forget that an effective 3-body force is already present at this point : it is sufficient to open the " bubble " in fig. 20b and let two further valence lines propagate.

GENERALIZED SELFCONSISTENCY.

If we compare fig. 20b to eq. (3.15), we see that core polarization is reminiscent of a process that will get cancelled by a kinetic energy term if we demand generalized selfconsistency, i.e. vanishing of the A_1 amplitudes. The analogy would be complete if we insisted on defining an optimum $|ref\rangle$ imposing vanishing of the A_1 and B_1 amplitudes. In that case degeneracy would be broken, the orbits could no longer be assumed to be spherical but we could operate on a non degenerate regime. This perspective is rather fascinating but we shall not explore it now, except to say that the $|ref\rangle$ determinant would no longer be a scalar but an intrinsic state from which the integral equations would generate a band of states of good angular momentum.

Be as it may, we have been reminded that the A_1 amplitudes exist and that we have to do something about them. In general A_n need not be a scalar, but A_1 must be a scalar and we shall request its vanishing, at least in an average sense. The idea was originally suggested in [ZU75] in a purely Hartree-Fock context, to define the optimum spherical field as a function of the number of particles. In our case it takes the following form : for a given $|ref\rangle$ enforce eq. (3.15), then argue as in section (4.III) that the resulting orbits will be very much the same whatever the choice of $|ref\rangle$ because of the insensitivity of the G-matrix to the jumps in valence space. Notice that this is not related to the REF-INDEPENDENCE RULE, which does not state that <u>all</u> the A_1 amplitudes can be cancelled simultaneously, which is what we are hoping for. Actually, this general cancellation can't even be imposed in a purely HF context. If we do spherical HF on configurations of the form $[N_1 \ldots N_{i1} n_1 \ldots n_m]$, nobody can guarantee that the resulting orbits will be independent of n_i. Nevertheless it is a good bet that they will depend very little.

CONNECTION WITH QT AND EQUATION (2.5). CHOICE OF MODEL SPACE.

Although our eqs. (3.26) and (3.27) are very similar to eqs. (2.21), and (2.10) to (2.12), the formulation of LCT is much less restrictive than the full block decoupling enforced through eq. (2.5). The reason is that the wave operator $\hat{A}$ cannot be taken to be independent of $|i\rangle$, as in eq. (3.26) if it must obey eq. (2.5), i.e. produce an effective interaction that yields a number of eigenstates equal to the dimensionality of the model space. When we come to think of it, this request is much too strong and somehow assumes an a priori choice of model space. What is done in eq. (2.5) is to obtain an $\hat{A}$ that is $|i\rangle$ dependent but that does not depend on the eigenstates, while in eq. (3.26) we do exactly the opposite. The new formulation is better in that we always have in mind working on a single eigenstate but it may well happen that the corresponding amplitudes A_k also produce good approximations to other eigenstates. If this is the case we shall know that our choice of model state makes sense. Furthermore, once we get to understand under what conditions a given integral equation can be treated as inhomogeneous, i.e. what has to be treated as A and what as B amplitudes, the model space will be defined by the equations, rather than imposed on them.

The REF-INDEPENDENCE RULE may provide clues in this respect. It certainly holds for the exact amplitudes A_k, but a bad choice of $|ref\rangle$ may force us to solve far more equations than we can cope with, or drive us to eigenstates we are not interested in. In the case of strict degeneracy in the model space, this is unlikely to happen and it is not too dangerous to conjecture that A_k will be eigenstate independent, that a rather low order approximation will suffice and that many eigenstates will make sense. When degeneracies are sufficiently broken we should be more careful in the choice of $|ref\rangle$ and consider the possibility of reducing the size of model space.

FOLDING : THE DIAGRAMMATIC RULE takes care of folding. All our diagrams are folded to start with.

CHOICE OR REFERENCE ENERGY :

Formally one can always ignore the selfconsistency requirement and work on a fixed basis throughout the model space. Then it becomes a very elementary exercise to separate from all diagonal matrix elements of the effective interaction the energy of the closed shell at the beginning of the space, provided a perturbative expansion is used. Unfortunately, by doing so the effect of the Pauli principle on the G-matrix is totally obscured and there is no way to retrieve the mechanism of density dependence (see discussion after eq. (3.29) and section (3.VI)). In a very basic sense, the Bloch Horowitz theory is misleading. We apologize for this act of " lèse-majesté ", but there is no point in referring the energies in ^{28}Si to the ground state of ^{16}O. This is a point of extreme subtlely that will be explained in lecture 4.

(3.V) RPA revisited and derived.

We shall study now the connection between the ground state properties and the excitation spectrum, in the RPA approximation.

If our model space consists of p-h jumps on the closed shell we can use an equation similar to (3.24) and look for exact solutions of the form

$$|E,k\rangle = \sum_{ph,\alpha\beta ij} \bar{X}_{ph} a^{+}_{p'} a_{h'} (1+ \hat{A}_{\alpha\beta ij} + \ldots)|0\rangle \quad , \tag{3.30}$$

to construct explicitly the secular equation we probe to the left with some $\langle ph|$ state, to get

$$\bar{X}_{p'h'}E = \bar{X}_{p'h'}(-\Sigma W_{\alpha\beta ij}A_{\alpha\beta ij} + \varepsilon_{p'h'} + E_o + A_{\alpha p'ih'}W_{\alpha p'ih'}) + \\ + \Sigma\bar{X}_{ph}(W_{ph'hp'} - A_{\alpha p'ih'}W_{\alpha pih}) \tag{3.31}$$

The first parenthesis includes the exact GS energy $(\Delta+E_o)$ and the Pauli compensation term as in eq. (3.13), while the second parenthesis contains V_{eff}. Except for the neglect of the $A_{(3)}$ amplitude, no approximations are involved so far. Writing $E= E_o + \Delta + \omega$, where ω is the excitation energy, doing some trivial relabelling and dropping the Pauli term we get

$$(\varepsilon_{ph} - \omega)\bar{X}_{ph} + \sum_{\alpha\beta ij} \bar{X}_{\beta j}(W_{pjh\beta} - A_{\alpha pih}W_{\alpha\beta ij}) = 0 \tag{3.32}$$

Therefore, if we have obtained the amplitudes $A_{\alpha\beta ij}$ in a GS calculation, we can set up the secular problem in the ph space. The interaction is manifestly non Hermitian and to define biorthogonality conditions we solve the " fictitious " eigenvalue problem for the Hermitian conjugate of V_{eff}, as in eq. (2.13)

$$\langle ph|V^{+}_{eff}|\overline{\beta j}\rangle = \langle\beta j|V_{eff}|\overline{ph}\rangle = W_{pjh\beta} - \sum_{\alpha i} A_{\alpha\beta ij}W_{\alpha pih} \tag{3.33}$$

Using the definition

$$Y_{\alpha i} = -\sum_{\beta j} X_{\beta j}A_{\alpha\beta ij} \tag{3.34}$$

we get the secular equation

$$(\varepsilon_{ph} - \omega)X_{ph} + \Sigma X_{\beta j}W_{pjh\beta} + \Sigma Y_{\alpha i}W_{\alpha pih} = 0 \tag{3.35}$$

and the biorthogonality condition

$$\sum_{ph} \bar{X}^{(\omega)}_{ph} X^{(\omega')}_{ph} = \delta(\omega\omega'), \tag{3.36}$$

which suggests we could try the following ersatz in eq. (3.34)

$$A_{\alpha\beta ij} = -\sum_{\omega} \bar{X}^{(\omega)}_{\alpha i}\, Y^{(\omega)}_{\beta j} = -\sum_{\omega} \bar{X}^{(\omega)}_{\beta j}\, Y^{(\omega)}_{\alpha i} \;. \tag{3.37}$$

The second equality is an imperative symmetry condition that $A_{\alpha\beta ij}$ must fulfill. Now we are going to use this ersatz to see to what extent it satisfies the $A_{\alpha\beta ij}$ equation (3.16), which we write as

$$\Big[-W_{\alpha\beta ij} + \varepsilon_{\alpha\beta ij}A_{\alpha\beta ij} + \Sigma A_{\alpha'\beta i'j}W_{\alpha i'i\alpha'} + \Sigma A_{\alpha\beta' ij'}W_{\beta j'j\beta'} - \Sigma A_{\alpha\delta i\ell}W_{\gamma\delta k\ell}A_{\beta\gamma jk}\Big] + \Big[\Sigma A_{\alpha'\beta ij'}W_{\alpha j'j\alpha'} + \Sigma A_{\alpha\beta' i'j}W_{\beta i'i\beta'} - \Sigma A_{\alpha\delta j\ell}W_{\gamma\delta k\ell}A_{\beta\gamma ik}\Big] + \text{other terms} = 0 \tag{3.38}$$

In the first (second) bracket we have grouped the ph rescatterings of the first (second) line in eq. (3.19) and the corresponding relinkings.
The other terms correspond to the pp and hh rescattering and the A^L_3 and A^L_4 contributions. Now we multiply by $X_{\alpha i}$, sum over αi, use eq. (3.37) in the first bracket and neglect the second bracket and other terms.

$$-\Sigma W_{\alpha\beta ij}X_{\alpha i} - (\varepsilon_{\beta j} + \omega)Y_{\beta j} + \Sigma(\varepsilon_{\alpha i} - \omega)X_{\alpha i}A_{\alpha\beta ij} + \Sigma X_{\alpha' i'}W_{\alpha' ii'\alpha}A_{\alpha\beta ij} - \Sigma Y_{\beta' j'}W_{\beta j'\beta'} + \Sigma Y_{\delta\ell}W_{\alpha\delta i\ell}A_{\alpha\beta ij} = 0 \quad , \tag{3.39}$$

where we have added and substracted ω in the single particle term of eq. (3.38).
We notice that $A_{\alpha\beta ij}$ is a common factor to terms that satisfy eq. (3.35) and we obtain the equation for $Y_{\beta j}$

$$(\varepsilon_{\beta j} + \omega)Y_{\beta j} + \Sigma Y_{\beta' j'}W_{\beta j'\beta'} + \Sigma X_{\alpha i}W_{\alpha\beta ij} = 0 \tag{3.40}$$

The coupled system (3.35) and (3.40) gives the Standard RPA equations for the excited states and our treatment makes clear what has been neglected in the general equation (3.16) to arrive at this result. The interesting aspect of these equations is that all reference to $A_{\alpha\beta ij}$ and to $\bar{X}_{ph}$ has been eliminated. These quantities can be retrieved by using one of the standard ways of defining RPA. Introduce operators $B^+_{\omega,k}$ and $B\omega,k$

$$B^+_{\omega,k} = \Sigma\big(X^{(\omega)}_{\alpha i}\, a^+_\alpha a_i - Y^{(\omega)}_{\beta j}\, a^+_j a_\beta\big)$$
$$B_{\omega,k} = \Sigma\big(Y^{(\omega)}_{\beta j}\, a^+_\beta a_j - X^{(\omega)}_{\alpha i}\, a^+_i a_\alpha\big) \tag{3.41}$$

and request

$$B_{\omega,k}|\bar{0}\rangle = 0 \qquad B^+_{\omega,k}|\bar{0}\rangle = |\omega,k\rangle \tag{3.42}$$

The second equation leads immediately to

$$\bar{X}^{(\omega)}_{\alpha i} = X^{(\omega)}_{\alpha i} - \Sigma Y^{(\omega)}_{\beta' j'} A_{\alpha\beta' i j'} \qquad (3.43)$$

by equating the coefficient of the ph term to that in eq. (3.30), while the vanishing of the ph coefficient in the first equation (3.42) yields eq. (3.34). By multiplying (3.43) by $Y^{(\omega)}_{\beta j}$ and summing over ω we get

$$A_{\alpha\beta i j} = \sum_{\omega} X^{(\omega)}_{\alpha i} Y^{(\omega)}_{\beta j} - \sum_{\omega} Y^{(\omega)}_{\beta j} \, Y^{(\omega)}_{\beta' j'} \, A_{\alpha\beta' i j'} \, , \qquad (3.44)$$

the integral equation defining $A_{\alpha\beta i j}$ [DA66]. Notice that the biorthogonality condition (3.36), using (3.43) and (3.34), translates into

$$\sum_{\alpha i} \bar{X}^{(\omega)}_{\alpha i} \, X^{(\omega')}_{\alpha i} = \sum_{\alpha i} X^{(\omega)}_{\alpha i} \, X^{(\omega')}_{\alpha i} - \Sigma Y^{(\omega)}_{\beta j} \, Y^{(\omega')}_{\beta j} = \delta_{\omega\omega'} \qquad (3.45)$$

Finally, we leave as an exercise the proof that the second equation in (3.37) is indeed satisfied.

It is characteristic of the RPA that the excitation spectrum is somewhat simpler to obtain than the ground state amplitudes $A_{\alpha\beta i j}$. The interest of our approach, is that it makes clear that RPA is simply a diagonalization of ph excitations acting on the correlated ground state, calculated under certain approximations that eq. (3.38) states unambiguously. It will be no doubt interesting to explore the way one can make use of the spirit of these approximations to devise truncation schemes for the general equations.

There are several other aspects of the preceeding derivations that may call for comments but we only draw attention on the very natural role of non Hermiticity in this problem.

(3.VI) Density dependence of the G-matrix.

In section (3.III) we have introduced the notion of screening for the hard core. Now we are going to show explicitly how a ladder summation among particle lines transforms the repulsive cores of the N-N potentials into a density dependence of the effective force. We shall use an unconventional approach to the G-matrix, based on the fact that prediagonalization of H in the 2p intermediate states reduces the ladder summation to 2^d order PT.

The first time the problem appears is in the " A_o " equation (3.2), which we rewrite

E = 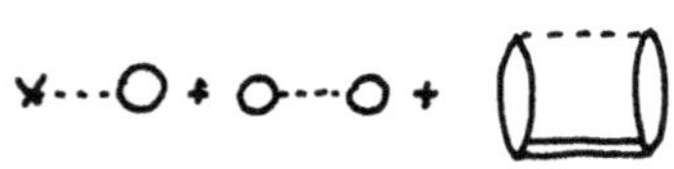 =

$$= \sum K_i + \sum_{i<j} W_{ijij} - \sum_{\substack{i<j \\ \alpha<\beta}} W_{ij\alpha\beta} A_{\alpha\beta i j} \, . \qquad (3.46)$$

We shall introduce the single hole energy in LOBT as

$$\bar{\varepsilon}_i = K_i + \sum_j \Big(\sum_{\alpha<\beta} W_{ijij} - W_{ij\alpha\beta} A_{\alpha\beta ij} \Big) = K_i + \sum_j \bar{W}_{ijij} = K_i + \bar{\bar{W}}_i \quad , \tag{3.47}$$

which serves as definition of the screened potential line $\bar{W}_{ijij}$ and screened single hole energy $\bar{\bar{W}}_i$.

The LOBT equation (3.23) for $A_{\alpha\beta ij}$ is

$$(\bar{\varepsilon}_{ij} - K_{\alpha\beta}) A_{\alpha\beta ij} = W_{\alpha\beta ij} - \sum W_{\alpha\beta\alpha'\beta'} A_{\alpha'\beta' ij} \equiv$$

$$(\bar{\varepsilon}_{ij} - \langle\bar{\alpha}\beta|H|\bar{\alpha}\beta\rangle) A_{\overline{\alpha\beta}ij} = W_{\bar{\alpha}\beta ij} \quad \text{, whenever}$$

$$\langle\overline{\alpha\beta}|H|\overline{\alpha\beta}'\rangle = \varepsilon_{\overline{\alpha\beta}} \delta_{\overline{\alpha\beta}\,\overline{\alpha\beta}'} \quad \text{, then}$$

$$A_{\overline{\alpha\beta}ij} = \frac{W_{\overline{\alpha\beta}ij}}{\bar{\varepsilon}_{ij} - \langle\overline{\alpha\beta}|H|\overline{\alpha\beta}\rangle} \tag{3.48}$$

In deriving these equations we have only used the obvious observation that rescattering among $|\alpha\beta\rangle$ states is suppressed if we choose a 2p basis $|\overline{\alpha\beta}\rangle$ that is diagonal in the interaction. Then the amplitude $A_{\overline{\alpha\beta}ij}$ in LOBT is given by 2^d order PT, and the quantity we are after, the screened potential matrix element becomes

$$\bar{W}_{ij} = W_{ij} - \sum_{\overline{\alpha\beta}} \frac{|W_{\overline{\alpha\beta}ij}|^2}{\bar{\varepsilon}_{ij} - \langle\overline{\alpha\beta}|H|\overline{\alpha\beta}\rangle} \quad . \tag{3.49}$$

By going to relative and center of mass coordinates

$$|ij\rangle \equiv |k\rangle|K\rangle \quad , \quad |\overline{\alpha\beta}\rangle \equiv |q\rangle|K\rangle \quad , \tag{3.50}$$

we can integrate out the CM states $|K\rangle$. The intermediate states $|q\rangle$ must be at the same time diagonal in H and orthogonal to the occupied states $|k\rangle$, for all values of k.

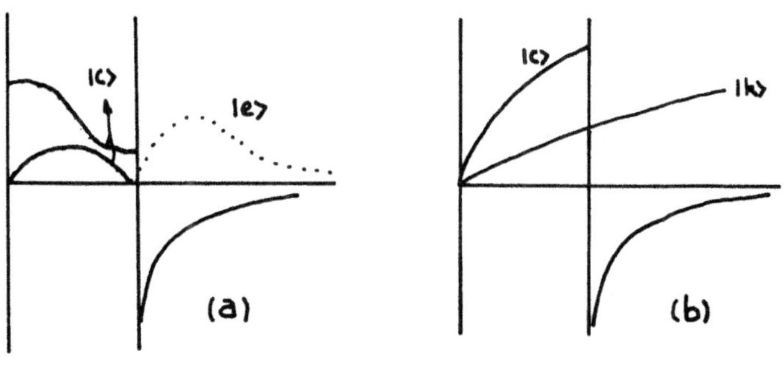

Fig. 21

If the potential has a hard core there is a natural separation of the Hilbert space in two regions, core and external

$$|q\rangle = |c\rangle \oplus |e\rangle \tag{3.51}$$

and the diagonalization can be carried out separately. Even if the core is not strictly infinite, but sufficiently strong, there is a point in distinguishing between the two type of states. The Pauli orthogonalization to the $|k\rangle$ states, gives

$$|c\rangle \longrightarrow |c\rangle - \sum_k |k\rangle\langle c|k\rangle = |\bar{c}\rangle$$

$$|\bar{\bar{c}}\rangle = (\langle\bar{c}|\bar{c}\rangle)^{-1/2}|\bar{c}\rangle = (1-\sum_k(\langle c|k\rangle|^2)^{-1/2}|\bar{c}\rangle \tag{3.52}$$

Now $|\bar{c}\rangle$ is no longer an eigenstate, assuming that originally $|c\rangle$ was diagonal with periodic boundary conditions as shown in fig. 21a. However, this is no problem. We can certainly diagonalize H again but we should better remember that our first priority is to eliminate the hard core and this can be done consistently by the following choices

1°) Split W_k, the energy in 1st order PT

$$W_k = W_k^R + W_k^c \quad , \tag{3.53}$$

where R stands for regular and denotes that part of the matrix element to which contributes the potential outside the core.

2°) In the 2d order term keep only core terms, i.e., disregard the regular part of the matrix elements and the intermediate states of type $|e\rangle$. Furthermore, for each W_k, choose $|c\rangle$ as shown in fig. 21b, so as to be that linear combination in c-space that reproduces the shape of $|k\rangle$ inside the core. All other core states are now strictly orthogonal to $|k\rangle$ and $|c\rangle$. Furthermore $K_c = K_k$. After Pauli orthogonalization, these properties are slightly perturbed but the effects introduced are of the same order that those neglected from the beginning.

As a consequence we are left with the following expressions

$$\bar{W}_{kk} = W_{kk}^R + \langle k|V^c|k\rangle - \frac{|\langle k|V^c|\bar{\bar{c}}\rangle|^2}{\bar{\bar{W}}_K - \langle\bar{\bar{c}}|V^c|\bar{\bar{c}}\rangle} \quad , \tag{3.54}$$

$$\langle k|V^c|k\rangle = L|\langle k|c\rangle|^2 \tag{3.55}$$

$$|\langle k|V^c|\bar{\bar{c}}\rangle|^2 = \left[1-\sum_{k'}|\langle c|k'\rangle|^2\right]^{-1}\left[\langle k|c\rangle - \sum\langle k|k'\rangle_c\langle k'|c\rangle\right]^2 L^2 =$$

$$= |\langle k|c\rangle|^2(1-\sum|\langle c|k\rangle|^2)L^2, \text{ since } \langle k|k'\rangle_c = \langle k|c\rangle\langle c|k'\rangle \tag{3.56}$$

$$<\bar{\bar{c}}|V^c|\bar{\bar{c}}> = \left[1-\sum|<c|k'>|^2\right]^{-1}<c-\sum<c|k'>k'|V^c|c-\sum<c|k'>k'>=$$

$$= L\left[1-\sum|<c|k'>|^2\right]^{-1}\left[1-\sum|<c|k'>|^2\right]^2 = L(1-\sum|<c|k'>|^2) \qquad (3.57)$$

$$\therefore\ \bar{W}_{kk} = W^R_{kk} + L|<c|k>|^2 - \frac{<c|k>^2(1-\Sigma|<c|k'>|^2)}{\bar{\bar{W}}_k - L(1-\Sigma|<c|k'>|^2)} L^2$$

$$= W^R_{kk} - \bar{\bar{W}}_k \frac{|<c|k>|^2}{1-\Sigma|<c|k'>|^2} \approx W^R_{kk} - \frac{\beta N\bar{W}_{kk}|<c|k>|^2}{1-\alpha N|<c|k>|^2} = \bar{W}_{kk}$$

$$\therefore\ \bar{W}_{kk} = (1-\alpha N|<c|k>|^2)\left(1+(\beta-\alpha)N|<c|k>|^2\right)^{-1} W^R_{kk} \qquad (3.58)$$

In deriving these equations we have called L the height of the core, which could be infinite or simply large. In eq. (3.58) we have used

$$\bar{\bar{W}}_k = \beta N\bar{W}_{kk}\ , \qquad \sum|<c|k'>|^2 = \alpha N|<c|k>|^2\ . \qquad (3.59)$$

Here N is the number of particles. If α and β are suitably chosen, these expressions are always valid. However, the physical argument is that these parameters are of order unity. In the first equality (see eq. (3.47)) it corresponds to replacing the two body " matrix element " $\bar{W}_{ijij}$ by an average $\bar{W}_i$. In the second equality we claim that the number of hole states $|k'>$ that have a sizable overlap with $|c>$ is large, i.e. of order N, (O(N)). If we write

$$|k> = (<\tilde{k}|\tilde{k}>)^{-1/2}|\tilde{k}> \quad , \quad |c> = (<\tilde{c}|\tilde{c}>)^{-1/2}|\tilde{c}> \qquad (3.60)$$

Then

$$|<c|k>|^2 = |<\tilde{c}|\tilde{k}>|^2(<\tilde{k}|\tilde{k}><\tilde{c}|\tilde{c}>)^{-1} = \gamma\frac{\Omega_c}{\Omega} \qquad (3.61)$$

where $\Omega(\Omega_c)$ is the volume of the system (core), and using $N/\Omega = \rho$, the density of the system, we can rewrite eq. (3.58) as

$$\bar{W}_{kk} = W^R_{kk}(1-a\rho)(1+b\rho)^{-1} \qquad (3.62)$$

We can safely assume $\beta>\alpha$, and therefore the denominator is strictly positive. For low densities the screened potential is essentially the regularized matrix element. For high densities the effective interaction becomes weaker or even repulsive and we have indeed a saturating mechanism.

The arguments in this section depend little on the assumption used throughout that we are dealing with an infinite system. This difference between infinite nuclear matter and finite nuclei is essentially one of boundary conditions. In the former case we use plane waves with box normalization and we can make an arbitrary choice of density . For finite nuclei an oscillator basis will leave the system free to choose its saturation point i.e. we are no longer entitled to fix the density : the only correct solution will correspond to the equilibrium value. This is where selfconsistency arguments come in, reminding us of the existence of eq. (3.15). Moreover, in a nucleus the second equation (3.59) is a bit dangerous : as α may have a non negligible N dependence for small N ($N^{-1/3}$ or $N^{-2/3}$!). The same objections hold for eq. (3.61). In a large system Ω_c and γ are O(1) while Ω is O(N). In small systems the distinction may need some subtlely.

Be as it may, we have shown that in some reasonable sense the effective force corresponds to a regularized N-N potential in which the hard core is eliminated and replaced by a multiplicative saturating expression we call density dependence.

It is probably worth insisting on the role of the Pauli orthogonalization, which is seen to be responsible for the whole mechanism we have described. Referring to eq. (3.29) we can easily understand that as we change the number of valence particle n, more Pauli excluded states appear. In view of the remarks we have made, we are led to expect some secular variation of the interaction as a function of n.

LECTURE 4. THE MONOPOLE SAGA.

(4.I) Definition of the monopole field. Its influence in Shell Model calculations.

In eq. (1.1) we have already mentioned the monopole field as the part of the Hamiltonian providing the simplest accurate estimate of the unperturbed energies for the configurations entering a Shell Model calculation.

From now on we assume that H is the effective Hamiltonian operating in a model space. We can split it quite generally in the following way

$$H = H_m + H_M$$

$$\mathrm{tr}\langle m|H_m|m\rangle = \mathrm{tr}\langle m|H|m\rangle = \frac{1}{D}\sum_i \langle m,i|H|m,i\rangle$$

$$\mathrm{tr}\langle H_M\rangle = 0 \qquad (4.1)$$

Here the traces are performed over fixed configurations m, i indicating the extra quantum numbers necessary to specify the D states of the configuration. Simpler as well as more complicated averages are possible, e.g. we could take traces over fixed number of particles to define an H_n (the scalar trace equivalent), or include the isospin (H_{nT}), or even construct H_{mT} in which we average at fixed isospin for each shell in a given configuration.

In [FR67] the interested reader may find a comprehensive introduction to the subject, which is at the heart of the spectral distribution methods developed by French and his school. For our purposes it will be sufficient to state without proof the form of H_m,

$$H_m = E_o + \sum_i n_i\varepsilon_i + \sum_{i<j} n_i\cdot n_j V_{ij}$$

$$n_i\cdot n_j = \left(1+\delta_{ij}\right)^{-1} n_i\left(n_j+\delta_{ij}\right)$$

$$V_{ij} = \left(N_i\cdot N_j\right)^{-1} \sum_{i_z,j_z} W_{ii_zjj_zii_zjj_z} = \left[\sum_{JT}(2J+1)(2T+1)\right]^{-1} \sum_{JT} W^{JT}_{ijij}(2J+1)(2T+1)\ , \qquad (4.2)$$

where the sums over JT are performed over all the Pauli allowed values. The quantum numbers i_z stand for the angular momentum and isospin projections for shell i.

What we call the monopole saga started from the following remark made by Pasquini and Zuker [PA76,PZ77] :

The Kuo Brown (KB) interaction in the fp shell [KB68] produces very good spectroscopy provided the following modifications are made

$$W^{JT}_{frfr}(KB) \rightarrow W^{JT}_{frfr}(KB)-(-)^T 300 \text{ keV}, \qquad r \equiv f_{5/2}p_{3/2}p_{1/2},\ f= f_{7/2} \qquad (4.3)$$

These changes affect the isovector centroïds as well as the scalar ones defined in (4.2), but by far the most spectacular effects are due to the latter. As an illustration we mention the 10 MeV upward shift in the 4p-4h states $f^{12}r^4$ with respect to the closed shell f^{16}. Modifications of the W_{ffff} and other matrix elements are also necessary but the general conclusion is that in the KB interaction H_m is completely wrong, while H_M is quite reasonable, i.e. the multipole (non monopole) part of the force.

It was also a remarkable fact that other realistic matrix elements [KLS69,OR72] derived from different forces and treating the G-matrix calculations differently, are strikingly similar to the KB set and exhibit the same monopole misbehaviour.

As H_m accounts entirely (by definition (4.1)) for the energy of the closed shell(s) at the end of the space, as well as for the single particle and single hole energies, it is quite reasonable to associate it with some global property of H, while H_M is more closely responsible for spectroscopic detail.

In view of this state of affairs, it was decided by Cortès and the author [CZ79,CO83,CZ84] to examine the situation in the p and sd shells where complete shell model calculations are possible. The monopole effect in a form similar to (4.3) is indeed present in these regions but it soon became evident that two body forces alone could not do and that three, later four and eventually five body effects were necessary. The number of necessary parameters becomes potentially enormous : H_m contains terms in V_{ijk}, V_{iij}, V_{iikk}, etc., a total of some 50 numbers (for forces up to rank 4) that there is little hope or point in trying to determine in a fit. As we shall show in eq.(4.12) the need of higher rank forces is guaranteed by sum rules that can be used to limit the number of free parameters. The saga aspect of the calculations was mainly due to the difficulty in finding convincing arguments to explain the apparently massive need of new many body effects. The hypothesis (rather the bet) that H_m is alone responsible for the problems of the effective interaction seemed quite radical at first : we felt that acceptable corrections should stop at the 3 body level and an enormous amount of time was wasted in the useless effort to eliminate other contributions. Strangely enough this was a matter of conceptual consistency.

The reason to insist on the importance of H_m was twofold :

i) realistic interactions behave quite decently for small number of particles. They rapidly deteriorate as n increases. Therefore it is natural to search the trouble in H_m, the only contribution to H that has explicit n dependence.

ii) As mentioned above, it is natural to relate H_m to global properties and there were very good reasons to believe that they were poorly treated in the early calculations. Therefore, we were in firm ground in demanding changes in H_m. However, the interest of these changes can't be purely phenomenological. We always thought of the calculations as a way to extract some insight into either the secular behaviour of the effective interaction or the necessity of bringing in true (as opposed to effective) higher rank forces. In other words : we insisted on the (eventual) need of calculating on some more fundamental basis whatever parameters were demanded empirically. Hence the refusal of accepting parameters about whose origin we did not have the slightest theoretical idea. This is were H_m becomes really important : if there is something we ought to be able to calculate properly it is the ground state of the closed shell nuclei. Since H_m contains precisely this information and propagates it from one end of the model space to the other it is to be expected that it should be the part of the effective interaction more amenable to calculation.

In retrospect it is quite easy to understand what was the theoretical misconception that turned out to be so time consuming : a poor understanding of the connection between density dependence and perturbation theory.

We were always prepared to accept and introduce density dependent effects. The work of Negele [NE70] and the subsequent development of Hartree-Fock calculations with density dependent forces leaves little doubt about their importance. Unfortunately, in their usual parametric forms (mainly Skyrme) these forces look too much like a rank 2+3 potential. On the other hand second order PT can produce such ranks (see eq. (2.20)) and it is tempting to use it as a guide. Even if we are prepared to take a coupled cluster view PT remains a useful guide : the G-matrix is essentially a second order calculation (as shown explicitly in (4.II)), a Bethe-Faddeev sum would be third order etc. Therefore one would naïvely associate rank 2 to the G-matrix, rank 3 to Bethe-Faddeev etc. But, as the exercise in (4.VI) shows the G-matrix is itself density dependent, and the mechanism responsible for it is not perturbative. In other words : density dependence will translate in a shell model space as a dependence on the number of particles, which may be quite complicated but which has nothing to do with higher orders in PT.

The formulation of LCT and the crude derivation of density dependence of the G-matrix in (4.VI) are consequences of the monopole saga and we believe they serve as good indication of the methodological validity of the idea of doing the calculations first and then the theory. Before we turn to the former, we would like to point out why an expression such as (3.62) explains why H_M is so good for the realistic forces. The medium range part of the N-N potentials contains most of their well understood features and most of the detailed information carried by their complicated parametrizations. At the same time it dominates the numerical values of the matrix elements of H_M. Eq. (3.62) conserves all the information and explains the dominance.

(4.II) Parametrization of the Monopole Field.

As pointed out in ii) of Lecture 1 (see fig. 3), exact calculations with two body forces will not saturate correctly. However we shall take the view that our effective interaction has been deduced from a Hamiltonian that does saturate correctly. LCT can easily accomodate non nucleonic effects. Knowing the exact results for two body forces, whatever we learn by enforcing exact saturation may shed light on these effects.

In determining the monopole field we shall demand

a) that the closed shell, the single particle and single hole nuclei at the ends of the model space be well reproduced (i.e. A=4,5,15 and 16 for the p shell, A=16,17,39,40 for the sd shell). These data are taken directly from experiment

b) that the two particle and two hole centroïds, i.e. the quantities V_{ij} in eq. (4.2) and the corresponding values for holes, be well reproduced. These data cannot be read directly from experiment but shell model calculations for a very small ($\leqslant 4$) number of particles and holes are very sensitive to their values and can be used to fix them accurately with very few exceptions.

The monopole field will be parametrized uniquely by requesting :

A) that the number of free parameters be equal to the quantities mentioned in a) and b) above.

B) That the forces entering the parametrization be of smallest possible rank.

As a preliminary step we write H_m in eq. (4.2) as

$$H_m = E_o + \varepsilon n + Wn\cdot n + \sum \hat{O}^{(1)}_{ji} N_j \left[\eta_j + (n-1)W_j\right] + \sum_{i\neq j} \hat{O}^{(2)}_{ij} \frac{N_i N_j}{2} W_{ij} =$$

$$= E_o + \varepsilon n + Wn\cdot n + \sum n_i \left[\eta_i + (n-1)W_i\right] + \sum n_i \cdot n_j W_{ij} \tag{4.4}$$

$$W = (N\cdot N)^{-1} \sum N_i \cdot N_j V_{ij} \qquad W_i = (N-2)^{-1}\left[\sum_j (N_j - \delta_{ij})V_{ij} - (N-1)W\right]$$

$$W_{ij} = V_{ij} - W_i - W_j - W \qquad \varepsilon = N^{-1} \sum N_i \varepsilon_i \qquad \eta_i = \varepsilon_i - \varepsilon$$

The operators $\hat{O}^{(k)}_{ij}$ are defined as

$$\hat{O}^{(1)}_{ij} = \left(\frac{n_i}{N_i} - \frac{n_j}{N_j}\right) = -\left(\frac{\bar{n}_i}{N_i} - \frac{\bar{n}_j}{N_j}\right)$$

$$\hat{O}^{(2)}_{ij} = \frac{n_i^{(2)}}{N_i^{(2)}} + \frac{n_j^{(2)}}{N_j^{(2)}} - \frac{2n_i n_j}{N_i N_j} = \frac{\bar{n}_i^{(2)}}{N_i^{(2)}} + \frac{\bar{n}_j^{(2)}}{N_j^{(2)}} - \frac{2\bar{n}_i \bar{n}_j}{N_i N_j} \tag{4.5}$$

$$n_i^{(k)} = n_i(n_i - 1)\dots(n_i - k + 1), \qquad \bar{n}_i = N_i - n_i \quad .$$

The centroïds W_i and W_{ij} obey the sum rules

$$\sum N_i W_i = 0 \qquad \sum_j (N_j - \delta_{ij}) W_{ij} = 0 \tag{4.6}$$

The parametrization (4.4) has been independently obtained by French and Kota (see prof. French's lectures). Notice that it gives a separation of H_m of the type

$$H_m = H_o + H_1 + H_2 \tag{4.7}$$

where H_o depends only on n, H_1 is a single particle field that depends linearly on n, while H_2 is of strict rank 2. From our point of view, the interest of these expressions comes from the high symmetry they exhibit under a p-h transformation ($n_i \to \bar{n}_i$) as shown in eq. (4.5). It is a remarkable property of $\hat{O}^{(k)}_{ij}$ that it is an operator of strict rank k for particles as well as for holes. (The generalization of $\hat{O}^{(1,2)}_{ij}$ to rank k is evident !). If now we write (4.4) in a hole representation we get

$$H_m = \bar{H}_o + \bar{H}_1 + \bar{H}_2 \quad ,$$

$$H_o = (E_o + N\varepsilon + N\cdot NW) - \bar{n}\left[\varepsilon + (N-1)W\right] + \bar{n}\cdot\bar{n}W$$

$$H_1 = -\sum \bar{n}_i \left\{\left[\eta_i + (N-2)W_i\right] - (\bar{n}-1)W_j\right\} \quad , \quad H_2 = \bar{n}_i \cdot \bar{n}_j W_{ij} \tag{4.8}$$

Each of the coefficients in this expression corresponds to a sum rule dictated by the two body nature of the force e.g.

$$\bar{E}_o = E_o + N\varepsilon + N\cdot NW, \qquad \bar{\varepsilon} = -[\varepsilon + (N-1)W]\ldots, \qquad \bar{W}_{ij} = W_{ij} \tag{4.9}$$

where $\bar{E}_o$, $\bar{\varepsilon}$, $\bar{W}_{ij}$ correspond to the energy of the closed shell, the hole centroïd and the two hole centroïd respectively. It is fair to say that all these sum rules are experimentally violated. To give an example : for $\bar{E}_o$ and $\bar{\varepsilon}$ to correspond to the observed values in ^{16}O and ^{15}O, the value of ε in A=5 should practically vanish. This implies no splitting between the $p_{3/2}$ and $p_{1/2}$ orbits. It is remarkable that this degeneracy emerges naturally from the Cohen and Kurath fits restricted to rank 2 forces [CK65]. Experimentally the separation $\varepsilon p_{3/2} - \varepsilon p_{1/2}$ exceeds 4 MeV.

To be able to obtain a parametrization that reproduces the observed values we write schematically

$$H_m = \bar{H}_o\bar{\rho}_o + \bar{H}_1\bar{\rho}_1 + \bar{H}_2\bar{\rho}_2 + H_o\rho_o + H_1\rho_1 + H_2\rho_2 \tag{4.10}$$

Here $\bar{\rho}_o$ for example is an operator that vanishes for n=0,1,2 and gives unity for $\bar{n}$= 0,1,2.

$$\bar{\rho}_o = \frac{10n^{(3)}}{N^{(3)}} - \frac{15n^{(4)}}{N^{(4)}} + \frac{6n^{(5)}}{N^{(5)}} \tag{4.11}$$

Actually, such an operator should affect the $\bar{E}_o$ term only. The term in $\bar{\varepsilon}$ vanishes automatically for $\bar{n}$= 0 and a simpler form is sufficient (and necessary to comply with the minimal rank requirement B above). Although the details are not uninteresting, they will be found elsewhere ([CZ84],[CO83]) and we only give the final form of the parametrization.

$$H_m = H_{(o)} + H_{(1)} + H_{(2)}$$

$$H_{(o)} = \frac{\bar{E}_o}{2} + \frac{\bar{E}_o}{2N}(n-\bar{n}) + \frac{1}{2(N=1)}(\varepsilon + \bar{\varepsilon})n\bar{n} + \frac{1}{2N^{(3)}}\left[2E_o - N(\varepsilon-\bar{\varepsilon})\right](n-\bar{n})n\bar{n}$$

$$+ \frac{1}{4(N-1)^{(3)}}\left[(N-1)(W+\bar{W})+2(\varepsilon+\bar{\varepsilon})\right]n^{(2)} + \frac{1}{4N^{(5)}}\left\{6\left[2E_o - N(\varepsilon-\bar{\varepsilon})\right] - N^{(2)}(W-\bar{W})\right\}(n-\bar{n})n^{(2)}n^{(2)}$$

$$H_{(1)} = \sum_i n_i\Big\{\eta_i + (n-1)W_i - \frac{(n-1)^{(2)}}{(N-2)^{(2)}}\left[3(\eta_i+\bar{\eta}_i)+(N-2)(2W_i+\bar{W}_i)\right] +$$

$$+ \frac{(n-1)^{(3)}}{(N-2)^{(3)}}\left[2(\eta_i+\bar{\eta}_i)+(N-2)(W_i+\bar{W}_i)\right]\Big\}$$

$$H_{(2)} = \sum_{i\leq j} n_i\cdot n_j\left[W_{ij} - \frac{n-2}{N-4}(W_{ij} - \bar{W}_{ij})\right] \tag{4.12}$$

If necessary H_m can be supplemented by

$$H_{(3)} = \sum_i \hat{O}_{i_1}^{(1)}(n-1)^{(2)}(\bar{n}-1)^{(2)}A_i + \sum_{i<j}\hat{O}_{ij}^{(2)}(n-2)(\bar{n}-2)\left[B_{ij}+(n-\bar{n})C_{ij}\right] + \sum_{i<j}\hat{O}_{ij}^{(3)}D_{ij} \tag{4.13}$$

which is an operator of at least rank 3 in p and h, that does affect the parameters defined in a) and b) above.

In eq. (4.12) we have used a mixed representation for $H_{(o)}$, which has the advantage of giving an explicit sum rule for each coefficient, e.g. at least rank 3 is necessary if $2E_o - N(\varepsilon-\bar{\varepsilon}) \neq 0$, etc. The uncertainties associated with the values of $\bar{\varepsilon}$, W and $\bar{W}$ are such that a rank 5 force is not imperative, but the data demand rank 4 at least.

The statement that once H_m is corrected, the KB and other realistic matrix elements produce quantitatively precise spectroscopy is borne out by the calculations. As H_m embodies the secular variation of the selfconsistent orbits it should be expected that the matrix elements in H_M also change. A clean prescription we adopt to account for this change, is to fix the size of the orbits at both ends of the model spaces using oscillator parameters $\hbar\omega$ that reproduce the observed radii. Then we interpolate linearly between these values. In the p shell this amounts to working with fixed sizes : ^{6}Li and ^{16}O have basically the same radii.

(4.III) Spectra and binding energies in the p and sd shells.

For the p shell we use bare G-matrix elements calculated using the Lee potential [KLS69]. In fig. 22 we show the results for ^{12}C. The cases labeled EMX correspond to different variants of H_m. When T appears in X it means that the isospin centroïds were allowed to change as well, producing variations that are not particularly significant for this nucleus. In EMT5 a rank 5 force is included while EMT4 does not have rank 5 but incorporates the rank 4 operators in (4.13). The main message is very simple : our calculations are as good as the Cohen and Kurath fit [CK65], column CK in the figure. This holds throughout the region, except for A= 5,6,7 where the CK calculations fail because of the single particle splitting (see discussion after eq. (4.9)), while our results are in good agreement with experiment. In the sd shell we use the standard KB matrix elements.

In fig. 23 we show the calculations for A= 22, which are by far the less satisfactory in the sd shell. The column Kuo+Mc corresponds to an earlier attempt with monopole corrections treated before the minimal parametrization (4.12) was found. In Kuo+H_m the parametrization is introduced but no attempt was made at optimizing the fit by playing with the uncertainties in the values of V_{ij} and $\bar{V}_{ij}$. The columns C.W (Chang-Wildenthal) and USD (Universal sd interaction)* represent the results with the best multiparametric fits available. In our calculations this nucleus was not included in the determination of the centroïds. Moreover, as there is still room for improvement in the treatment of H_m, we can conclude that the monopole correction

---

*We thank Dr. Wildenthal for making available his matrix elements before publication.

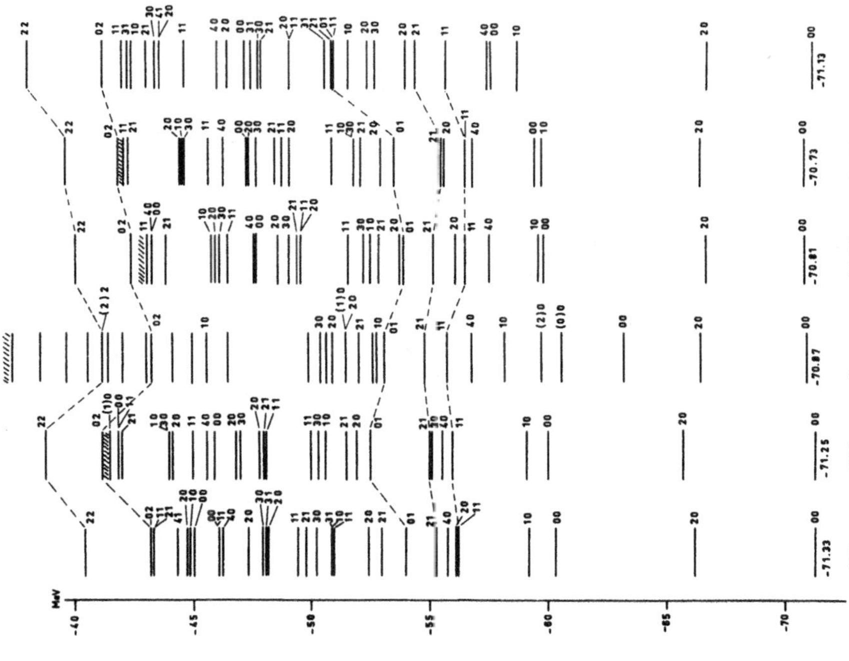
c12
EM
EMT
Exp
EMT5
EMT4
C.K.
MeV

Fig. 22

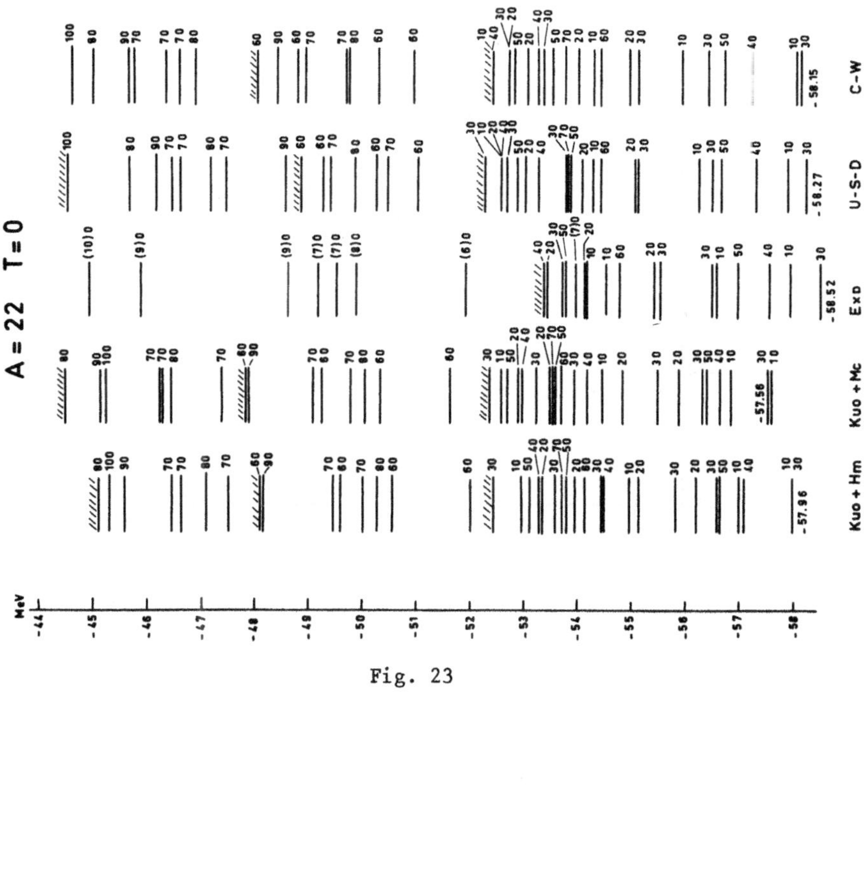
A = 22 T = 0
Kuo + Hm
Kuo + Mc
Exp
U-S-D
C-W
MeV

Fig. 23

is practically sufficient to make " realistic " matrix elements competitive with the numerical fits.

We now turn to the binding energies. In fig. 24 we plot the energies of the lowest states for each isospin T, with respect to ^{4}He, for p shell nuclei (A=4+M, M=0,1,11,12 are fitted and give perfect agreement by definition). The agreement between experimental and calculated values is seen to be extremely good.

By far the worse discrepancy occurs for the GS of ^{8}Be (~1.5 MeV), which is reassuming : this nucleus is unbounded and made of two well separated α particles. The smallness of the system makes a simple shell model description viable (see discussion in iii) following fig. 4 in Lecture 1) and if there is a surprise it is that such a description works at all. It should be noticed that the T=1 and 2 states in mass 8 are perfectly given by the calculations. Incidentally : for the sd shell the agreement between calculated and observed binding energies is even better : the GS of ^{22}Na shown in fig. 23 gives the worse discrepancy.

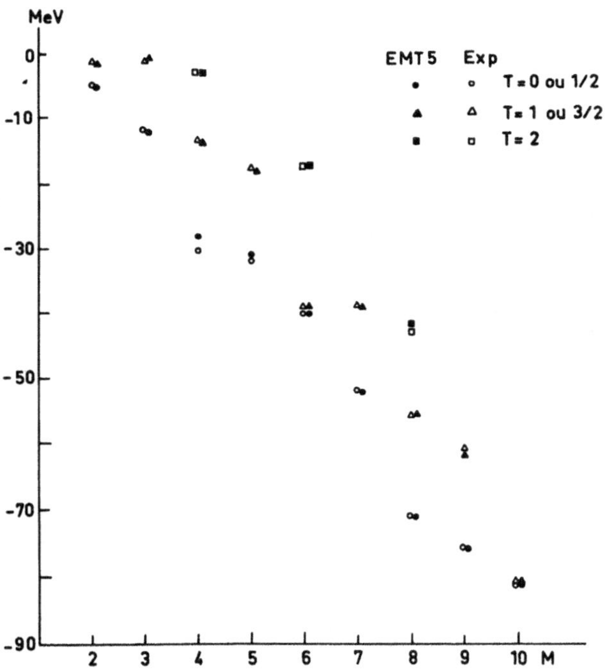

Fig. 24

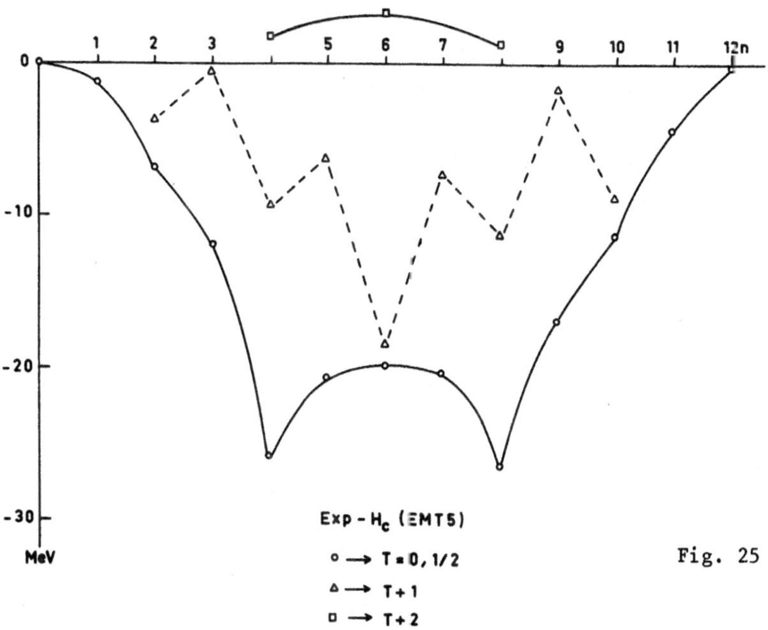
1
2
3
5
6
7
9
10
11
12n
0
-10
-20
-30
MeV
Exp - H_c (EMT5)
○ → T = 0, 1/2
△ → T + 1
□ → T + 2

Fig. 25

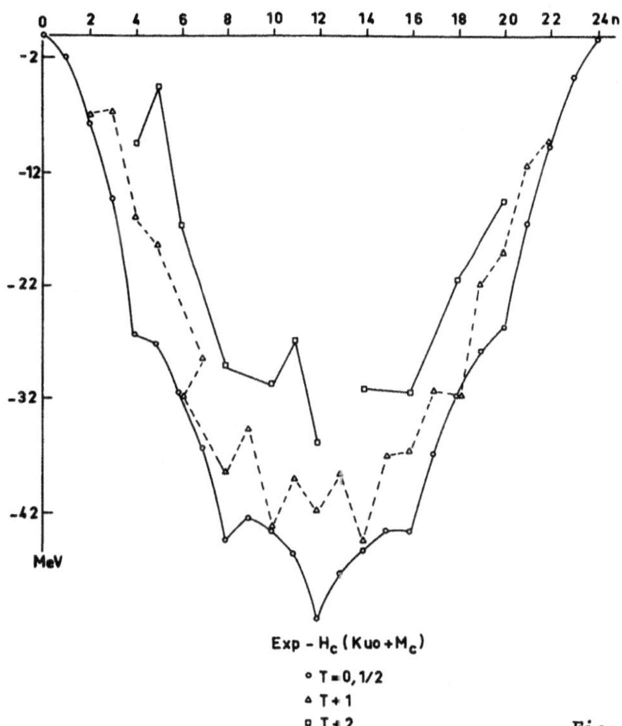
0
2
4
6
8
10
12
14
16
18
20
22
24n
-2
-12
-22
-32
-42
MeV
Exp - H_c (Kuo + M_c)
○ T = 0, 1/2
△ T + 1
□ T + 2

Fig. 26

The most interesting feature of fig. 24 however, is its apparent lack of clear systematic trend. If instead of referring the energies to the GS of ^{4}He (a perfectly arbitrary procedure as emphasized in the last paragraph of (3.IV)), we refer them to $H_{(o)}$ we obtain the striking regularity of fig. 25 (H_c in the figure is $H_{(o)}$ in eq. (4.12)). For the sd shell the corresponding result is shown in fig. 26. Notice in particular the possibility of " predicting " to within some acceptable accuracy the missing point for T= 5/2 in A=29.

(4.IV) Brief remarks on density dependence.

Let us examine the term $H_{(o)}$ in eq. (4.12). Direct indication of the need of at least 3-body forces is furnished by the coefficient $2\bar{E}_o - N(\varepsilon-\bar{\varepsilon})$. If we use the experimental energies to fix its value and disregard the need of higher rank terms we obtain the following results for the binding energies of ^{16}O (with respect to ^{4}He) and ^{40}Ca (with respect to ^{16}O)

$$E(^{16}O)= -140 \text{ MeV}= -155 \ (1+2 \text{ body})+ 15 \ (3 \text{ body})$$
$$E(^{40}Ca)= - 240 \text{ MeV}= -300 \ (1+2 \text{ body})+ 60 \ (3 \text{ body}) \quad (4.14)$$

These numbers represent purely nuclear contributions : the Coulomb energies have been substracted.

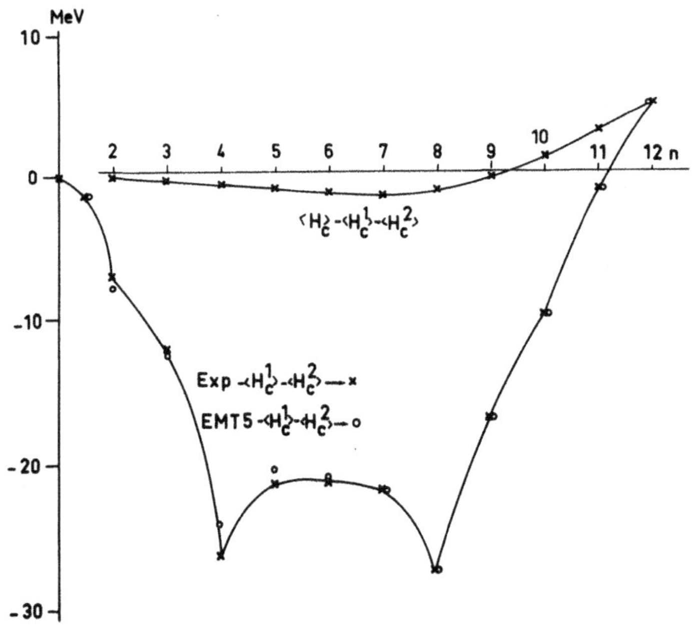

Fig. 27

By the time H_m has been constructed in full, the effects of ranks higher than 2 show a much smoother behaviour as can be seen in fig. 27 which is similar to fig. 25 except that the reference line is no longer $H_{(o)}$ but the rank 1+2 contributions to it. The sum for ranks 3+4+5 is plotted separately and we can see that by the time we reach ^{16}O it adds up to a mere 5 MeV after being slightly negative up to A=13. In other words : the contributions of different ranks are individually large but they cancel each other and the overall effect is a very smooth modulation of the base line. If we try to understand this behaviour in terms of perturbation theory we are in trouble since the expansion does not show any indication that it is converging. Perhaps the best way to understand what is happening comes from the USD form chosen by Wildenthal

$$W_{rstu}(n)= W_{rstu} \left(\frac{18}{18+n}\right)^{0.3} \qquad n= 0,22 \tag{4.15}$$

which is not far from what eq. (3.62) suggests. Here it is useful to remember the remarks made after this equation derived, for an infinite system. For one thing ρ may have to be written as something subtler than N/Ω (or A/Ω). For another, the matrix elements of W itself have an A dependence, which is not simply A^{-1} as in nuclear matter. Be as it may, the modulation in (4.15) is quite plausible, but in spite of it smoothness it is not an analytic function of n. However, the fact that it cannot be expanded in series does not mean it can't be expanded in powers of n : it is a discrete function defined at 23 points and the most general polynomial of rank 23 will reproduce it perfectly. In practice one needs rank 6. However, the USD centroïds for particles and holes can be incorporated in a minimal parametrization of type (4.12) and it is very unlikely that we shall be able to distinguish spectroscopically between different forms of the density, or n-dependence.

Probably the most interesting clues emerging from these calculations are related to what we call non nucleonic effects. The data seem to be demanding a modulation of the two body force that may be parametrized in terms of higher rank forces but there is no indication that genuine (effective or true !) three and higher body forces are necessary. Exact calculations fail to give good saturation with two body N-N potentials, but the spectroscopic evidence we have shown, indicates that the medium range behaviour of these potentials is very adequate, provided we enforce good saturation. It should be added here, that forces not related to the N-N phase shifts may give good spectroscopy in special cases, but they can compete neither in detailed nor overall behaviour with the realistic potentials.

From these remarks it is very tempting to conclude that what is probably necessary and sufficient to ask of the non nucleonic degrees of freedom is not many body forces but a density dependent mechanism that very gently modulates the two body force to produce correct saturation and a good monopole field.

REFERENCES

[BA75] B. Barrett ed. Lecture Notes in Physics 40 (1975), Springer Verlag.

[BBP63] H. Bethe, B. Brandow and A. Petschek, Phys. Rev. 129 (1963) 225.

[BR67] B. Brandow, Rev. Mod. Phys. 39 (1967) 771.

[BR83] G.E. Brown and M. Rho, Physics Today (Feb. 1983) 24.

[CK65] S. Cohen and D. Kurath, Nucl. Phys. 73 (1965) 1.

[CO83] A. Cortès, Ph.D Thesis, CRN Strasbourg 1983.

[CZ79] A. Cortès and A.P. Zuker, Phys. Lett. 84B (1979) 25.

[CZ84] A. Cortès and A.P. Zuker, to be published

[DA66] J. Da Providência, Phys. Lett. 21 (1966) 668.

[EM81] K. Emrich, Nucl. Phys. A351 (1981) 379.

[EZ79] J.G. Zabolitzky and W. Ey, Nucl. Phys. A328 (1979) 507.

[EZ81] K. Emrich and J.G. Zabolitzky, Nucl. Phys. A351 (1981) 439.

[FR67] J.B. French in Nuclear Structure, A. Hossain ed., North Holland 1967.

[GI63] J.N. Ginocchio and J.B. French, Phys. Lett. 7 (1963) 37.

[IA79] F. Iachello in Lecture Notes in Physics 119 (1979) Bertsh & Kurath eds., Springer Verlag.

[KB66] T.T.S. Kuo and G.E. Brown, Nucl. Phys. 85 (1966) 40.

[KB68] T.T.S. Kuo and G.E. Brown, Nucl. Phys. A114 (1968) 241.

[KL69] S. Kahana, H.C. Lee and K. Scott, Phys. Rev. 180 (1969) 956.

[KLZ78] H. Kümmel, K.H. Lührmann and J.G. Zabolitzky, Phys. Rep. 36C (1978) 1.

[KU82] W. Kutzelnigg, J. Chem. Phys. 77 (1982) 3081.

[LI78] J. Lindgren, Int. J. Quantum Chem. Symposium 12 (1978) 33.

[MC64] J.D. McCullen, B.F. Bayman and L. Zamick, Phys. Rev. 134 (1964) 515.

[NE70] J. Negele, Phys. Rev. C1 (1970) 1260.

[OF76] R. Offermann, W. Ey and H. Kümmel, Nucl. Phys. A273 (1976) 349.

[OR72] G. Oberlechner and J. Richert, Nucl. Phys. A191 (1972) 577.

[PA76] E. Pasquini, Ph.D thesis, CRN Strasbourg 1976.

[PZ77] E. Pasquini and A.P. Zuker, Florence Conference 1977. Ed. P. Blasi and R. Ricci (Editrice Compositrice Bologna)

[PZ81a] A. Poves and A.P. Zuker, Phys. Rep. 71 (1981) 141.

[PZ81b] A. Poves and A.P. Zuker, Phys. Rep. 70 (1981) 235.

[VSW73] J.P. Vary, P.U. Sauer and C.W. Wong, Phys. Rev. C7 (1973) 1776.

[ZU75] A.P. Zuker, Czech. J. Phys. B25 (1975) 311.

[WI83] B.H. Wildenthal, private communication.

SOME PROGRESS TOWARDS " UNIVERSAL " EFFECTIVE INTERACTIONS

J.M.G. GÓMEZ

Departamento de Fisica Atómica y Nuclear

Facultad de Ciencias, Universidad de Salamanca

Salamanca, Spain

1. Introduction

The approximation methods introduced to treat the nuclear many-body problem usually imply that the appropriate nuclear force is an effective interaction, different from the free nucleon-nucleon interaction. An effective interaction is thus intimately related to a given nuclear model and its scope is generally confined to the description of a limited number of nuclei or nuclear states. However, in recent years there has been some progress towards " universal " effective nucleon-nucleon interactions, in the sense that they may be reasonably suitable to describe bulk properties of nuclear ground states throughout the periodic table and also properties of excited states.

In the shell model, the effective interaction is a natural consequence of the severe truncation of configuration space required in practical calculations. In the usual approach, an inert closed-shell core is assumed and the active valence nucleons are restricted to very few orbits, for example a major shell. Starting from a realistic free nucleon-nucleon interaction, the effective two-body matrix elements appropriate for the truncated space can be derived by the reaction matrix method used by Kuo and Brown[1)]. A less fundamental, but practical approach when the number of two-body matrix elements is not too large, is to treat them as free parameters which are varied to give best fit to a selected set of nuclear data. A typical example is the successful calculations of Cohen and Kurath[2)] in the p shell. Direct transfer reactions are also used to determine selected two-body matrix elements in some nuclei near closed shells. The most favourable case is when certain nuclear states can be identified with a pure two-particle configuration outside closed shells, i.e. states of the type $|j_1 j_2 J\rangle$, but most often the two-particle configuration is fragmented among a multiplet of nuclear levels, and then the knowledge of spectroscopic factors is required for extracting the two-body matrix elements of the residual interaction. Obviously the same treatment can be utilized to extract two-hole or particle-hole matrix elements from appropriate nuclei. A review of this method has been given by Schiffer and True[3)] and more recently by Daehnick[4)]. In section 2 we briefly comment some outstanding systematics of these empirical matrix elements troughout the periodic table.

Finally let us consider the widely used approach of phenomenological nuclear potentials. In order to restrict the parameters of the force to a reasonable number, simple shapes are usually assumed for the radial dependence of central and noncentral terms. Typical examples are a Yukawa shape, or a combination of two Yukawa shapes of different ranges, or two Gaussian shapes. Although noncentral terms play an important role in some nuclear properties, surprinsingly good results are often obtained using central forces only, in combination with a simple one-body or two-body spin-orbit term. A particularly simple and successful central force is the surface delta interaction, on the modified surface delta interaction[5,6].

A quatitatively different kind of phenomenological forces are the density-dependent interactions used in Hartree-Fock calculations. The work of Negele and Vautherin[7] has shown that starting from a realistic free nucleon-nucleon interaction, an effective interaction containing a density dependence can be derived by application of the local-density approximation and the density-matrix expansion. Because the exchange term greately complicates the numerical selfconsistent calculations, zero-range density dependent forces have been proposed and successful results have been obtained for many bulk properties of spherical nuclei. Well-known forces of this kind are for example the Skyrme-type interactions[8,9]. These schematic forces have proved to be very useful in Hartree-Fock calculations, but they obviously have limitations. For example, the usual (linear) dependence on the density, leads to a too small surface diffuseness in the charge densities when it is requested that the calculated mean single particle spacing be the observed ones. In the spirit of universality, the best choice would then be a force like SIII[9], representing a compromise between best level density and best surface diffuseness.

In later work, many modifications and extensions of the original Skyrme forces have been proposed in order to improve the description of different nuclear properties, such as excitation energies of the isoscalar giant monopole resonances[10,11] and giant magnetic states[12], Coulomb displacement energies[13], stability of the Hartree-Fock minimum in odd-mass and odd-odd nuclei[14,15] etc.

Among zero-range forces of the Skyrme type, an interaction of the " universal " kind has been proposed by Warroquier et al[16], suitable for both ground states and excited states of nuclei throughout the periodic table. This interaction includes a combination, $(1-X_3)V_o + X_3W_o$, of a two-body density-dependent term V_o and the original Skyrme three-body term W_o. These two terms give identical contributions to the Hartree-Fock energy of closed shell nuclei, but their antisymmetric two-body matrix elements are different. Thus after all the other parameters of the force had been determined from bulk properties of closed shell nuclei, where the value of X_3 is irrelevent, this parameter was adjusted to reproduce the low-lying levels of nuclei with two identical nucleons outside closed shells. The spectra of ^{18}O, ^{42}Ca, ^{50}Ca, ^{58}Ni and ^{134}Te were calculated and it was found that a good fit of the experimental levels can be obtained with a constant value of X_3 throughout the mass table, provided that : (i) the two-body matrix elements are renormalized to take into account the 3p1h and 4p2h core excitations not contained in the truncated 2p configuration space, and (ii) empirical single-particle energies are used instead of the self-consistent Hartree-Fock single-particle energies.

It is becoming increasingly clear that a unified description of bulk and spectroscopic properties of nuclei is desirable. In Zuker's lectures this point is abundantly stressed and we want to present some calculations in which this request is fulfilled under seemingly unfavourable conditions.

From the point of view of Hartree-Fock calculations the light nuclei in the 1p shell are usually considered " exotic " and out of reach. However, in section 3 we shall describe calculations in this region that given evidence that the Gogny D1 force[17)] and a variant of it, the PKN force[18)] which include the necessary density dependence for successful Hartree-Fock calculations, can also be used as effective interactions giving fair results in open shell nuclei.

2. Systematics of empirical two-body matrix elements.

Two-body matrix elements of the residual interaction in the shell model can be deduced from the experimental energy levels of selected nuclei when the spectroscopic factors in one-nucleon transfer reactions are known. Let us assume the usual decomposition of H as the sum of a one-body term H_o and the two-body residual interaction V,

$$H = H_o + \sum_{i<j} V(\vec{r}_i, \vec{r}_j), \tag{1}$$

and denote by ψ_k the eigenstates of H and by $\phi_{o\alpha}$ the eigenstates of H_o. The matrix elements of H and V in the basis $\phi_{o\alpha}$ are related by

$$H_{\alpha\beta} = E_{o\alpha}\, \delta_{\alpha\beta} + V_{\alpha\beta}. \tag{2}$$

The basis functions $\phi_{o\alpha}$ can be expanded in terms of a complete set of eigenstates ψ_k in the first member of this equation, leading to

$$V_{\alpha\beta} = \sum_k \langle\phi_{o\alpha}|\psi_k\rangle E_k \langle\psi_k|\phi_{o\beta}\rangle - E_{o\alpha}\delta_{\alpha\beta}. \tag{3}$$

Thus we see that to extract matrix elements from experimental levels one needs to know the overlap integrals. A particular case where they can be determined from experimental data is a one-nucleon transfer reaction on a target well described as a single particle (or hole) on a closed shell core. The experimental spectroscopic factor gives the square of the overlap integral and from eq. (3) one obtains the two-body diagonal matrix elements,

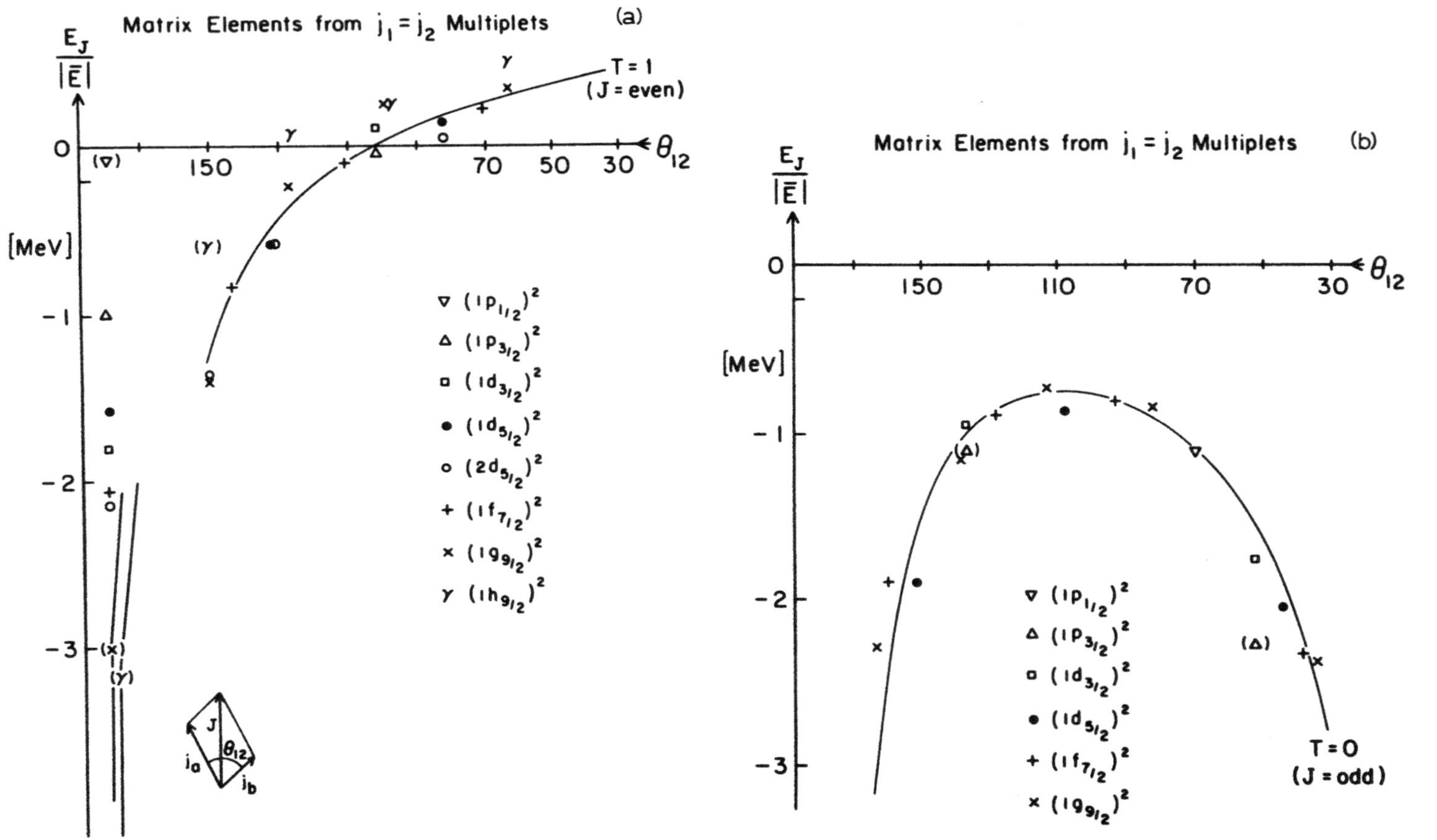

Fig. 1. Empirical two-body matrix elements E_J divided by an average energy scaling factor $|\bar{E}|$.

$$\langle j_1 j_2 J|V|j_1 j_2 J\rangle = \sum_k S_k(j_1 j_2 J) E_k(j_1 j_2 J) - E_o(j_1 j_2) \ . \qquad (4)$$

Here j_1 and j_2 are the angular momenta of the single particle in the target and the transfered nucleon, respectively, and k runs over the populated levels. The unperturbed centroïd energy E_o of the $j_1 j_2$ multiplet can be obtained from the experimental binding energies and single-particle energies.

Using this method, a large number of empirical matrix elements have been determined and interesting regularities have been observed[3,4]. The monopole strength of the residual interaction is given by

$$\bar{E}(j_1 j_2 T) = \frac{\sum_J (2J+1) V(j_1 j_2 JT)}{\sum_J (2J+1)} \ , \qquad (5)$$

where $V(j_1 j_2 JT)$ are the two-body diagonal matrix elements and for equivalent particles, $j_1 = j_2$, the sum runs only over the allowed values of J. The empirical $\bar{E}(T)$ values have a remarkably smooth dependence on the mass number A, and a good fit is obtained[4] with

$$\bar{E}(T=0) = -31.5 \quad A^{-0.75} \ , \qquad (6)$$

$$\bar{E}(T=1) = -2.25 \quad A^{-0.75} \ . \qquad (7)$$

It should be noted that this behaviour applies to the multiplets with $j_1 = j_2$ as well as those with $j_1 \neq j_2$, although this point is better established for T=0 than for T=1, due to the larger relative error bars of the latter.

A very interesting feature of the matrix elements is their similarity when they are divided by a suitable linear conbination $\bar{E}$ of the monopole energies of eq. (5), which acts as scaling factor, and are plotted against the classical angle θ_{12} between the orientation of j_1 and j_2 coupled to J, in the vector coupling model of angular momenta[3]. This angle is a measure of the overlap of the orbital wave functions for the two nucleons.

Fig. 1 (from ref.[4]) illustrates the striking similarity of the matrix elements from $j_1 = j_2$ multiplets, over a large region of the periodic table.

These examples and other regularities of the empirical matrix elements suggest a " universal " character of the nuclear residual interaction.

3. Spectra of 1p shell nuclei with a density dependent interaction.

Phenomenological interactions of the type used in Hartree-Fock calculations are usually expected to give a reasonably good description of nuclear bulk properties throughout the periodic table rather than details of the structure of nuclei in a particular region. On the other hand, the opposite behaviour is expected of effective interactions utilized in shell model calculations.

A particularly successful interaction in Hartree-Fock calculations is the finite-range force D1 of Gogny[17)], both for ground states and for particle-hole states[19)] in closed-shell nuclei. To test the validity of this interaction in open-shell configurations, we have performed shell model calculations for several light nuclei in the full configuration space of the 1p shell. We also present calculations with the PKN foce[18)] which has the same analytical structure but different parameters.

The form of the potential is given by

$$V = \sum_{k=1,2} \sum_{i<j} \exp\left(-r_{ij}^2/\mu_k^2\right)\left(W_k + B_k P_{ij}^{\sigma} - H_k P_{ij}^{\tau} - M_k P_{ij}^{\sigma}\right)$$

$$+ \xi \sum_i \vec{\ell}_i \cdot \vec{s}_i + t_o \sum_{i<j} \rho^{\gamma}\left(\frac{\vec{r}_i + \vec{r}_j}{2}\right) \delta\left(\vec{r}_{ij}\right)\left(1 + x_o P_{ij}^{\sigma}\right) . \qquad (8)$$

For simplicity in the calculations, the original two-body zero-range spin-orbit term of the D1 force has been replaced here by a one-body term, but we expect that the results will not be too sensitive to this choice for the p shell nuclei. The two-body potential is central, consisting of a superposition of two Gaussians and a zero-range density-dependent term.

Table 1.

Parameters of the interactions PKN and D1. Range μ in fm and strengths in MeV (t_o in MeV. fm^4).

	μ	W	B	H	M	γ	t_o	x_o	ξ
PKN	0.7 1.2	-232.3 -45.6	115.3 -40.1	144.7 -51.9	-193.7 -44.5	$\frac{1}{3}$	1350	0.1	4
D1	0.7 1.2	-402.4 -21.3	-100 -11.77	-496.2 37.27	-23.56 -68.81	$\frac{1}{3}$	1350	1	4

The parameters of the interactions are given in table 1. The force PKN was derived from D1 with the purpose of improving the description of ground states in the p shell, having in mind applications based on sum rule techniques. However, some restrictions were imposed in order to preserve properties of the force in N=Z closed shell nuclei. The Hartree-Fock energy contribution of the Gaussian part in these nuclei depends on two combinations of the parameters only, M+W and 5M- 2(B-H). We have chosen for our force the values of D1 for these combinations, and also for γ, t_o and μ_k. As a further constraint on our effective interaction we required that the position and strength of the electric giant dipole resonance in ^{16}O be correct. The remaining freedom in the parameters was utilized to obtain ground state energies and wave functions as close as possible to those given by Cohen and Kurath[20] in their (8-16) 2BME fit to the effective interaction in the p shell. Furthermore we imposed the correct single particle splitting in A=15. For A=5 the observed result was obtained, without further parameters, a result that can be obtained only through density dependence. For the radial part, harmonic oscillator wave functions were used, with realistic values of the size parameter deduced for each nucleus from the experimental charge radius.

Table 2

Matrix elements $\langle j_1 j_2 JT|V|j_3 j_4 JT\rangle$ for the 1p shell (MeV)

$2j_1 2j_2$, $2j_3 2j_4$	JT	Cohen-Kurath	PKN (^{16}O)	D1 (^{16}O)
3 3 , 3 3	01	-3.19	-4.79	-3.76
	21	-0.17	-1.11	-1.05
	10	-3.58	-3.78	-4.07
	30	-7.23	-5.80	-5.61
3 3 , 3 1	21	-1.92	-1.66	-1.49
	10	3.55	3.45	4.35
3 3 , 1 1	01	-4.86	-3.43	-2.66
	10	1.56	2.21	2.66
3 1 , 3 1	11	0.92	0.06	0.00
	21	-0.96	-2.28	-2.11
	10	-6.22	-6.27	-7.13
	20	-4.00	-5.32	-5.62
3 1 , 1 1	10	1.69	0.62	0.62
1 1 , 1 1	01	-0.26	-2.36	-1.88
	10	-4.15	-4.32	-4.47

In table 2 the 15 two-body matrix elements of the p shell for the D1 and PKN interactions in ^{16}O are given, and the (8-16) 2BME set of Cohen and Kurath[20] is included for comparison. The similarity of the three sets is quite evident, and perhaps surprising. We recall that in the well-known calculations of Cohen and Kurath the two-body matrix elements and the single-particle energies are treated as completely free parameters with no a priori conditions, and they are determined from a fit to selected low-lying levels of p shell nuclei. On the other hand the force D1 is an effective density-dependent interaction determined from static properties of nuclei throughout the periodic table ($A \geqslant 16$), and adjusted to give an acceptable description of pairing effects.

It should be noted that while the (8-16) 2BME matrix elements have the same values for all the p shell nuclei, the D1 and PKN matrix elements are different for each nucleus, due to the density dependence of the interaction and the realistic sizes of radial wave functions. The matrix elements are largest in ^{10}B, with differences up to $\sim$25 % from the values in ^{16}O.

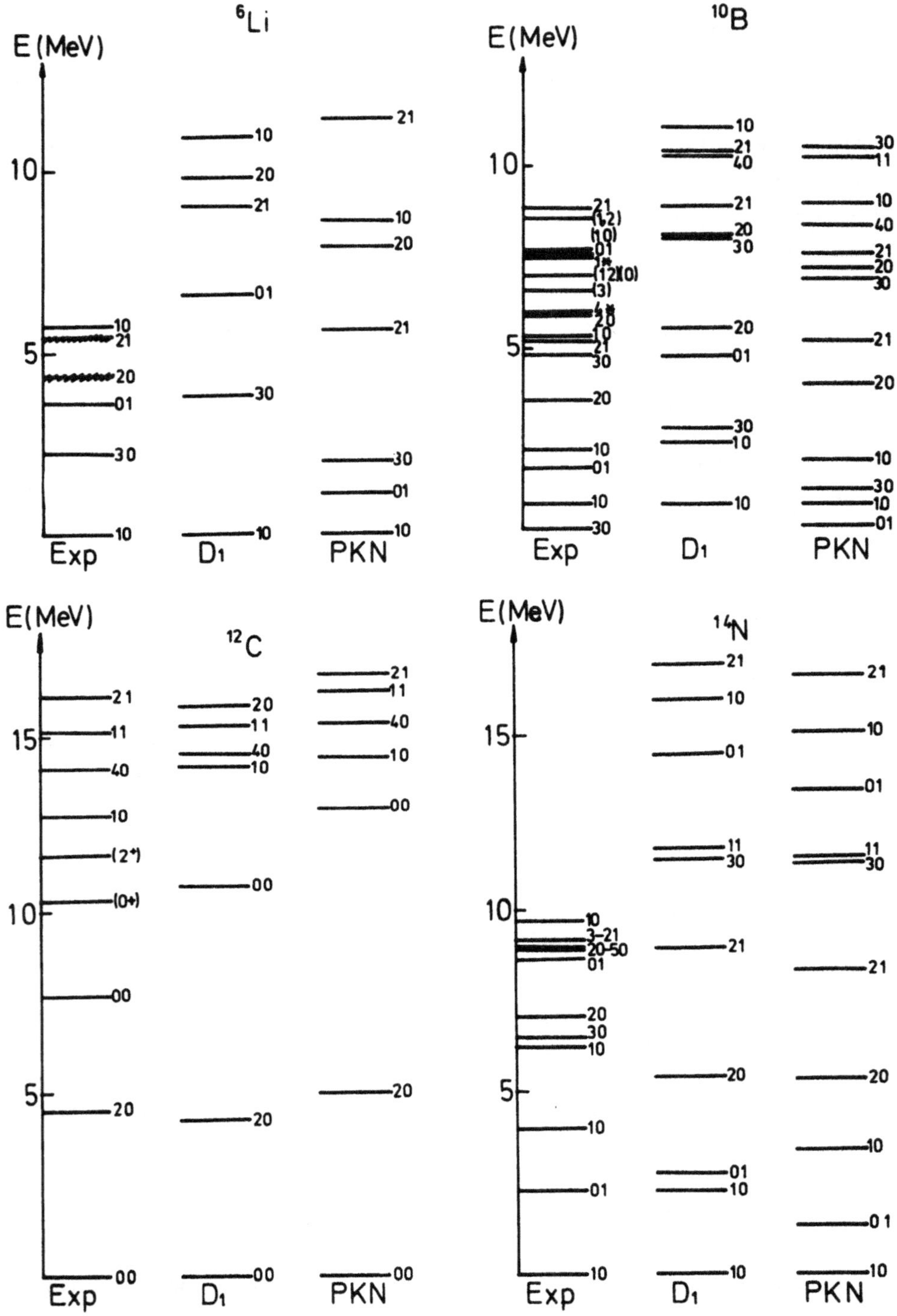

Fig. 2. Level schemes computed for forces D1 and PKN compared to experiment.

Table 3

Ground-state shell-model wave functions generated by the Cohen-Kurath, D1 and PKN forces.
The components are labeled by the number of nucleons, angular momentum, isospin and seniority in $p_{3/2}(nJTv)$ and $p_{1/2}(n'J'T')$. Ω is the overlap with the Cohen-Kurath wave function.

Nucleus	n	J	T	v	,	n'	J'	T'	CK	D1	PKN
$^{10}B(3^+0)$	2	3	0	2	,	4	0	0	-0.1144	-0.0432	-0.0636
	3	$\frac{5}{2}$	$\frac{1}{2}$	3	,	3	$\frac{1}{2}$	$\frac{1}{2}$	0.0616	0.1107	0.0608
	3	$\frac{7}{2}$	$\frac{1}{2}$	3	,	3	$\frac{1}{2}$	$\frac{1}{2}$	-0.1145	-0.0892	-0.0788
	4	2	0	2	,	2	1	0	-0.0066	0.0774	-0.0072
	4	2	0	4	,	2	1	0	0.0992	0.1297	0.1111
	4	3	1	2	,	2	0	1	-0.4168	-0.3138	-0.3564
	4	4	0	4	,	2	1	0	-0.1231	-0.1243	-0.1142
	5	$\frac{5}{2}$	$\frac{1}{2}$	3	,	1	$\frac{1}{2}$	$\frac{1}{2}$	0.2586	0.4426	0.2890
	5	$\frac{7}{2}$	$\frac{1}{2}$	3	,	1	$\frac{1}{2}$	$\frac{1}{2}$	-0.3533	-0.3577	-0.3264
	6	3	0	2	,	0	0	0	0.7613	0.7192	0.8022
										Ω=0.9688	Ω=0.9945
$^{12}C(0^+0)$	4	0	0	0	,	4	0	0	-0.3131	-0.2594	-0.2603
	5	$\frac{1}{2}$	$\frac{1}{2}$	3	,	3	$\frac{1}{2}$	$\frac{1}{2}$	-0.2686	-0.3638	-0.2631
	6	0	1	0	,	2	0	1	0.6122	0.6095	0.6182
	6	1	0	2	,	2	1	0	-0.2360	-0.3469	-0.2631
	8	0	0	0	,	0	0	0	-0.6319	-0.5554	-0.6416
										Ω=0.9849	Ω=0.9981
$^{14}N(1^+0)$	6	1	0	2	,	4	0	0	0.0270	0.2402	0.1683
	7	$\frac{3}{2}$	$\frac{1}{2}$	1	,	3	$\frac{1}{2}$	$\frac{1}{2}$	0.3127	-0.0815	0.0404
	8	0	0	0	,	2	1	0	0.9495	0.9673	0.9849
										Ω=0.8994	Ω=0.9523

In table 3 the ground state wave functions of ^{10}B, ^{12}C and ^{14}N obtained with these interactions are given. Clearly they are very similar, particuarly in ^{10}B and ^{12}C. In fig. 2 the computed level schemes are compared to experiment. A reasonable agreement is observed.

We conclude that a finite-range density-dependent effective interaction of the Gogny's type is capable of describing a large number of static and dynamical nuclear properties throughout the periodic table, including open-shell nuclei. Hopefully it may provide clues for the definition of some " universal " effective force.

References.

1. T.T.S. Kuo and G.E. Brown, Nucl. Phys. 85 (1966) 40 ; Nucl. Phys. A114 (1968) 241.
2. S. Cohen and D. Kurath, Nucl. Phys. 73 (1965) 1.
3. J.P. Schiffer and W.W. True, Rev. Mod. Phys. 48 (1976) 191.
4. W.W. Daehnick, Phys. Rep. 96 (1983) 317.
5. P.W.M. Glaudemans, P.J. Brussaard and B.H. Wildenthal, Nucl. Phys. A102 (1967) 593.
6. P.J. Brussaard and P.W.M. Glaudemans, Shell-Model Applications in Nuclear Spectroscopy (North-Holland, Amsterdam, 1977)
7. J.W. Negele and D. Vautherin, Phys. Rev. C5 (1972) 1472 ; Phys. Rev. C11 (1975) 1031.
8. D. Vautherin and D.M. Brink, Phys. Rev. C5 (1972) 626.
9. M. Beiner, H. Flocard, Nguyen Van Giai and P. Quentin, Nucl. Phys. A238 (1975) 29.
10. H.S. Köhler, Nucl. Phys. A258 (1976) 301.
11. H. Krivine, J. Treiner and O. Bohigas, Nucl. Phys. A336 (1980) 155.
12. Nguyen Van Giai and H. Sagawa, Phys. Lett. 106B (1981) 379.
13. J.M.G. Gómez and J. Martorell, Nucl. Phys. A410 (1983) 475.
14. B.D. Chang, Phys. Lett. 56B (1975) 205.
15. K.H. Passler, Nucl. Phys. A257 (1976) 253.
16. M. Warroquier, K. Heide and G. Wenes, Nucl. Phys. A404 (1983) 269 ; Nucl. Phys. A404 (1983) 298.
17. D. Gogny, in Nuclear Self-Consistent Fields, Eds. G. Ripka and M. Porneuf (North-Holland, Amsterdam, 1975) p. 335.
18. J. Navarro, J. Bernabeu, J.M.G. Gómez and J. Martorell, Nucl. Phys. A375 (1982) 361.
19. J.P. Blaizot and D. Gogny, Nucl. Phys. A284 (1977) 429.

ON SOME PROPERTIES OF THE WIGNER DISTRIBUTION FUNCTION OF SPHERICAL NUCLEI

J. Martorell and E. Moya de Guerra

Dept. Fisica Atomica i Nuclear, Universitat de Palma de Mallorca, Spain

Dept. Fisica Atomica y Nuclear, Universidad de Extremadura, Badajoz, Spain

The study of heavy ion reactions and of collective excitations such as giant resonances, via methods derived from the Hartree Fock approximation, has renewed the interest on the properties of the Wigner transform of the one body density matrix [(1)]. The Wigner transform is particularly useful to clarify the dynamics of such processes [(2-7)]. In addition, extensive use of semiclassical methods to describe both static and dynamical nuclear properties, has also shown the interest of the Wigner transform as a starting point for discussing this kind of approximations [(8-9)].

In this context we want to discuss some simple properties of the ground state Wigner transform of the doubly magic nuclei, obtained using the Density Dependent Hartree Fock approximation with realistic effective interactions. First we want to discuss what are its most salient features as a function of **p**, and to what extent they can be reproduced by simpler models : Thomas Fermi and Harmonic Oscillator. Afterwards, following the line suggested in ref. 10, we will discuss the ability of the so called "Density Matrix Expansions" (DME), as opposed to semiclassical approaches, to reproduce also the most important features of the DDHF Wigner transform. More extensive details of our work will be given in a forthcoming publication [(11)].

Following ref. (8) we define the Wigner transform as :

$$f(\bar{R},\bar{p}) = \int d\bar{s}\, e^{-i\bar{p}.\bar{s}}\, \rho(\bar{R}+\bar{s}/2,\bar{R}-\bar{s}/2) \qquad (1)$$

where

$$\rho(\bar{r}_1,\bar{r}_2) = \sum_{\sigma} \rho(\bar{r}_1\sigma,\bar{r}_2\sigma) \qquad (2)$$

and $\bar{R} = (\bar{r}_1+\bar{r}_2)/2$, $\bar{s} = \bar{r}_1-\bar{r}_2$. For Slater determinants:

$$\rho(\bar{r}_1,\bar{r}_2) = \sum_{i=1}^{A} \sum_{\sigma} \varphi_i^*(\bar{r}_1,\sigma)\, \varphi_i(\bar{r}_2,\sigma) \qquad (3)$$

where the φ_i are the single particle wavefunctions(the index i symbolizes the quantum numbers : $n_i l_i j_i m_i$).

As already discussed by Wigner [(1)], the classical analog of the Wigner transform gives the momentum distribution at point $\bar{R}$. In the quantal case this is no more true, and in fact f is not even positive definite. However it retains some useful properties, i.e. : the mass density is given by

$$\rho(\bar{R}) = \int d\bar{p} \quad f(\bar{R},\bar{p}) \tag{4}$$

and the momentum distribution by

$$n(\bar{p}) = \int d\bar{R} \; f(\bar{R},\bar{p}) \tag{5}$$

in addition, the local current can be written as

$$\bar{j}(\bar{R}) = \int d\bar{p} \quad \bar{p} \quad f(\bar{R},\bar{p}) \tag{6}$$

whereas the kinetic energy density τ is directly related to the integral with the second power of $\bar{p}$

$$\tau = \int d\bar{p} \quad \bar{p}^2 \; f(\bar{R},\bar{p}) \quad + \quad \frac{1}{4} \, \bar{\nabla}^2 \rho(\bar{R}) \tag{7}$$

(we use everywhere units with $\hbar = 1$). This shows that, to get intuitive understanding of the Wigner transform,it will be sometimes useful to resort to the classical analog and to speak of f as a "local momentum distribution". It is also useful to remember that in the Thomas Fermi approximation, where locally the nucleons are treated as a Fermi gas :

$$f_{TF}(\bar{R},\bar{p}) = 4 \, \theta \, (p_F - p) \tag{8}$$

with $\quad p_F = (3\pi^2 \rho(\bar{R})/2)^{1/3}$.

For J=0, spin saturated nuclei, it is easy to show

that ρ depends only on R^2, s^2 and $(\bar{R}.\bar{s})^2$, and similarly that f is a function only of R^2, p^2 and $(\bar{R}.\bar{p})^2$. Clearly in this case $\mathcal{J}$, and any other integrals involving odd powers of $\bar{p}$, vanish. Since the DDHF wavefunctions are known only numerically, the Fourier transform in eq. (1) must be computed also numerically. The authors of ref. (2) performed such calculation in configuration space. Alternatively, we preferred to decompose the φ_i in a harmonic oscillator basis and then take advantage of the fact that for these functions the Fourier transforms are analytic. Our final expression is then:

$$f_{DDHF}(\bar{R},\bar{p}) = 8\sqrt{\pi}\, e^{-(p^2b^2 + R^2b^2)} \sum_{n_i l_i j_i} \sum_{n_1 \geq n_2} \left(1 - \frac{1}{2}\delta_{n_1 n_2}\right)$$

$$\cdot\, a^{n_i}_{n_1 l_i j_i}\, a^{n_i}_{n_2 l_i j_i} \frac{(2j_i+1)}{(2l_i+1)^{1/2}} \sum_{l\ \mathrm{even}} \sum_{N \geq n} \left(1 - \frac{1}{2}\delta_{N,n}\right)$$

$$\cdot\, (-1)^{l_i + n + l/2} \left[\frac{n!\, N!\, (2l+1)}{\Gamma(n+l+3/2)\,\Gamma(N+l+3/2)}\right]^{1/2} \langle nl, Nl, 0 | n_1 l_i, n_2 l_i, 0\rangle$$

$$\cdot\, P_l(\cos\theta)\,(2Rp)^l \left[L_n^{l+1/2}(2p^2b^2)\, L_N^{l+1/2}(2R^2/b^2) + (-1)^{n_1+n_2} L_n^{l+1/2}(2R^2/b^2)\, L_N^{l+1/2}(2p^2b^2)\right] \qquad (9)$$

where θ is the angle between $\bar{R}$ and $\bar{p}$, and the a's are the coefficients of the expansion of the φ_i into the harmonic oscillator basis of oscillator parameter b. This expression generalizes eq.(28) of ref.12, which applies only to harmonic oscillator wavefunctions, to any single particle wavefunctions expanded in an oscillator basis. For the calculation of the DDHF wavefunctions, we used the realistic effective force G0 of Sprung and Banerjee [(13)] derived from the nuclear matter G-matrix corresponding to the Reid potential, and performed DDHF calculations as those described in refs. 14 and 15, where the Local Density Approximation (LDA) is invoked to apply the nuclear matter results to finite nuclei.

It is well known from previous work (see e.g. refs.12 and 16 and references therein) that the Wigner transform as a function of $p=|\vec{p}|$ presents strong oscillations for the harmonic oscillator wavefunctions. It is then interesting, as a consistency check, to find wether these oscillations are still present in the Wigner transforms corresponding to DDHF wavefunctions, because one of the basic assumptions of the LDA is that locally in finite nuclei the momentum distribution ressembles that of nuclear matter, and therefore should be flat and have a sharp Fermi surface. The same assumption is also frequently made in setting up approximations for collective vibrations, see e.g. ref. 17 .

The DDHF results for ^{16}O, ^{40}Ca and ^{208}Pb are shown in fig.1, where they are compared with the Thomas Fermi approximation for f, and with the harmonic oscillator (H.O.).By the latter we mean harmonic oscillator wavefunctions for the levels that are occupied in the DDHF calculations, with a parameter b such that the r.m.s. mass radius is the same as that given by DDHF (analytic expressions for these and other nuclei are given in Appendix 1).

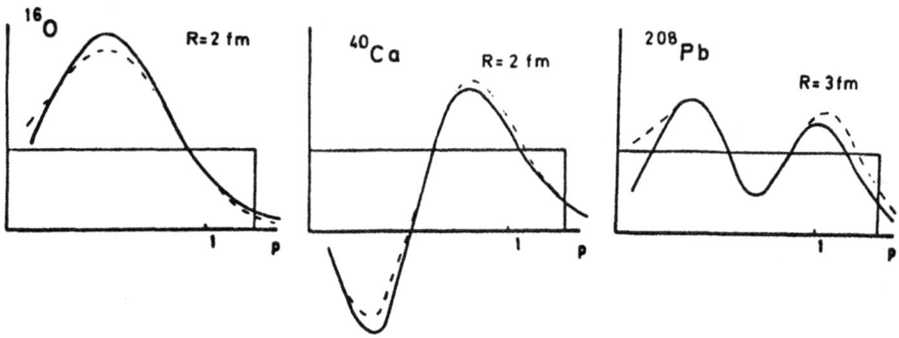

Fig.1 The DDHF Wigner transform(continuous line) compared to H.O. (dashed line) and Thomas Fermi. ($\theta = 0$).

In fig.1 one clearly sees that qualitatively the DDHF results do not differ from the oscillator ones, while there is a big difference with the Thomas Fermi approximation. To discuss more appropriately the latter difference, we show in fig.2 again the ^{208}Pb case but with $p^2 f(\bar{R},\bar{p})$ plotted instead of f, to take into account that for integral properties like those in eqs.(5-7) these are the two quantities to compare. It is then apparent that Thomas Fermi gives correctly the average trends of f, but does not account for the relatively small oscillations at p below p_F and substitutes the smooth decrease of f around p_F by a step function. The qualitative agreement with the harmonic oscillator model shows that both oscillations and surface difusseness do reflect the finiteness of the system and in particular also one of its manifestations : the shell structure. Both aspects have been extensively discussed in the literature (see ref.16 and references therein), and we want to stress here only that they are also present in the DDHF Wigner transform.

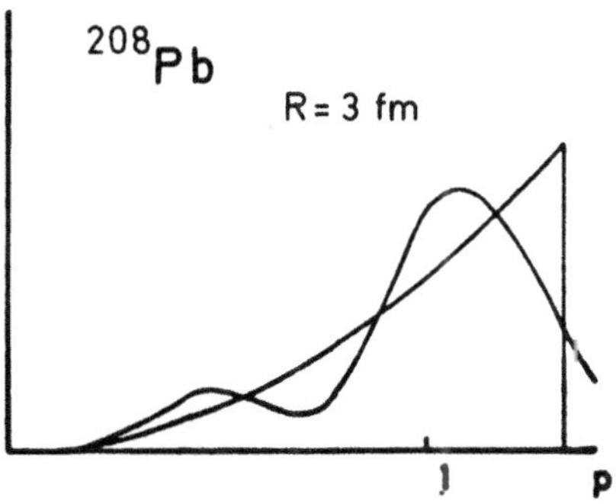

Fig. 2. $p^2 f(R,p)$ at θ=0, DDHF v.s. Thomas Fermi.

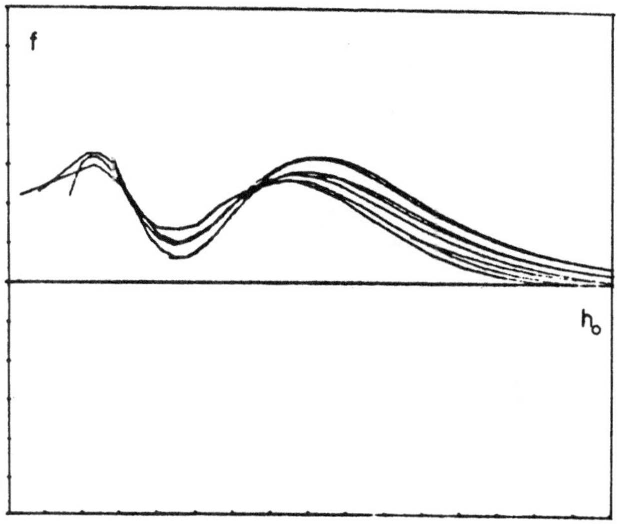

Fig. 3. f_{DDHF} v.s. h_o, at θ=0

The close analogy with the harmonic oscillator results is also illustrated by fig.3 where in the x-axis the variable is $h_0 = 2(p^2 b^2 + R^2/b^2)$. For pure harmonic oscillator wavefunctions one would obtain a single line (see Apperdix 1), here it is seen that although each fixed R=1,2,...8 fm leads to a different line,these are very close to each other, so that qualitatively the behaviour is the same as that shown by the harmonic oscillator.Another consequence of this result is that it clearly shows that the oscillations are not connected to local features around each value of R, but to the finite size of the nucleus, and correspondingly to shell effects. This can be confirmed analytically in two simple model examples :

A) Instead of the DDHF model and occupations one uses the harmonic oscillator model with its own occupation numbers. In this way the magic numbers become (closure of main oscillator shells) : A= 4, 20, 40, 80,140,.. .The analytic expression for f_{HO} is well known:

$$f_{HO}(\bar{R},\bar{p}) = \sum_{K=0}^{N} (-1)^{K}\, 32\, e^{-h_0/2}\, L_{K}^{2}(h_0) \qquad (10)$$

where the L_{K}^{2} are the generalized Laguerre polynomials :

$$L_{n}^{\alpha}(x) = \sum_{m=0}^{n} (-1)^{m} \binom{n+\alpha}{n-m} \frac{x^{m}}{m!} \qquad (11)$$

Therefore each shell contributes to f a polynomial of degree K times a decreasing exponential. Since the number of oscillations of f is directly given by the degree of the polynomial, it is clear that the shell effects are responsible for the oscillations. It is also known that these are quite suppressed when Strutinski averaging is performed on the occupations (see ref.16). An even more drastic averaging corresponding to continuously decreasing occupations is characteristic of the model introduced in nuclear physics by Bhaduri and Sprung (18) :

B) Again the wavefunctions are pure harmonic oscillator, but the occupations of each shell are given by the law t^{K}, where t is determined fixing the total number of nucleons: $t=1-(4/A)^{1/3}$.

For this model:

$$f_{BS}(\bar{R},\bar{p}) = \sum_{K=0}^{\infty} (-1)^K \; 32 \; e^{-h_0/2} \; L_K^2(h_0) \; t^K \tag{12}$$

, and the series in K can be summed analytically :

$$f_{BS}(\bar{R},\bar{p}) = \frac{32}{(1+t)^3} \; e^{-h_0[(1-t)/(1+t)]/2} \tag{13}$$

which shows clearly that only a monotonic decrease remains, all oscillations having been washed out by the drastic smoothing of the occupations.

Moments of the Wigner Transform and the DME

It was pointed out by Campi and Bouyssy[10] that a better understanding of the Density Matrix Expansion proposed by Negele and Vautherin[19] for the one body density matrix, could be gained by studying the corresponding approximation for the Wigner transform. The DME can then be viewed as an approximate expansion of $\rho(\bar{R},\bar{s})$ such that each of its terms is separable in the two variables R and s. To derive it, one can proceed as follows :

1) $\rho(\bar{R},\bar{s})$ is approximated by its angle averaged value, i.e. writing

$$\rho(\bar{R},\bar{s}) = \exp[\, i\bar{s}.\bar{p} \,] \; \rho(\bar{R}) \tag{14}$$

where $\hat{\bar{p}} = -i \; (\bar{\nabla}_1 - \bar{\nabla}_2)/2$, and $\rho(\bar{R}) = \rho(\bar{R},\bar{s}=0)$, one gets

$$\rho(\bar{R},s) = \frac{1}{4\pi} \int d\Omega_s \; \exp[\, i\bar{s}.\bar{p} \,] \; \rho(\bar{R}) = j_0(s\hat{p}) \; (\bar{R}) \tag{15}$$

2) next the expansion

$$j_o(ab) = \sum_{n=0}^{\infty} (4n+3)\ (-1)^n \frac{j_{2n+1}(ak)}{ak} \frac{P_{2n+1}(b/k)}{b/k} \tag{16}$$

where k is arbitrary, and meaningful only when $-1 \leqslant b/k \leqslant 1$, is used to make the ansatz :

$$\rho_{DME}(\bar{R},s) = \sum_{n=0}^{\infty} (4n+3)\ (-1)^n \frac{j_{2n+1}(sk)}{sk} \frac{P_{2n+1}(\hat{p}/k)}{\hat{p}/k} \rho(\bar{R}) \tag{17}$$

where

$$\frac{P_{2n+1}(\hat{p}/k)}{\hat{p}/k} \rho(\bar{R}) = C_{2n+1} \frac{M_{2n}}{k^{2n}} + C_{2n-1} \frac{M_{2n-2}}{k^{2n-2}} + \ldots \tag{18}$$

and the C_i are the coefficients of the 2n+1 Legendre polynomial, while

$$M_k = \int p^k f(\bar{R},\bar{p})\ d\bar{p} \tag{19}$$

Notice that now a condition like $-1 \leqslant \hat{p}/k \leqslant 1$ is meaningless, so that the validity of the ansatz has to be checked by direct comparison of ρ_{DME} with the original $\rho(\bar{R},\bar{s})$. This is done in fig.4 and it is seen that the results are quite satisfactory.

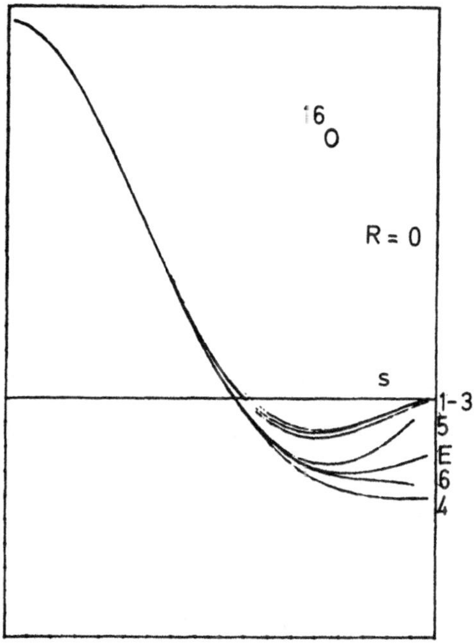

Fig. 4. : ^{16}O with H.O. wave-functions, $\rho_{DME}(R,s)$, n_{max} =0 to 6, v.s. the exact $\rho(R,s)$.

In practice, only a few terms of the expansion are retained : a) the simplest choice is to fix $k=p_F(\bar{R})$ and keep only the first term : this is the well known Thomas Fermi approximation. b) Negele and Vautherin's[19] choice was to choose $k = p_F(\bar{R})$ and keep only the first two terms of the expansion, c) Campi and Bouyssy's[10] choice was to keep also two terms, but choosing $k(\bar{R})$ such that the second term vanishes for each $\bar{R}$. In this way only one term contributes and one obtains a modified Slater-Thomas-Fermi type of approximation with $k \neq p_F(\bar{R})$.

Let us now look at the corresponding approximations for $f(\bar{R},\bar{p})$. The transform of ρ_{DME} gives :

$$f_{DME}(\bar{R},p) = \frac{(2\pi)^3}{4\pi k^3} \sum_{n=0}^{\infty} (4n+3) \frac{P_{2n+1}(p/k)}{p/k} \frac{P_{2n+1}(\hat{p}/k)}{\hat{p}/k} \rho(\bar{R})\, \theta(p-k) \tag{20}$$

so that each term is multiplied by a step function. The first consequence of this result is therefore that $\dot{f}_{DME}$ is a distribution, discontinuous at p=k , and zero for any p> k . Clearly then the shape of f_{DME} as a function of p will have nothing to do with that of f_{DDHF}. Explicit results are shown in fig.5 ,for different values of n_{max} , the value of n at which the series is truncated. It is seen that increasing n_{max} increases both the number and the amplitude of the oscillations. For any finite k the results are similar, and for finite n_{max} there is no convergence at all towards f_{DDHF}.

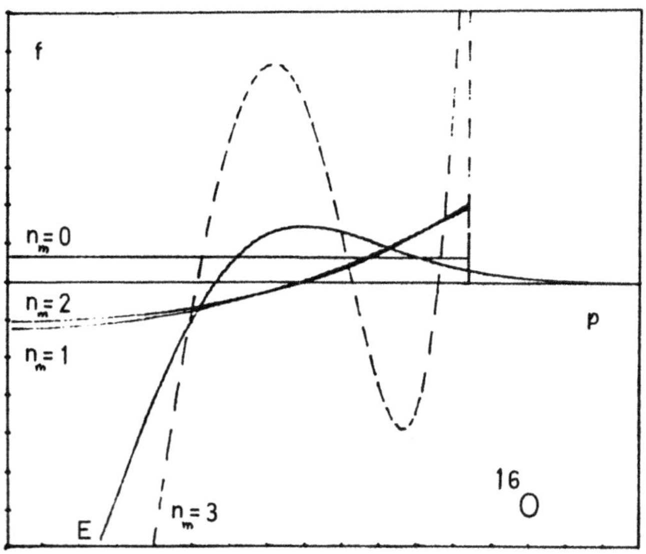

Fig. 5. : ^{16}O H.O.wave - functions, f_{DME} v.s. the exact f.

The interpretation of this somewhat disappointing result comes from the following property of f_{DME} (details given in ref. 11) : The quantities M_k (eq.19) have the property :

$$M_k\,(DME) = M_k\,(DDHF) \tag{21}$$

for all k when the DME expansion is not truncated, and for all $k \leqslant n_{max}$ when the DME is truncated at order n_{max}. This is a remarkable result: while f_{DME} is a distribution and thus cannot reproduce the shape of f_{DDHF}, it is a good approximation in the sense of the distributions because all integrals over p involving that function will be accurately given by the DME approximation to f. In that sense therefore the DME is superior to semiclassical approximations which as is well known do not reproduce exactly ρ, τ and the other quantities related to successive moments of f. The price to pay is of course that f_{DME} cannot be used to describe f as a function in p-space.

Appendix I

The Wigner transforms for harmonic oscillator single particle wavefunctions have the following analytic expressions :

$$f(^{16}O) = 32\,(-2 + h_0)\, e^{-h_0/2}$$

$$f(^{40}Ca) = 32\,(4 - 3h_0 + h_0^2/2)\, e^{-h_0/2}$$

$$f(^{48}Ca) = 16\left[4 - 2h_0 + \frac{1}{5}h_0^2 + \frac{4}{7!!}h_0^3 + d\left(-\frac{8}{5} + \frac{8}{35}h_0\right)\right] e^{-h_0/2}$$

$$f(^{90}Zr) = 16\left[-7 + \frac{22}{3}h_0 - 2h_0^2 + \frac{1}{7}h_0^3 + \frac{5}{9!!}h_0^4 + d\left(4 - \frac{8}{7}h_0 + \frac{4}{63}h_0^2\right) + \frac{2}{63}d^2\right] e^{-h_0/2}$$

$$f\left({}^{208}Pb\right) = 16\, e^{-h_0/2} \Big[-2 + 5h_0 - \frac{7}{2} h_0^2 + \frac{15}{14} h_0^3 - \frac{65}{504} h_0^4$$

$$+ \frac{405}{8 \cdot 11!!} h_0^5 + \frac{7}{13!!} h_0^6 + d \left(6 - \frac{32}{7} h_0 + \frac{20}{21} h_0^2 - \frac{16}{231} h_0^3 +\right.$$

$$\left. + \frac{210}{13!!} h_0^4 \right) + d^2 \left(\frac{10}{21} - \frac{8}{77} h_0 + \frac{2}{429} h_0^2 \right) + d^3 \frac{140}{13!!} \Big]$$

$$h_0 = 2\left(p^2 b^2 + R^2/b^2 \right) \quad , \quad d = 4 R^2 p^2 \sin^2 \theta$$

References

1 . E.P. Wigner, Phys. Rev. 40(1932) 749.

2 . M. Prakash et al. , Nucl. Phys. A385(1982) 483.

3 . J. Hufner, M.C. Nemes , Phys. Rev. C23(1981) 2538.

4 . H.K. Tang et al, Phys. Lett. 101B(1981) 10.

5 . D.M. Brink, M. di Toro, Nucl. Phys. A372(1981) 151.

6 . K. Goeke and J. Speth, Ann. Rev. Nucl. Sci. 32(1982) 65.

7 . B.K. Jennigs and A.D. Jackson, Phys. Rep. 66(1980) 141.

8 . P. Ring and P. Schuck, "The Nuclear Many Body Problem", Springer Verlag 1980.

9 . B.K. Jennings, Phys. Lett. 74B(1978) 13, Phys. Lett.96B (1980) 1.

10. X. Campi and A. Bouyssy, Phys. Lett. 73B(1978)263 , Nukleonika 24(1979) 1.

11. J. Martorell and E. Moya de Guerra, to be published.

12. S. Shlomo and M. Prakash, Nucl. Phys. A357(1981) 157.

13. D.W.L. Sprung and P.K. Banerjee,Nucl. Phys. A168(1971)273.

14. X. Campi and D.W.L. Sprung, Nucl. Phys. A194(1972) 401.

15. D.W.L. Sprung et al., Nucl. Phys. A253(1975) 1.

16. M. Prakash, S. Shlomo and V.M. Kolomietz, Nucl. Phys. A370 (1981) 30.

17. G.F. Bertsch, Nucl. Phys. A249(1975) 253.

18. R.K. Bhaduri and D.W.L. Sprung, Nucl. Phys. A297(1978) 365.

19. J. Negele and D. Vautherin, Phys. Rev. C5(1972) 1472.

ATDHF: ITS SCOPE AND LIMITATIONS

E. Moya de Guerra
Dpto. de Física Atómica y Nuclear. Facultad de Ciencias
Universidad de Extremadura
BADAJOZ (Spain).

ABSTRACT.

ATDHF formulations for large amplitude collective nuclear modes are reviewed together with the conditions of validity that establish the range of applicability of the theory. Application to soluble models is presented to clarify the meaning of different validity conditions and equations of path. Extensions of the standard ATDHF formalism are also discussed.

I. Introduction.

The adiabatic limit of time dependent Hartree Fock theory (or ATDHF) was initially formulated in two different ways by Villars[1] and by Baranger and Veneroni[2]. The first is a state vector formulation while the second is a density matrix formulation, their relationship has been extensively discussed in the literature[3-4], and I shall in this talk concentrate on the first. The purpose of this formulation was to provide a microscopic basis suitable for the description of large amplitude, low velocity, collective modes of the nucleus, based on our experience that the HF (or BCS) theory (used with a proper effective interaction) gives an excellent account of many nuclear properties. The point is then, that using as constraint the hipothesis that at any deformation the nuclear states can be approximated by Slater determinants, this formalism allows to derive a collective hamiltonian whose mass tensor and potential energy are self-consistently calculated on a microscopic basis.

From the pioneering work of Bohr and Mottelson up to our days there is ample evidence that the nuclear many body Schrödinger equation

$$H \Psi_n = E_n \Psi_n$$

admits one -(or few-) body solutions of the form

$$H_c\left(-i \frac{\partial}{\partial q_i}, q_i\right) f_n(q_i) = E_n f_n(q_i)$$

where H_c is quadratic in the momenta. This is clearly the case for collective vibrational modes, for collective rotations of strongly deformed nuclei, and presumibly for the fission and fusion modes at low energies, where the rate at which nuclear deformation changes on time is small as compared to single particle velocities.

As said before, the purpose of ATDHF is to construct microscopically such collective hamiltonians. To this end one starts from the minimum action principle

$$\delta I_{12} / \delta \phi(t) = 0 \quad , \quad I_{12} = \int_{t_1}^{t_2} dt \; \langle \phi(t) | H - i \frac{\partial}{\partial t} | \phi(t) \rangle$$

using as ansatz Slater determinants parametrized in terms of one, or a few, collective coordinates (q_i) and their canonical conjugate momenta (p_i),

$$\phi(t) = \phi(q_i, p_i) = \exp.(i \Sigma p_i Q_i) \; \phi(q_i)$$

The variational condition with this trial wave function leads to classical equations of motion for the q_i, p_i variables, subject to a set of self-consistency conditions(1-5). This set of self-consistency conditions has to be solved in order to determine the potential energy and inertial tensor of the classical hamiltonian, and hence the classical dynamics. Once the classical hamiltonian is knowm one has to proceed to its quantization in order to obtain the desired description of the actual quantized collective nuclear motion.

The quantization problem has been addressed from different points of view by Goeke and Reinhard(3) and by myself(6). I shall not discuss this question here, let me just say that Goeke and Reinhard give a prescription for the case of just one collective coordinate while I was mainly interested in the case of several coordinates. As far as I could check both prescriptions coincide for the case of zero point energy corrections due to vibrational modes, further than that it remains a question that requires further investigation.

Most of the work on ATDHF, both formal and numerical(1,3,4,7), have dealt with just one collective coordinate (I shall refer to this as standard ATDHF), using the very notion of pure adiabatic motion as completely decoupled. It is then clear, at least qualitatively, what the limitations of the standard ATDHF are. It should only be valid for the description of (1) low velocity collective modes (2) in the regions where the coupling to intrinsic excitations (or other possible collective modes) can be neglected. What it is not so clear is how to put this conditions of validity in a quantitative form. Different workers in the field give different prescriptions for the condition of validity concerning point (1). In particular Goeke and his collaborators claim(3-7) that implementing their validity condition is essential to get from the set of self-consistency conditions of Villars the collective path as the path of minimal potential energy, while Mukherjee and Pal [4] invoque that to this end one more variational condition is to be added (see section III).

As stated before it seems that there should be two validity conditions, one for the small velocity and one for the small coupling, after all these are two different physical concepts and one may find situations in which collective modes are decoupled and fast or slow and coupled. However the second condition is never explicitely considered as such, in standard ATDHF.

It is important to realize that there is nothing in the ATDHF formalism that prevent us from extending it to the problem of slow coupled modes. After all if we want to describe a process such as fission or fusion we know that dissipation phenomena occur (and are important[8]) and there is no way to occount for this if we consider these modes as decoupled. So, it seems to me that studying the coupling between modes within the ATDHF formalism is crucial for both, 1) clearing up the validity conditions of the standard ATDHF, and 2) enlarging the scope of the standard ATDHF formulation to make it applicable to the study of friction or damping phenomena.

In what follows I shall then outline some remarks concerning these two points on the basis of a generalization of ATDHF that I did with Villars several years ago. Starting from the minimal action principle we generalized[5] the ATDHF theory for the "fission mode"[*] (parametrized in terms of one single collective coordinate) so as to describe simultaneously intrinsic excitations. To simplify the study of the coupling betwen the fission mode and the intrinsic excitations we did not include rotational degrees of freedom (not to be forgotten in any realistic calculation!) and treated the intrinsic excitations as RPA modes at any given deformation. We found a set of self-consistency conditions to determine the fission path and the coupling at any point along the path[5]:

$$\begin{aligned}
&\langle \delta\phi(q_1)| H - \lambda Q_1 |\phi(q_1)\rangle = 0 \\
&\langle \delta\phi(q_1)| [H, iQ_i] - P_i / M_i^0 |\phi(q_1)\rangle = 0 \\
&\langle \delta\phi(q_1)| [H, -iP_i] - K_i Q_i - K_{ij} Q_j |\phi(q_1)\rangle = 0 \qquad \begin{cases} i=1, & j=K \\ i=K, & j=1 \end{cases} \\
&\langle \phi(q_1)| [Q_i, -iP_j] |\phi(q_1)\rangle = \delta_{ij}
\end{aligned} \tag{1}$$

(*) Here and in what follows "fission mode" is used to denote a genuine large amplitude collective mode such as fission of fusion.

where we call q_1 the fission mode coordinate, q_k ($k \geq 2$) the coordinates for the bound motion perpendicular to the fission path. The restoring force and inertial parameters are functions of q_1 (see ref.(5) for details), and the constraining parameter λ is given by

$$\lambda = \langle \phi(q_1) | [H, -i P_1] | \phi(q_1) \rangle = d V_0(q_1) / d q_1 \tag{2}$$

The corresponding classical hamiltonian contains three parts,

$$\mathcal{H}(q_i, p_i) = \mathcal{H}_a(q_1, p_1) + \mathcal{H}_v(q_k, p_k) + \mathcal{H}_{coup.} \tag{3}$$

the pure adiabatic motion in the fission direction,

$$\mathcal{H}_a = P_1^2 / 2 M_1^0 + V_0(q_1) \tag{4}$$

the vibrational motion in the perpendicular directions,

$$\mathcal{H}_v = \frac{1}{2} \sum_{k \geq 2} \left(\frac{P_k^2}{M_k^0} + K_k q_k^2 \right) \tag{5}$$

with vibrational frequencies depending on deformation, and a explicit coupling term

$$\mathcal{H}_{coup} = \frac{1}{2} P_1^2 \sum_k \frac{q_k}{M_1^{(k)}} + P_1 \sum_{k,l} \frac{q_k P_l}{M_{1l}^{(k)}} \tag{6}$$

whose main part comes from the dependence on the bound coordinates of the inertia along the fission mode.

Keeping this equations in mind, I shall address the two points mentioned above. First (section II) I shall make a few remarks on the second point, then (section III) I shall discuss the first point, that of validity conditions, using as illustration two-dimensional models.

II. ATDHF and damping.

Concerning the point of extending ATDHF to describe friction or damping, there are several questions that come to mind by looking at eqs. (3-6). Supose we have solved the set of self-consistency conditions (1), and therefore we know what the classical hamiltonian (3) is at any point along the fission path, is this hamiltonian self-consistently derived usefull for describing friction or damping?. At least, we see that the main feature of the way in which the coupling comes (dependence of the collective inertia on the bound amplitudes and of the vibrational frequencies on deformation) is in agreement with the observation of Swiatecki and Bjornholm [8] that the friction in the fission process is highly dependent on deformation and hence on the total inertia associated to the fission mode.

The second remark to be considered is that at any given deformation many of the RPA modes will be of the 1p-h type for which there is not a classical analogue. This implies that the vibrational variables in eq. (3) have to be considered quantum mechanically even at the accessible classical regions in the fission mode, where this mode can be considered classically. This means that instead of the completely classical hamiltonian (3) we must use the hamiltonian[9]

$$\hat{\mathcal{H}}(c_k, c_k^+; q_1, p_1) = \mathcal{H}_0(q_1, p_1) + \sum_k \omega_k c_k^+ c_k + \frac{1}{2}\sum_k F_k (c_k^+ + c_k) + \frac{1}{2}\sum_{k,l} G_{kl}\, i(c_l^+ c_k^+ - c_l c_k + c_l^+ c_k - c_k^+ c_l) \quad (7)$$

in which the bound motion is fully quantized. The first part of this hamiltonian

$$\mathcal{H}_0(q_1, p_1) = \mathcal{H}_a + \frac{1}{2}\sum_k \left(\omega_k - \frac{\langle P_k^2\rangle}{M_k^0} - K_k \langle Q_k^2\rangle - K_{1k}\langle Q_k Q_1\rangle\right) - \frac{p_1}{M_1^0}\sum_k \left\langle \frac{\partial P_k}{\partial q_1} Q_k \right\rangle \quad (8)$$

represents the classical hamiltonian for the frozen motion in the fission direction, i.e., for the path along which the vibrational modes remain in their ground state. It contains, appart from the renormalization of the potential energy due to zero point motion, an imaginary term that is linear in the fission mode velocity.

We also see that if we neglect the coupling betwen the bound modes (which is responsible for the two phonon creation and anihilation part of eq. (7)) the remaining hamiltonian has the usual form of that for a system that consists of a subsystem and a thermal reservoir. The thermal reservoir being a reservoir of phonons and the coupling betwen the reservoir and the subsystem being linear in the phonon creation and anihilation operators. This kind of systems have been extensively discussed in the literature[10], essentially what is to be done is to use the method of ensamble averages to derive the desired time irreversible equations from the time reversible ones.

Complete quantization of this hamiltonian is obviously required to describe purely quantum mechanical phenomena such as tunneling. But the main conclusion is that one sees that this hamiltonian can be used as the starting point for models of damped collective motion, as those based on linear response theory[11] with the advantage that, since all the parameters are derived self-consistently at any point inthe path, one would be able to elucidate the regions at which the intrinsic excitations can be treated globally as a heat bath, and to single out the modes that being highly collective must be excluded from that treatment at a given deformation.

III. Validity conditions. The decoupled collective path.

Let us now go back to the first point mentioned in the introduction. Namely, what are the conditions for the validity of the standard ATDHF equations[1],

$$\text{I} \quad \langle \delta\phi(q)| H - \lambda Q |\phi(q)\rangle = 0$$

$$\text{II} \quad \langle \delta\phi(q)| [H, iQ] - P/M |\phi(q)\rangle = 0$$

$$\text{III} \quad \langle \phi(q)| [Q, -iP] |\phi(q)\rangle = 1$$

with

$$P\,\phi(q) = i \frac{\partial}{\partial q}\phi(q) \quad ,$$

to define the collective path for slow decoupled motion.

The small velocity condition imposed by Villars[1] is that

$$\langle \phi(q)| p^2 Q^2 |\phi(q)\rangle \ll C \tag{9}$$

where C is a collectivity index representing the average number of p-h excitations contributing to the collective mode. We shall see in subsection III.2 that this condition is indeed essential.

On the other hand, Mukherjee and Pal[4] say that in order to get from the set of self-consistency conditions of Villars (eqs. I-III) the collective path as the path of minimal potential energy, an additional self-consistency condition is to be satisfied, namely

$$\text{IV} \quad \langle \delta\phi(q)| [[H, iQ], iQ] - \frac{\partial M^{-1}}{\partial q} Q + 2 M^{-1} \frac{\partial Q}{\partial q} |\phi(q)\rangle = 0$$

or

$$\text{IV}' \quad \langle \delta\phi(q)| [[H, iQ], iQ]\lambda^2 + \frac{2\lambda}{M}[H, -iP] - 2\lambda\omega_0(q) Q |\phi(q)\rangle = 0$$

where

$$\omega_0(q) = \frac{1}{2\lambda}\frac{d}{dq}\left(\frac{\lambda^2}{M}\right)$$

This fourth condition of Mukherjee and Pal (M-P) arises from the terms quadratic in p^2 in the expansion of the variational equation

$$\langle \delta\phi| i\frac{\partial}{\partial t} - H |\phi\rangle = 0$$

with

$$\phi = \phi(q, p) = \exp.(ipQ)\,\phi(q) \quad , \quad Q = Q(q) \tag{10}$$

Using the Generator Coordinate method Goeke and Reinhard (G-R) obtain instead the validity condition[(3-7)]

$$p^2 \| (\partial_p Q)_{ph} \| \ll 1 \tag{11}$$

with $(\partial_p Q)_{ph}$ determined from

$$[H, \partial_p Q]_{ph} = \frac{2i}{M} (\partial_q Q)_{ph} - i[[H, Q], Q]_{ph} - i Q \partial_q M^{-1} \tag{12}$$

where A_{ph} means the p-h part of the operator A with respect to $\phi(q)$ and orthogonal to the path.

The origin of the G-R condition can be easily understood in the language of the usual derivation of ATDHF, as coming from the fact that the expansion (10) differs to order p^2 from the exact expansion by at most a new one p-h operator. Indeed to second order in p^2 one must have that

$$\phi(q, p) = \phi(q) + p \left(\partial_p \phi(q, p)\right)_{p=0} + \frac{1}{2} p^2 \left(\frac{\partial^2}{\partial p^2} \phi(q, p)\right)_{p=0}$$

and since $\phi(q, p)$ is a Slater determinant we can allways write

$$\phi(q, p) = \mathcal{U}(q, p)\, \phi(q)$$

where $\phi(q)$ is a Slater determinant independent of p and U is a unitary operator,

$$\mathcal{U}(q, p) = \exp.(i p Q')$$

Q' being a one p-h hermitian time even operator that may depend on p as well as on q. Then setting

$$\left(\partial_p \mathcal{U}(q, p)\right)_{p=0} = i\, Q'(p=0) \equiv i\, Q$$

the difference betwen the standard expansion, $U_o(q, p) = e^{ipQ}$ (see eq. (10)), and the exact expansion comes, to order p^2, from the difference

$$\left(\frac{\partial^2}{\partial p^2} \mathcal{U}(q, p)\right)_{p=0} - \left(\frac{\partial^2}{\partial p^2} \mathcal{U}_o(q, p)\right)_{p=0} = i \left.\frac{\partial Q'}{\partial p}\right|_{p=0}$$

This new operator $\partial Q'/\partial p|_{p=0}$ is the operator $(\partial_p Q)$ Kept by Goeke and Reinhard. It can also be easily seen that if one includes this new operator, the fourth condition of M-P is modified into eq. (12) of G-R. Therefore the validity condition of G-R is a weaker condition than that of M-P.

On the other hand it can be verified that with the usual election of the operator Q as a hermitian, time-even, one particle-hole operator, the fourth condition of Mukherjee and Pal follows in part as an identity from the set of conditions I-III and does not necessarilly add anything new to that set.

To clarify the meaning of different conditions and equations for the one-dimensional decoupled path it is worth applying them to soluble models that are two-dimensional in the semiclassical limit. We shall latter on discuss two specific models, the landscape model[12] and the SU(3) model[13], but first let us consider a general two-dimensional classical hamiltonian of the form

$$\mathcal{H} = \sum_{i,j=1}^{2} \frac{p_i p_j}{2 M_{ij}} + V(x_1, x_2) \tag{13}$$

with

$$M_{ij} = M_{ij}(x_1, x_2) = M_{ji}$$

The problem is to find a canonical transformation of the coordinates x_1, x_2 (and their canonical conjugate momenta p_1, p_2) into a new set of coordinates q_1, q_2 (and momenta Π_1, Π_2) such that, if q_1 represents the large amplitude collective coordinate, the hamiltonian (13) can be written in terms of this new coordinates as

$$\mathcal{H} = \frac{\Pi_1^2}{\mathcal{M}_1} + \mathcal{V}_0(q_1) + \frac{1}{2}\left(\frac{\Pi_2^2}{\mathcal{M}_2} + K_2 q_2^2\right) + \mathcal{H}_{coup.} \tag{13'}$$

where $\mathcal{H}_{coupl.}$ is of the form of eq. (6).

If such a transformation exists the hamiltonian (13') can be found from the solution of the set of self-consistency conditions (1). At any given point q_1 of the collective path this set of self-consistency conditions reads

$$\begin{pmatrix} x_1^{(2)} & x_2^{(2)} \end{pmatrix} \begin{pmatrix} V_1 \\ V_2 \end{pmatrix} = 0 \tag{14}$$

$$\begin{pmatrix} x_1^{(1)} & x_2^{(1)} \end{pmatrix} \begin{pmatrix} V_1 \\ V_2 \end{pmatrix} = \lambda \tag{15}$$

$$\lambda^2 / \mathcal{M}_1 = \begin{pmatrix} V_1 & V_2 \end{pmatrix} \begin{pmatrix} M_{11}^{-1} & M_{12}^{-1} \\ M_{12}^{-1} & M_{22}^{-1} \end{pmatrix} \begin{pmatrix} V_1 \\ V_2 \end{pmatrix} \tag{16}$$

with

$$\lambda = x_1^{(1)} (V_1 + \alpha V_2) \tag{15'}$$

$$\alpha = \frac{x_2^{(1)}}{x_1^{(1)}} = (V_1 M_{12}^{-1} + V_2 M_{22}^{-1}) / (V_1 M_{11}^{-1} + V_2 M_{12}^{-1}) \tag{17}$$

In addition for $x_1^{(2)}$, $x_2^{(2)} \neq 0$ we have

$$(x_1^{(2)})^2 / \mathcal{M}_2 = (\alpha^2 M_{11}^{-1} + M_{22}^{-1} - 2\alpha M_{12}^{-1}) / (\alpha + V_1/V_2)^2 \tag{18}$$

$$K_i = \begin{pmatrix} x_1^{(i)} & x_2^{(i)} \end{pmatrix} \begin{pmatrix} V_{11} & V_{12} \\ V_{12} & V_{22} \end{pmatrix} \begin{pmatrix} x_1^{(i)} \\ x_2^{(i)} \end{pmatrix} \quad , \; i = 1,2 \tag{19}$$

$$K_{ij} = (x_1^{(i)} \quad x_2^{(i)}) \begin{pmatrix} V_{11} & V_{12} \\ V_{12} & V_{22} \end{pmatrix} \begin{pmatrix} x_1^{(j)} \\ x_2^{(j)} \end{pmatrix} = -\frac{1}{2} (x_1^{(12)} \quad x_2^{(12)}) \begin{pmatrix} V_1 \\ V_2 \end{pmatrix} \qquad i \neq j \tag{20}$$

where we use the notation

$$x_i^{(j)} = \partial x_i / \partial q_j \big|_{q_2=0} \qquad x_i^{(12)} = \partial^2 x_i / \partial q_1 \partial q_2 \big|_{q_2=0}$$
$$V_i = \partial V / \partial x_i \qquad V_{ij} = \partial^2 V / \partial x_i \partial x_j \tag{21}$$

That set of eqs. guarantees that at any point q_1, $q_2 = 0$ in the collective path

$$\partial \mathcal{V}(q_1, q_2)/\partial q_2 = \partial^2 \mathcal{V}(q_1, q_2)/\partial q_1 \partial q_2 = \partial^2 \mathcal{H}/\partial \pi_1 \partial \pi_2 = 0$$

which are the essential properties of the hamiltonian (13'). All the parameters in this hamiltonian can be determined from eqs. (14) to (20) in terms of $x_1^{(1)}$ and $x_1^{(2)}$, but those could be fixed by the normalization condition

$$(x_1^{(i)})^2 + (x_2^{(i)})^2 = 1 \quad , \quad i = 1, 2$$

or any other convenient normalization. This is equivalent to choosing the more convenient step lengths Δq_1, Δq_2 in the numerical solution of differential equations.

It is important to notice at this point that the self-consistency conditions (I-III) give rise to eqs. (14) to (17), and that solving the complete set of eqs. (1) (i.e., eqs. (14) to (20) for dimension 2) is only necessary to determine the collective path and the coupling along the path.

However we can also use the set (1) to look for the decoupled collective path. To this end it seems natural to require that if we start from any given point x_1°, x_2° in the collective path (i.e., q_1°, $q_2^\circ=0$, as for instance de H-F minimum) the next point in the x_1, x_2 plane should lie in the collective path (this implies that $x_1^{(2)}=x_2^{(2)}=x_1^{(12)}=x_2^{(12)}=0$ at any point x_1° x_2°) and that the collective mode is at any fixed point the mode of minimal restoring force and maximal inertia (i.e., the lowest frequency local mode). This prescription for the decoupled collective path is analogous to that of Rowe and Basserman[14] and defines the collective path as the most stable path. With this prescription we get, instead of eqs. (18) to (20), the equation

$$x_2^{(1)} / x_1^{(1)} = (C_L - A_{11}) / A_{12} \tag{22}$$

where C_L is the lowest eigenvalue of

$$|A - C| = 0 \tag{23}$$

and the matrix A is

$$A = \begin{pmatrix} M_{11}^{-1} & M_{12}^{-1} \\ M_{12}^{-1} & M_{22}^{-1} \end{pmatrix} \begin{pmatrix} V_{11} & V_{12} \\ V_{12} & V_{22} \end{pmatrix} \tag{24}$$

Equation (22) together with eq. (17) define the line $q_2=0$ (or $x_1=x_1(x_2)$ in the x_1, x_2 plane) as the line of minimal potential provided the eigenvalues C of eq. (23) are real. This requires that $A_{12}\, A_{21} \geq 0$, $|A_{12}\, A_{21}| < (\frac{A_{11}-A_{22}}{2})^2$ if $A_{12}\, A_{21} < 0$, that can be seen as the conditions for the existence of the decoupled collective path given by eqs. (17) and (22), or

$$\frac{V_1 M_{12}^{-1} + V_2 M_{22}^{-1}}{V_1 M_{11}^{-1} + V_2 M_{12}^{-1}} = \frac{A_{22}-A_{11}}{2\,A_{12}} - \left(\left(\frac{A_{11}-A_{22}}{2\,A_{12}}\right)^2 + \frac{A_{21}}{A_{12}}\right)^{1/2} \tag{25}$$

Analogous considerations and equations follow for the case of N-dimensional problems. In what follows we shall apply the above equations to two specific models where we can also see what is the meaning of the different validity conditions presented at the beginning of this section.

III.1 The potential landscape model.

The potential landscape model[12-15] is a pure classical model hamiltonian of the form (13) with constant diagonal inertia tensor ($M_{11}=M_{22}=m$, $M_{12}^{-1}=0$), and potential energy given by

$$V(x_1,x_2) = \frac{1}{2} m\,\omega_2^2\, x_2^2 + \frac{1}{2} m\,\omega_1^2\, x_1^2 + \beta\, x_1\, x_2^2 \tag{26}$$

This potential has its minimum at $x_1=x_2=0$ and a saddle point at $x_1=-\frac{m\omega_2^2}{2\beta}$
$x_2^2 = m^2\,\omega_1^2\,\omega_2^2/(2\beta^2)$.
For this model eq. (25) reads

$$V_2/V_1 = (V_{22}-V_{11})/2V_{12} - \left[(V_{11}-V_{22})^2/(2V_{12})^2 + 1\right]^{1/2} \tag{27}$$

which gives the line $x_2^2=x_2^2(x_1)$ in the x_1, x_2 plane:

$$x_2^2 = -\frac{m}{\beta}\,\omega_1^2\, x_1 + \frac{\alpha}{4\beta^2}\left\{m\,\omega_1^2 + \alpha - \left[(m\omega_1^2+\alpha)^2 - 16\, m\,\beta\,\omega_1^2\, x_1\right]^{1/2}\right\} \tag{27'}$$

where

$$\alpha = m\,\omega_2^2 + 2\,\beta\, x_1$$

It can indeed be verified that this line passes trough the minimum and saddle points and follows the bottom of the valley of the potential (26). For numerical applications and graphical representations see refs. (12, 15).

For this model the condition of Mukherjee and Pal[4] reads

$$(V_1 V_{11} + V_2 V_{12}) / (V_1 V_{12} + V_2 V_{22}) = V_1 / V_2$$

It is easy to check that the line (27') satisfies this condition. However this condition is satisfied by the equation (see eqs. (17), (22))

$$x_2^{(1)} / x_1^{(1)} = V_2 / V_1 = (C - V_{11}) / V_{12}$$

independently of wether C is the larger or the smaller eigenvalue of eq. (23) and therefore is not a good stabilizing condition. Analogous remark apply to the weaker condition of Goeke and Reinhard. But one can see that if one solves eq. (28) for $x_2^2 = x_2^2(x_1)$ imposing the condition that it passes through the H-F minimum one recovers eq. (27'), while this is not the case when the Goeke Reinhard condition is implemented. Indeed in ref. (15) it is stressed that the G-R path is not identical to that given by eq. (27').

III.2 The SU(3) model.

A better illustration of the application of ATDHF is obtained in the SU(3) model. This is an exactly soluble, three level, schematic shell model. It is analogous to the SU(2) Lipkin model, with the advantage over this that in the semiclassical (TDHF) limit the dimensionality is two, instead of one, and therefore serves as a true test of the different sets of variational equations and validity conditions.

The model is defined[13-16] as a system of N fermions distributed over three, N-fold degenerate, energy levels with energies $e_0 < e_1 < e_2$. The two body interaction is such that it scatters any pair of particles from one level (i) to another (j) leaving their substates (m) unchanged. The model hamiltonian can then be written in terms of the generators of an SU(3) algebra

$$H = \sum_{i=0}^{2} \epsilon_i K_{ii} + \frac{1}{2} \sum_{i<j} V_{ij} (K_{ij}^2 + K_{ji}^2) \tag{28}$$

where, in terms of the fermion creation and anihilation operators,

$$K_{ij} = \sum_{m=1}^{N} a_{im}^+ a_{jm} = K_{ji}^+ \qquad , j, i = 0, 1, 2$$

It is then a simple matter to verifie that the operators K_{ij} satisfy the conmutation relations of an SU(3) algebra with $\hat{N} = \sum k_{ii}$ being the invariant of the group.

In the absence of interaction the ground state of the system is the state

$$|N, 0, 0\rangle \equiv |0\rangle \tag{29}$$

The hamiltonian H does not connect different representations and therefore, any of its eigenstates can be written in terms of the basis states of the representation $[N,0,0]$ whose generating state is the state $|0\rangle$ defined above.

The TDHF and ATDHF approximations to this model have been extensively studied. I shall here only comment what the results pertinent to the present discussion are, skiping over lengthy details that can be found in references (13, 16, 17). The more general Slater determinant for this model can be parametrized in terms of four angles $(\theta_1,\theta_2;\varphi_1,\varphi_2)$ as

$$\Psi(\theta_i;\varphi_i)=\exp.[i\varphi_1\{(G_{10}+G_{01})\cos\varphi_2+(G_{20}+G_{02})\sin\varphi_2\}]\;\phi(\theta_i) \tag{30}$$

where $\phi(\theta_1,\theta_2)$ represents the "time even" Slater determinant

$$\phi(\theta_i)=\exp.[\theta_2(K_{21}-K_{12})]\;\exp.[\theta_1(K_{10}-K_{01})]\,|0\rangle\equiv\mathcal{U}_2\,\mathcal{U}_1|0\rangle \tag{31}$$

and

$$G_{ij}=\mathcal{U}_2\,\mathcal{U}_1\,K_{ij}\,\mathcal{U}_1^+\,\mathcal{U}_2^+$$

With θ_1,θ_2 playing the role of coordinates (x_1,x_2) the TDHF variational principle

$$\langle\delta\psi|H-i\frac{\partial}{\partial t}|\psi\rangle=0$$

gives rise to the equations of motion(16)

$$\dot{p}_i=-\frac{\partial\mathcal{H}}{\partial\theta_i}\quad,\quad\dot{\theta}_i=\frac{\partial\mathcal{H}}{\partial p_i} \tag{32}$$

with

$$\mathcal{H}=\langle\psi|H|\psi\rangle \tag{33}$$

$$p_1=N\sin 2\varphi_1\;\cos\varphi_2\qquad,\;p_2=N\sin 2\varphi_1\sin\varphi_2\sin\theta_1 \tag{34}$$

In the ATDHF approach the time dependent Slater determinant is parametrized in this case as

$$\Psi_a(\theta_i,p_i)=\exp.(i\sum_i p_i Q_i)\,\phi(\theta_i)=(1+i\sum p_i Q_i+\cdots)\,\phi(\theta_i) \tag{35}$$

with the static determinants $\phi(\theta_i)$ and velocity phase to be determined from the set of self-consistency conditions

$$\text{I}'\qquad\langle\delta\phi|H-\sum_i\frac{\partial V}{\partial\theta_i}Q_i|\phi\rangle=0$$

$$\text{II}'\qquad\langle\delta\phi|[H,iQ_i]-\sum_j P_j/M_{ij}|\phi\rangle=0$$

$$\text{III}'\qquad\langle\phi|[Q_i,-iP_j]|\phi\rangle=\delta_{i,j}$$

with $p_i\,\phi = i\,\partial\phi/\partial\theta_i$ (i=1,2), and ATDHF equations of motion

$$\dot{\theta}_i = \partial \mathcal{H}_a/\partial p_i \qquad , \quad \dot{p}_i = -\,\partial\mathcal{H}_a/\partial\theta_i \tag{36}$$

where

$$\mathcal{H}_a = \langle\Psi_a|H|\Psi_a\rangle = \sum_{i,j}\left(\frac{p_i\,p_j}{2M_{ij}}\right) + V(\theta_i) \tag{37}$$

Choosing the parameters of the model hamiltonian (28) as

$$e_0 = -1 \ , \ e_1 = 0 \quad , \ e_2 = n \geq 1$$

$$-V_{01} = -V_{02} = \chi/(N-1) \quad , \chi > 1 \qquad , V_{12} = 0$$

the solution of eqs. I'-III' gives the following result for the inertia tensor and constraining operators[16]

$$M_{11}^{-1} = \frac{1}{2N}\left\{\cos 2\theta_1\,(1+n\sin^2\theta_2) + \chi(1+\sin^2 2\theta_1)\right\}$$

$$M_{12}^{-1} = n\ \sin 2\theta_2\,\cos\theta_1/(4N\sin\theta_1) \tag{38}$$

$$M_{22}^{-1} = \frac{1}{2N\sin^2\theta_1}\left\{\cos^2\theta_1\,(1+\chi(1+2\sin^2\theta_1)) + n\,(\cos^2\theta_2 - \sin^2\theta_1\sin^2\theta_2)\right\}$$

$$Q_1 = (G_{10}+G_{01})/(2N) \qquad , \ Q_2 = (G_{20}+G_{02})/(2N\sin\theta_1) \tag{39}$$

The potential energy is given by

$$V(\theta_i) = -N\left\{1-\sin^2\theta_1[(1+n\sin^2\theta_2) - \chi\cos^2\theta_1]\right\} \tag{40}$$

We can now ask ourselves what is the range of applicability of ATDHF, i.e., under which conditions the ATDHF equations of motion (36) describe the same dynamics as the TDHF eqs. (32). The validity condition of Villars (see eq. (9)) reads

$$\langle\phi|(\sum_i p_i\,Q_i)^2|\phi\rangle \ll N \tag{9'}$$

or

$$(p_1^2 + p_2^2/\sin^2\theta_1) \ll 4N^2 \tag{41}$$

But looking at the exact expressions of p_1, p_2 in the semiclassical limit (see eqs. (34)) we see that this gives the condition $\sin^2 2\varphi_1 \ll 4$, or $\varphi_1^2 \ll 1$. In ref. (16) it was indeed verified that it is in this limit that the TDHF equations of motion (see eqs. (32) to (34)) reduce to the ATDHF eqs. of motion (36).

This means that the set of self-consistency conditions I'-III' together with the validity condition of Villars give in an unambiguous way the slow motion limit of TDHF, and that they should only be applied in the regions where this validity condition is satisfied. Provided this condition is satisfied we can use the ATDHF equations (1) in the form explained at the beginning of this section to look for the decoupled one-dimensional collective path.

The H-F minimum is, in this case, the point $\theta_1 = \frac{1}{2}\arccos 1/\chi$, $\theta_2 = 0$, and at this point the RPA frequencies are $\omega_1 = 2(\chi^2 - 1)$, $\omega_2 = n(n+\chi+1)$. For $n > \chi - 1$ the lowest frequency at the H-F minimum is ω_1 and the decoupled collective path is easily found as the line θ_1, $\theta_2 = 0$, connecting the H-F minimum and the saddle point $\theta_1 = \pi/2$, $\theta_2 = 0$. For $n+1 < \chi$ the lowest frequency mode at the H-F minimum is the mode 2 and the equation for the path is a rather complicated equation, given by eq. (25), that connects the H-F minimum and the saddle point $\theta_1 = \frac{1}{2}\arccos 2/\chi$, $\theta_2 = \pi/2$.

As said before this specific quantum mechanical model can be exactly solved. It would therefore be interesting to compare the exact solutions of the Schrödinger equation with those of quantized ATDHF, this has so far not being investigated and deserves further work.

Finally I would like to comment briefly on the slow motion limitation of ATDHF, as expressed by the condition (9) of Villars. We should bare in mind that this limitation comes from the fact that we wish to obtain finally a collective quantum hamiltonian quadratic in the momenta (for which the corresponding classical energy is of order p^2). However we all know that, for instance, in the rotational model of Bohr and Mottelson, higher than first order terms in J(J+1) have to be included in the parametrization of the rotational energy in order to account for the higher levels of the experimental rotational spectra. In the ATDHF scheme this amounts to keep terms of order p^4 in the expansion of $\mathcal{H} = \langle\phi(q,p)|H|\phi(q,p)\rangle$. This can be done within the ATDHF formalism by modifying the phase velocity $[i\, p\, Q]$ (see eq. (10)) into $[i\, p\, (Q + pQ' + p^2 Q'')]$, where Q' and Q" are hermitian, one particle-hole, time-odd and time-even operators, respectively. This extension of the standard ATDHF equations has been investigated by Toukan and Villars[18], and gives rise to additional variational conditions (for the Q', Q" operators) with the validity condition (9) replaced by a new condition quartic in p. But for most practical purposes getting a good hamiltonian quadratic in the momenta is the main issue.

REFERENCES.

1. F. Villars, Nucl. Phys. A285 (1977) 269.

2. M. Baranger and M. Vénéroni, Ann. Phys. (N.Y.) 114 (1978) 123.
M.J. Giannoni, Ann. Phys. 131 (1981) 356.

3. K. Goeke and P.G. Reinhard, Ann. Phys. 112 (1978) 328; Phys. Rev. C20 (1979) 1546.

4. A.K. Mukherjee and M.K. Pal, Nucl. Phys. 373 (1982) 289; Proc. of the Int. Symposium on "Time-Dependent Hartree-Fock and Beyond" (Springer-Verlag, 1982) 358.

5. E. Moya de Guerra and F. Villars, Nucl. Phys. A285 (1977) 297.

6. E. Moya de Guerra, Phys. Rev. Lett. 48 (1982) 922.

7. K. Goeke and P.G. Reinhard, Ann. of Phys. 124 (1980) 249; Nucl. Phys. A312 (1978) 121; Phys. Lett. B69 (1977) 17.
P.G. Reinhard, J.A. Maruhn and K. Goeke, Phys. Rev. Lett. 44 (1980) 1740.
J.N. Urbano, K. Goeke and P.G. Reinhard, Nucl. Phys. A370 (1981) 329.
K. Goeke, contribution to this Workshop and refs. therein.

8. W.J. Swiatecki and S. Bjørnholm, Phys. Rep. 4 (1972) 325.
W.J. Swiatecki, Nucl. Phys. A376 (1982) 275.

9. E. Moya de Guerra, in Proc. of the Int. Symposium on "Time-Dependent Hartree-Fock and Beyond" (Springer-Verlag, 1982) 372.

10. R.F. Fox, Phys. Rep. 48 (1978) 179, and refs. therein.

11. H. Hofmann and P.J. Siemens, Nucl. Phys. A257 (1976).
H.A. Weidenmuller, Progr. in Part. and Nucl. Phys. 3 (1980) 49.

12. A. Hayashi and S. Iwasaki, Progr. Theor. Phys. 63 (1980) 1063.

13. S.Y. Li, A. Klein and D.R. Dreizler, J. Math. Phys. 11 (1970) 975.

14. D.J. Rowe and R. Bassermann, Can. J. Phys. 54 (1976) 1941.

15. K. Goeke, P.G. Reinhard and D.J. Rowe, Nucl. Phys. A359 (1981) 408.

16. E. Moya de Guerra and F. Villars, Nucl. Phys. A298 (1978) 109.

17. R.D. Williams and S.E. Koonin, Nucl. Phys. A391 (1982) 72.
A.K. Mukherjee and M.K. Pal, Phys. Lett. B100 (1981) 457.

18. A. Toukan, Ph. D. Thesis (M.I.T., 1977), unpublished.

DRESSED STATES, NUCLEAR CORRELATIONS AND QUENCHING

I. CHAVES and A. POVES
Departamento de Física Teórica, Universidad Autónoma
Cantoblanco, MADRID-34, Spain

The role of non-nucleonic degrees of freedom in nuclear properties -mainly M1 and Gamow-Teller transitions- is nowadays a hot point in nuclear physics (1). There has been quite a few calculations trying to understand M1 scattering strengths in terms of $\Delta(1236)$ admixtures in the nuclear wavefunctions. The purpose of this talk is to point out the extreme relevance of a correct understanding of the purely nuclear structure problem before any claim for non-nucleonic effects. It is suspicious enough to realize that the nuclear calculations taken as a basis to establish the existence of quenching of the above quoted transitions are in general too simplified. We shall show how the improvement of the nuclear structure description can change the view in the case of M1 scattering strengths of Calcium isotopes.

Let's review briefly the situation; the experimental (2) data for M1 transitions $Ca(0^+)$ (ground state) $\rightarrow$ $Ca(1^+)$ (excited states) are:

^{42}Ca	^{44}Ca	^{46}Ca	^{48}Ca	
1.2	2.1	-	5.4	($\mu_n^{\,2}$)

The predictions of the extreme shell model, i.e., a description of the states as $|0^+\rangle = |1f_{7/2}{}^n\rangle$ and $|1^+\rangle = |1f_{7/2}{}^{n-1}\,1f_{5/2}\rangle$, are:

^{42}Ca	^{44}Ca	^{46}Ca	^{48}Ca	
3.0	6.0	9.0	12.0	($\mu_n^{\,2}$)

Clearly the discrepancies are very large. The pertinent question is: what is hidden under the label $(1f_{7/2}{}^n)$ of the shell model?. The answer is well known; a complex process of renormalization. Starting from the shell model bare states, the true physical state can be written as:

$$|0^+\text{phys}\rangle = |0^+\rangle + \langle J|\hat{V}|0^+\rangle\,|J\rangle \qquad (1)$$

where $\langle J|\hat{V}|0^+\rangle$ is an operator which garantees the decoupling between the model state $|0^+\rangle$ and the remaining Hilbert space $|J\rangle$. On that basis one can demostrate in effective interaction theory (3), that:

$$\langle 0^+\text{phys}\ |H|0^+\text{phys}\rangle = \langle 0^+\ |H + H_R|0^+\rangle$$

H_R being in principle an state dependent many body operator. Nevertheless H_R turns out to be such as to be mocked by an state independent two body operator over very small mass regions; that's why the shell model works. The crucial point now is that the same procedure has to be done for transitions; simple bare states force heavily renormalized transition operators. This has not been done except for E2 effective charges.

Using again wave function (1) it can be easily shown that.

$$\langle 1^+\text{phys}|\vec{\sigma}\cdot\vec{\tau}\ |0^+\text{phys}\rangle = \langle 1^+|\vec{\sigma}\cdot\vec{\tau} + \vec{\Sigma}\ |0^+\rangle$$

where $\vec{\Sigma}$ can be a complicated effective transition operator.

The importance of these effects has been demonstrated by the calculation of McGrory and Wildenthal of M1 scattering strengths of Ca-isotopes (4). They use for the 0^+ and 1^+ states, the wave functions resulting from the diagonalization of an effective interaction in the model space spanned by the orbits ($1f_{7/2}$ $2p_{3/2}$ $2p_{1/2}$ $1f_{5/2}$), without truncations. The effective interaction was adjusted to reproduce the energy levels of Ca-isotopes. The results they get are:

^{42}Ca	^{44}Ca	^{46}Ca	^{48}Ca	
1.9	4.0	6.4	8.9	($\mu_n{}^2$)

by taking these correlations into account one half of the discrepancy is gone. The configurations included are a small part of those coming from the complete nuclear correlations, therefore one is led to think that nuclear mechanisms are crucial to understand the quenching.

It is well known that (fp) -shell correlations are not the only ones relevant in the Calcium region. The importance of core excitations -mainly 2p-2h from the sd-shell- has been demostrated by spectroscopic (5) as well as charge radii isotope shift studies (6). We had old results(7) concerning the ^{41}Ca magnetic moment that encouraged us to go on in calculating M1-scattering stregths.

The calculation proceeds as follows:

a) The model space includes two orbits, $1d_{3/2}$ and $1f_{7/2}$

b) The configurations we retain are the basic ones $1f_{7/2}{}^{n}$ plus $(1d_{3/2})^{-2}$ $1f_{7/2}{}^{n+2}$ (2p-2h configurations) and $(1d_{3/2})^{-4}$ $1f_{7/2}{}^{n+4}$ (4p-4h configurations). In some cases, we truncate in seniority to keep the calculations in reasonable dimensionalities; when so it will be specified in the text.

c) In all cases we use the effective interaction built up by Zuker in his study of the core excited states in the Calcium Region.

d) Once the diagonalizations made we compute the transition matrix elements among the relevant states, getting the following results:

i) INCLUDING UP TO 2P-2H EXCITATIONS (B(M1); $0^+ \rightarrow 1^+$)

^{42}Ca	^{44}Ca *	^{46}Ca*	^{48}Ca	
2.85	5.87	8.71	11.18	($\mu_n{}^2$)

(in the starred nuclei the calculation is made with minimum seniority).

ii) INCLUDING UP TO 4P-4H EXCITATIONS (B(M1); $0^+ \rightarrow 1^+$)

^{42}Ca *	^{44}Ca*	^{46}Ca*	^{48}Ca	
2.81	5.82	8.62	10.95	($\mu_n{}^2$)

Simple perturbation theory arguments show that we are faced to a reduction in strength due to norm effects, that's why the seniority truncation erodes a bit the effect.

Combining our results with McGrory and Wildenthal's, the situation is the following;

	^{42}Ca	^{44}Ca	^{46}Ca	^{48}Ca	
TH	1.7	3.8	6.0	7.8	
EXP	1.2	2.1	-	5.4	($\mu_n{}^2$)

We shall focuss the discussion from now on in the ^{48}Ca case where the theoretical calculation is most accurate;

a) The experimental B(M1) strength is a 70% of the theoretical prediction (to be compared with the 33% of initial ones).

b) It is well known that M1 transitions are very sensitive to fine details of the wave function -or in other language, effective M1 operators should be rather fancy-.

c) We have left out secular mechanisms of quenching coming probably from the renormalization associated to the hard-core. These could be responsible for the discrepancy in the ^{41}Ca spectroscopic factor (.7 to be compared with a theoretical prediction of .9).

d) Consistently, non nucleonic effects, if present, are at most 10-15% of the effect.

Finally one can look a bit closer to our results to understand how the quenching appears through the inclusion of excited configurations which are not connected with the leading one by the M1 operator. Let Ψ be the result of our calculation for the 0^+ state. We can project Ψ on the 0p-0h, 2p-2h and 4p-4h spaces;

$$P_0\Psi = \Psi_0 \quad P_2\Psi = \Psi_2 \quad P_4\Psi = \Psi_4 \quad \text{, then}$$
$$\Psi = a\Psi_0 + b\Psi_2 + c\Psi_4$$

Similarly, the 1^+ wave function ϕ can be expressed as:

$$\phi = a'\phi_0 + b'\phi_2 + c'\phi_4$$

The transition matrix element is then:

$$\langle\phi|M1|\Psi\rangle = aa'\langle\phi_0|M1|\Psi_0\rangle + bb'\langle\phi_2|M1|\Psi_2\rangle + cc'\langle\phi_4|M1|\Psi_4\rangle$$

from our calculation, $\langle\Psi_2|M1|\phi_2\rangle \simeq \langle\Psi_4|M1|\phi_4\rangle$ and $a \simeq a'$, and to a very good approximation.

$$\langle\phi|M1|\Psi\rangle = \langle\phi_0|M1|\Psi_0\rangle - b^2 (\langle\phi_0|M1|\Psi_0\rangle - \langle\phi_2|M1|\Psi_2\rangle) \quad (2)$$

Therefore, the reduction of the matrix element, depends on; i) the amount of mixing and ii) the transition properties of the projection of the state on the excited configurations. The states we consider give a rather large $\langle\phi_2|M1|\Psi_2\rangle = 3.0\mu_n$ compared with $\langle\phi_0|M1|\Psi_0\rangle = \sqrt{12.0}\mu_n$, but expression (2) clearly suggests that configurations of very small amplitude, having $\langle\phi_c|M1|\Psi_c\rangle$ very small can contribute to the quenching.

REFERENCES

1. C. GAARDE, Nucl. Phys. A396 (1983) 127c
   W. WEISE, Nucl. Phys. A396 (1983) 373c

2. A. RICHTER, Nucl. Phys. A374 (1982) 177c

3. A. POVES and A. ZUKER, Phys. Reports 71 (1981) 141

4. J.B. McGRORY and B.H. WILDENTAL, Phys. Lett. 103B (1981) 173

5. A. ZUKER, Proc. Int. Conf. "$1f_{7/2}$ nuclei" (Ed. R. Ricci), Ed. Compositori, Bologna 1971.

6. E. CAURIER and A. POVES, Nucl. Phys. A385 (1982) 407

7. L. CHAVES, unpublished work.

ATDHF CALCULATION ON THE ^{4}He-^{16}O SYSTEM

D. Provoost[+], F. Grümmer, K. Goeke[++]
Institut für Kernphysik
Kernforschungsanlage Jülich
D-5170 Jülich, West Germany

The quantized ATHF theory[1)] leads to the following differential equation for the maximum decoupled collective path of the lowest large amplitude collective mode:

$$\frac{d}{dq}|\phi_q\rangle = c(q)[H,H_{ph}]_{ph}|\phi_q\rangle \tag{1}$$

with q being a coordinate which labels the Slater determinants along the path. The index "ph" indicates the 1p-1h part of the operator with respect to $|\phi_q\rangle$. Working out eq. (1) in terms of the single-particle wave functions we get

$$\frac{\partial}{\partial q}|\varphi_q^{(\alpha)}\rangle = c(q)(1-\rho_o(q))\{W_1(q)+W_o(q)(1-2\rho_o(q))W_o(q)\}|\varphi_q^{(\alpha)}\rangle \tag{2}$$

$$\alpha = 1\ldots A$$

where the s.p. density matrix is given by

$$\rho_o(q) = \sum_{\alpha=1}^{A} |\varphi_q^{(\alpha)}\rangle\langle\varphi_q^{(\alpha)}| \tag{3}$$

The Hartree-Fock Hamiltonian W_o reads

$$W_o(q) = T + Tr\{\rho_o(q)v\} \tag{4}$$

with the two-body interaction v, and the W_1 is given by

$$W_1(q) = Tr\{v[W_o,\rho_o]\} \tag{5}$$

More details of the theory are given in Ref. 2.

The calculations are performed with the BKN force[3)] and a modified interaction[4)]. With these forces the Hartree-Fock Hamiltonian becomes

+ also: Limburgs Universitair Centrum, B-3610 Diepenbeek, Belgium
++ also: Institut für Theoretische Kernphysik, Universität Bonn, D-5300 Bonn, West Germany

$$W_o(\bar{r},\bar{r}') = \delta(\bar{r}-\bar{r}')\{-\frac{\hbar^2}{2m}\nabla^2 + U(\bar{r})\} \tag{6}$$

$$U(\bar{r}) = \frac{3}{4}t_o\,\rho_o(\bar{r}) + \frac{3}{16}t_3\,\rho_o(\bar{r})\rho_o^{\alpha}(\bar{r})$$
$$+ V_o\int\rho_o(\bar{r}')\frac{\exp\{-\frac{|\bar{r}-\bar{r}'|}{a}\}}{\frac{|\bar{r}-\bar{r}'|}{a}}d^3\bar{r} + \frac{e^2}{4}\int\frac{\rho_o(\bar{r}')}{|\bar{r}-\bar{r}'|}d^3\bar{r}' \tag{7}$$

The parameters are given in Table 1.

	BKN	modified interaction
α	1.0	0.25
t_o	-497.726	0.00
t_3	17270.0	14316.2
V_o	-363.043	-5998.66
a	0.4598	0.2786

Table 1: Parameter values of the used interactions.

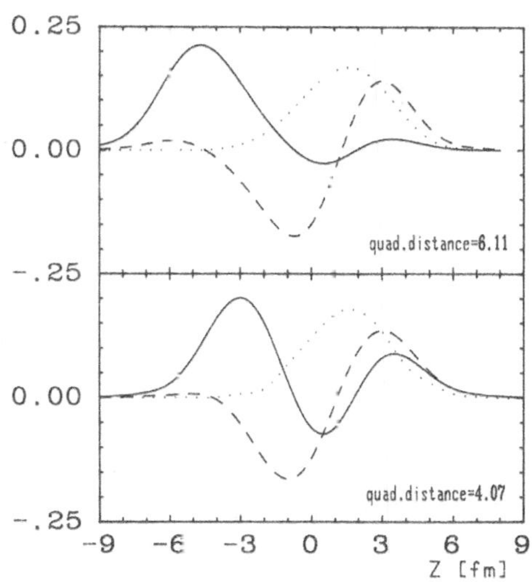

Fig. 1: Single particle wave functions along the axis of collision of the α-^{16}O system for two different configurations. At a cluster distance of 6.11 fm and at the HF point of ^{20}Ne (= quad. distance 4.07 fm) respectively. The s.p. wave functions have asymptotically the following symmetries:

—— α s_x, s_y, s_z
... ^{16}O s_x, s_y, s_z
--- ^{16}O s_x, s_y, p_z

Eq. (2) is solved in a gridbox of dimension 16x16x32 fm^3 with a spacing of 1 fm in every direction. Furthermore quartet symmetry is assumed so that protons and neutrons are treated on the same level with an effective charge e/2. Whenever a Fourier transformation is made to the momentum space the momenta which give rise to an energy higher than 100 MeV are neglected.

In Fig. 1 we show the single particle wave functions along the z-axis for two configurations. In the first configuration the α-particle overlaps slightly with the ^{16}O so that we still recognize the asymptotic wave functions: the α s-wave function and the ^{16}O s and p_z wave functions. The second configuration is close to the HF state of ^{20}Ne and one sees how the original α s-wave is deformed.

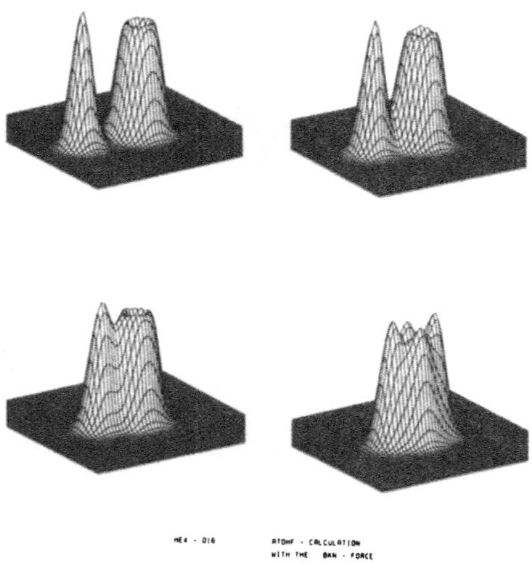

Fig. 2: Particle densities in the (x,z) plane for the α-^{16}O system along the path at quadrupole distances 7.4, 6.2, 4.7 and 4.07 fm respectively.

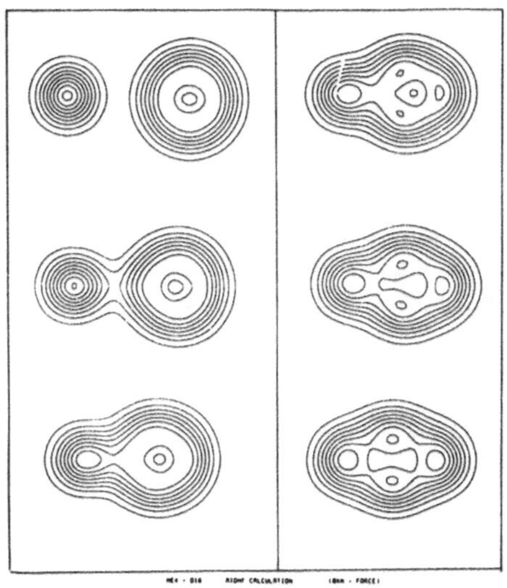

Fig. 3: Particle densities in the (x,z) plane for the α-^{16}O system along the path at quadrupole distances 7.4, 6.2, 4.6, 4.25, 4.15 and 4.07 fm respectively.

Three-dimensional plots of the particle densities in the (x,z) plane along the path are shown in Fig. 2. In the first picture the α-particle is still separated from the ^{16}O. The big depth in the middle of the ^{16}O is due to the deficiencies of the used BKN force. In the next pictures the α-particle starts to overlap more and more with the ^{16}O until finally they both completely loose their own identity to form a symmetric ^{20}Ne HF ground state. That this state is really symmetric might be clearer from a two-dimensional plot in Fig. 3. The middle of the ^{16}O should again be considered as a depth.

The determinantal wave functions used so far are not parity eigenstates. We therefore make a projection on good parity.

$$|\phi^{\pm}(q)\rangle = N(|\phi(q)\rangle \pm |\phi(-q)\rangle) \qquad (8)$$

where N is a normalization factor.

The parity projected energy curves are given by

$$E^{\pm}(q) = \langle\phi^{\pm}(q)|H|\phi^{\pm}(q)\rangle \qquad (9)$$

An analogous definition holds for the unprojected one.

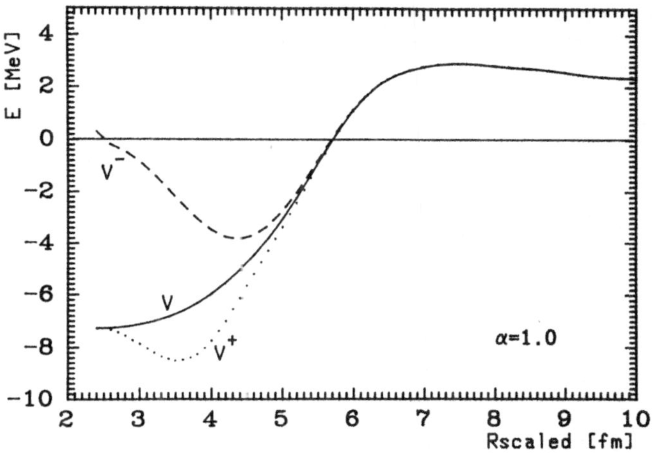

Fig. 4: Classical potential V anc energy curves for the wave functions with positive parity V^+ and negative parity V^- as a function of the coordinate R_{scaled} (see eq. (18) for definition) for the BKN force (α = 1.0).

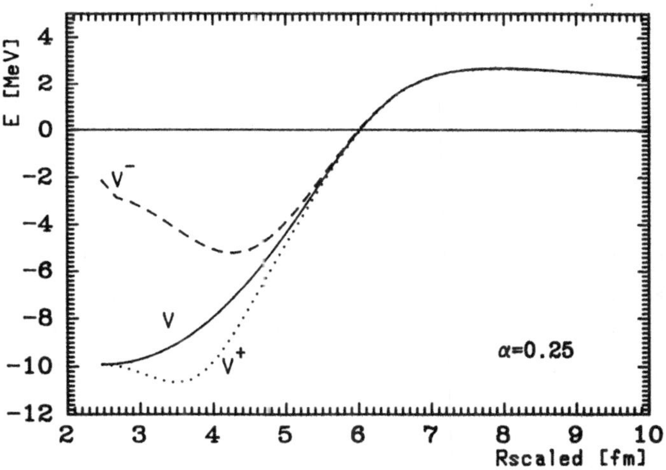

Fig. 5: Classical potential V and energy curves for the wave functions with positive parity V^+ and negative parity V^- as a function of the coordinate R_{scaled} (see eq. (18) for definition) for the modified force (α = 0.25).

In Fig. 4 and Fig. 5 the energies for both interactions, respectively, are plotted using as collective coordinate q the distance of the fragments R_{scaled}, a detailed definition of which will follow. From the actual binding energies those of the separate fragments are subtracted. The static results are summarized in Table 2.

force	$E_\alpha+E_{^{16}O}$	E_{HF}	E^+_{HF}	E^-_{HF}	barrier
α = 1.0	-133.38	-140.64	-141.87	-137.19	2.92
α = 0.25	-141.64	-151.55	-152.31	-146.87	2.72

Table 2: Energies in MeV of: a separate α-^{16}O system, the HF ground state, the positive parity HF ground state, the intrinsic state of the negative parity bands, the barrier height for the two different interactions used.

In the next step of the quantized ATDHF a collective Schrödinger equation is extracted from the collective path $\{|\phi_q\rangle\}$ in the Gaussian overlap approximation:

$$[-\frac{1}{4}\{\frac{d^2}{dq^2}\frac{\hbar^2}{2M}+\frac{d}{dq}\frac{\hbar^2}{M}\frac{d}{dq}+\frac{\hbar^2}{2M}\frac{d^2}{dq^2}\} + V(q)-Z(q)+\frac{\hbar^2}{2\theta}\ell(\ell+1)]g_\ell(q) = E\,g_\ell(q) \tag{10}$$

where the mass parameter M is given by

$$M^{-1}(q) = \frac{1}{\hbar^2}\langle\phi_q|[Q,[H,Q]]|\phi_q\rangle \tag{11}$$

and

$$Q = (\frac{\partial V}{\partial q})^{-1} H_{ph} \tag{12}$$

Here the classical potential V(q) is the expectation value of the Hamiltonian

$$V(q) = \langle\phi_q|H|\phi_q\rangle \tag{13}$$

and the Z(q) are the quantum corrections which take into account the kinetic energy along the path and the zero point energies for translational and rotational motion. For further details see Refs. 2 and 5. So far the centrifugal term $\frac{\hbar^2}{2\theta}\ell(\ell+1)$ has been added to the eq. (10). An angular momentum projection will be done later to improve the results.

A noteworthy point now is the free choice of the labelling parameter q of the collective path $\{|\phi_q\rangle\}$. If one wants to obtain a mass parameter which is asymptotically equal to the reduced mass then the coordinate should be chosen in such a way that as long as both nuclei do not overlap it is equal to the cluster distance.

For the following one has to note that two spherical particles at a distance R from each other show a quadrupole moment of the total system given by

$$Q_2 = 2\mu R^2 \tag{14}$$

if the center of mass is placed at the origin, μ being the reduced mass. Hence we define in analogy the so-called quadrupole distance

$$R_2 = \sqrt{\frac{\langle Q_2 \rangle}{2\mu}} \tag{15}$$

where $\langle Q_2 \rangle$ is the total quadrupole moment of the α-^{16}O system. In the asymptotic region R_2 agrees with the actual cluster distance.

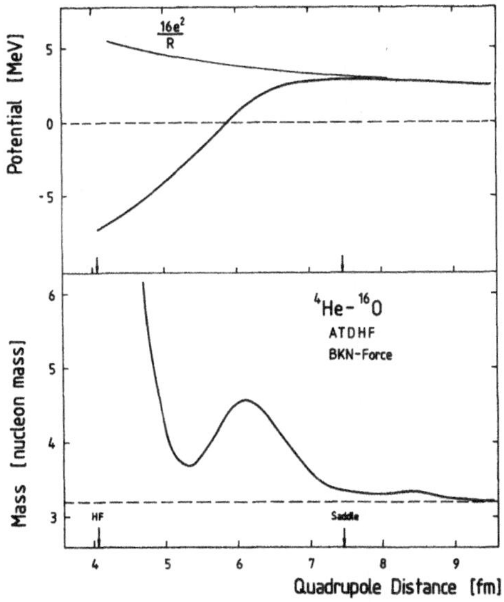

Fig. 6: Collective mass M and potential V for α-$^{16}O \rightarrow {}^{20}Ne$ as a function of the quadrupole distance using the BKN interaction.

Since the ^{20}Ne HF ground state is deformed the corresponding value of the R_2 coordinate is 4.07 fm. The classical potential V and mass parameter M as functions of this R_2 coordinate are given in Fig. 6.

It turns out that the mass parameter starts to diverge in coming closer to the HF point. Indeed from Fig. 3 it is clear that the quadrupole moment of the three last systems is more or less constant although the internal structure changes from an asymmetric to a symmetric one. So in this region of the path the R_2 coordinate does not distinguish very well the Slater determinants although they correspond to a rather different structure of the system. Hence the $M(R_2)$ is bound to raise steeply.

The octupole moment might be an alternative choice for the collective coordinate since it becomes zero for symmetric systems. The octupole moment for a 2 particle system with the center of mass in the origin and for separate particles with no net quadrupole and octupole moment, is given by

$$Q_3 = -2\mu \frac{A_2-A_1}{A_2+A_1} R^3 \tag{16}$$

where R is the cluster distance, A_1 and A_2 the masses of the two separate particles and μ the reduced mass. So we define another coordinate R_3 called octupole distance

$$R_3 = \sqrt[3]{-\frac{A_1+A_2}{2\mu(A_2-A_1)}\langle Q_3\rangle} \tag{17}$$

which is in the asymptotic region again equal to the cluster distance. The potential and mass parameter as functions of R_3 are shown in Fig. 7.

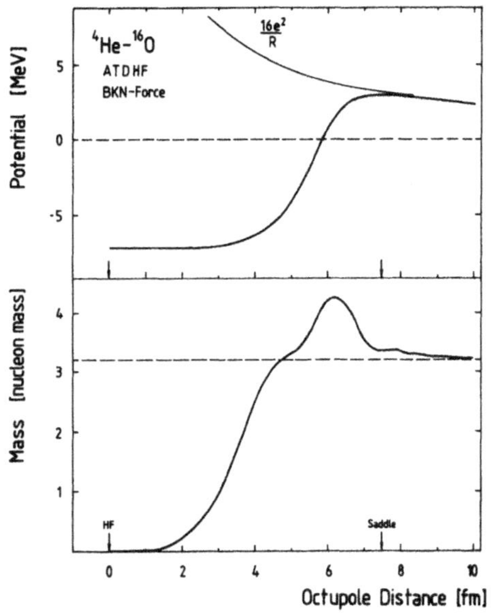

Fig. 7: Collctive mass M and potential V for α-^{16}O↔^{20}Ne as a function of the octupole distance using the BKN interaction.

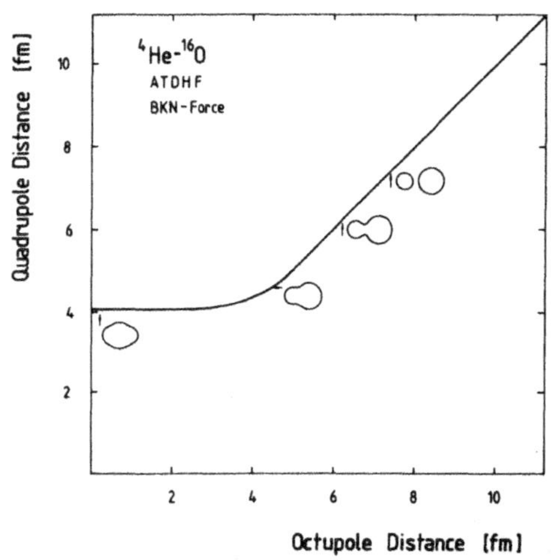

Fig. 8: Quadrupole distance R_2 as a function of the octupole distance R_3 for the α-^{16}O↔^{20}Ne system using the BKN interaction.

One sees now that the region near the HF point gets expanded so that the potential shows a flat minimum. The divergency of the mass is also overcome and the mass parameter goes now even to small values in the neighbourhood of HF.

The comparison between the two choices R_2 and R_3 is made in Fig. 8 which clearly illustrates the suitability of the R_3 coordinate near the HF point.

However it should clearly be pointed out that the physical information contained in the quantities V, M and Z is identical no matter which coordinate is actually used, if one ignores for a moment the technical difficulties encountered in handling a diverging mass. Hence it is justified to transform to a new and convenient coordinate, R_{scaled}, in which the mass parameter is constant. This is achieved by e.g. the following transformation:

$$dR_{scaled} = dR_3 \sqrt{\frac{M(R_3)}{\mu}} \qquad (18)$$

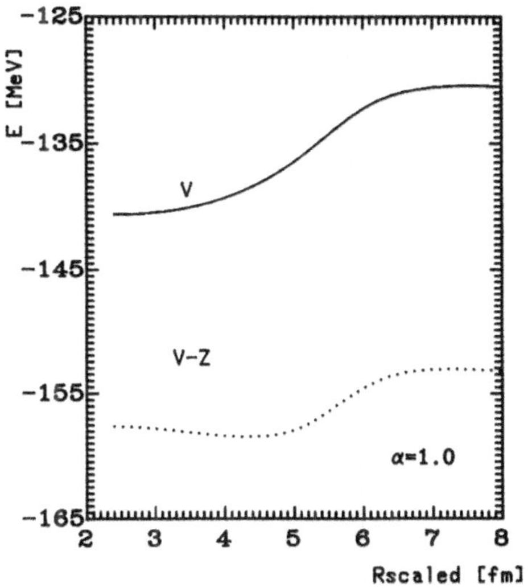

Fig. 9: Potential V and quantum corrected potential V-Z in absolute values for α-^{16}O-^{20}Ne using the BKN interaction.

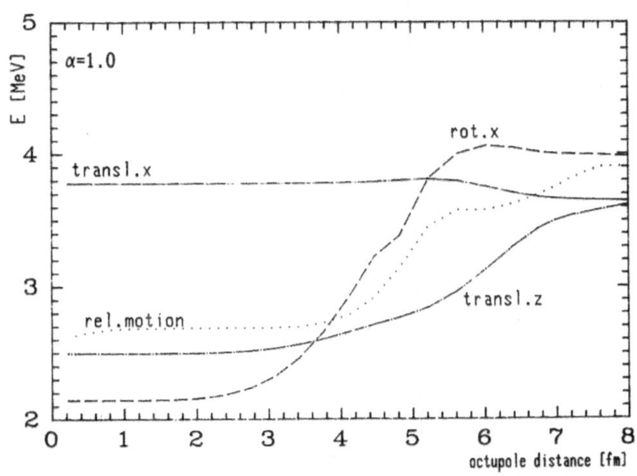

Fig. 10: The quantum corrections (zero point energies) for the total α+^{16}O-^{20}Ne system. Plotted are the corrections for relative motion, for rotation around the x-axis to be counted twice, for translation along the x-axis to be counted twice and for translation along the z-axis, i.e. the collision axis.

The potential V and the quantum corrected V-Z are shown in absolute values as a function of this new coordinate in Fig. 9 and the different contributions to the Z are shown in Fig. 10.

After subtracting the asymptotic binding energy the V-Z is inserted into the Schrödinger equation, eq. (10), and solved for each partial wave. This gives for the force with α = 0.25 a bound state for ℓ=0. All other states are resonances which are rather narrow. Since the BKN potential V-Z is not deep enough to give bound states all states obtained are resonances.

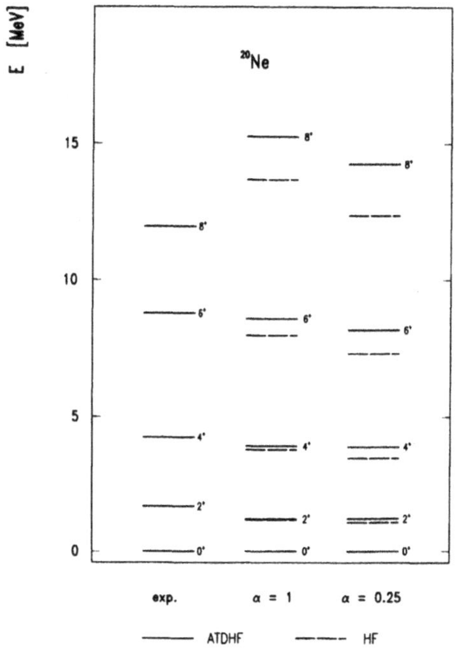

Fig. 11: Ground state rotational band of ^{20}Ne. Experimental values, ATDHF calculation with BKN force (α=1) and with modified force (α = 0.25) are compared. The results obtained with ^{20}Ne HF ground state plus adding a centrifugal term are also given.

They are compared with the experimental data in Fig. 11. The 2^+-, 4^+- and 6^+-states are in good agreement with experiment. The 8^+-state however is known to have no longer a rotational structure, so we cannot expect to find agreement with experiment since we added by hand a centrifugal term proportional to $\ell(\ell+1)$.

The results obtained with positive parity projection of the classical potential V are very similar and therefore not shown here. For comparison the spectra obtained by adding a centrifugal term to the ^{20}Ne HF ground state are also given. The similarity to the ATDHF results shows that the ^{20}Ne HF ground state has a big overlap with the ATDHF one.

In some preliminary calculations also the elastic α-^{16}O scattering process has been investigated with rather encouraging results. In Figs. 12 and 13 the elastic

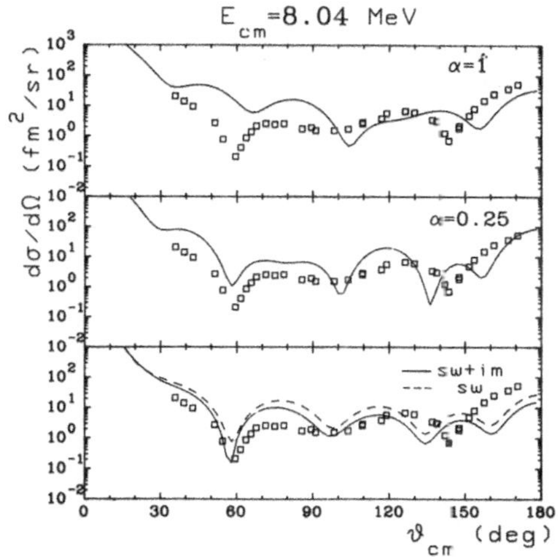

Fig. 12: Differential scattering cross sections for the elastic scattering of α on ^{16}O for a CM bombarding energy of 8.04 MeV. The experimental data are taken from Ref. 6 and compared with the BKN and modified force calculations (α = 1.0 and α = 0.25 respectively) and with the Woods-Saxon fit to the experimental data once with and once without imaginary part.

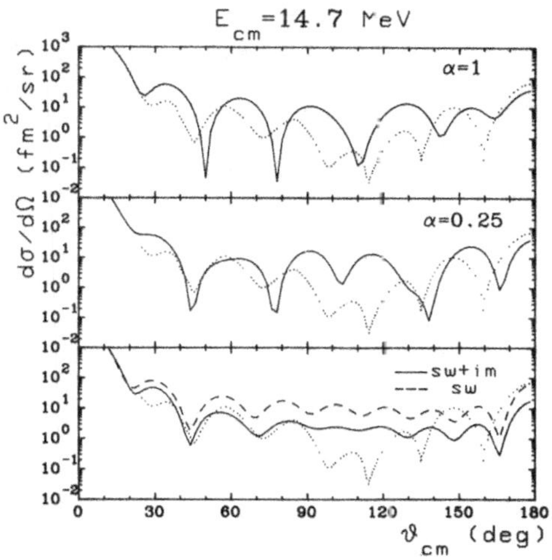

Fig. 13: Differential scattering cross sections for the elastic scattering of α on ^{16}O for a CM bombarding energy of 14.7 MeV. The experimental data --- are taken from Ref. 7 and compared with the BKN and modified force calculations (α = 1.0 and α = 0.25 respectively) and with the Woods-Saxon fit to the experimental data once with and once without imaginary part.

differential scattering cross sections are plotted for a CM bombarding energy of 8.04 and 14.7 MeV respectively. The upper part shows the comparison experiment-BKN calculations, the middle part experiment-modified force calculations and the lower part compares experiment with the Saxon-Woods fit to the experimental data. The results for the lower energy are fairly good especially if one compares the calculated cross sections with the Saxon-Woods fit (without imaginary part) to the data.

The above discussed investigations show that it is indeed possible to describe realistically the low energy α-^{16}O system, starting completely from a variational ansatz as formulated in the quantized ATDHF theory. We have demonstrated that in fully three-dimensional calculations using well known effective interactions the rotational ground state spectrum of ^{20}Ne and the elastic α-scattering data can be reproduced without adjusting any free parameters. Further investigations, in particular focussed on angular momentum properties and channel coupling, are in progress.

We thank M. Bouten, Limburgs Universitair Centrum, Belgium, for useful discussions and continuous support.

1) K. Goeke et al., these proceedings.
2) K. Goeke, F. Grümmer, and P.-G. Reinhard, Ann. Phys. 150 (1983) 504.
3) P. Bonche, S. Koonin, and J.W. Negele, Phys. Rev. C13 (1976) 1226.
4) P.-G. Reinhard, J. Friedrich, K. Goeke, F. Grümmer, D.H.E. Gross, to be published.
5) J.N. Urbano, K. Goeke, and P.-G. Reinhard, Nucl. Phys. A370 (1981) 329.
6) M.K. Mehta, W.E. Hunt, and R.H. Davis, Phys. Rev. 160 (1967) 791.
7) J.C. Corelli et al., Phys. Rev. 116 (1959) 1184.

Lecture Notes in Physics

Selected Issues from

Lecture Notes in Mathematics

GPSR Compliance

The European Union's (EU) General Product Safety Regulation (GPSR) is a set of rules that requires consumer products to be safe and our obligations to ensure this.

If you have any concerns about our products, you can contact us on ProductSafety@springernature.com

In case Publisher is established outside the EU, the EU authorized representative is:

Springer Nature Customer Service Center GmbH
Europaplatz 3
69115 Heidelberg, Germany

Batch number: 10704763

Printed by Printforce, the Netherlands